NATURAL HISTORY
UNIVERSAL LIBRARY

U0215821

西方博物学大系

主编：江晓原

A HISTORY OF BRITISH FISHES

不列颠鱼类志

[英] 威廉·亚雷尔 著

华东师范大学出版社

图书在版编目（CIP）数据

不列颠鱼类志 = A history of British fishes：英文 /
（英）威廉·亚雷尔（William Yarrell）著. — 上海：华东
师范大学出版社, 2018
（寰宇文献）
ISBN 978-7-5675-8078-7

Ⅰ. ①不… Ⅱ. ①威… Ⅲ. ①鱼类–水产志–英国–
英文 Ⅳ. ①Q959.408

中国版本图书馆CIP数据核字(2018)第180512号

不列颠鱼类志
A history of British fishes
（英）威廉·亚雷尔（William Yarrell）

特约策划　黄曙辉　徐　辰
责任编辑　庞　坚
特约编辑　许　倩
装帧设计　刘怡霖

出版发行　华东师范大学出版社
社　　址　上海市中山北路3663号　邮编 200062
网　　址　www.ecnupress.com.cn
电　　话　021-60821666　行政传真　021-62572105
客服电话　021-62865537
门市（邮购）电话　021-62869887
地　　址　上海市中山北路3663号华东师范大学校内先锋路口
网　　店　http://hdsdcbs.tmall.com/

印刷者　虎彩印艺股份有限公司
开　　本　787×1092　16开
印　　张　72.75
版　　次　2018年8月第1版
印　　次　2018年8月第1次
书　　号　ISBN 978-7-5675-8078-7
定　　价　1198.00元（精装全二册）

出版人　王　焰

（如发现本版图书有印订质量问题，请寄回本社客服中心调换或电话021-62865537联系）

总　目

《西方博物学大系》总序

江晓原

　　《西方博物学大系》收录博物学著作超过一百种，时间跨度为 15 世纪至 1919 年，作者分布在 16 个国家，写作语种有英语、法语、拉丁语、德语、弗莱芒语等，涉及对象包括植物、昆虫、软体动物、两栖动物、爬行动物、哺乳动物、鸟类和人类等，西方博物学史上的经典著作大备于此编。

中西方"博物"传统及观念之异同

　　今天中文里的"博物学"一词，学者们认为对应的英语词汇是 Natural History，考其本义，在中国传统文化中并无现成对应词汇。在中国传统文化中原有"博物"一词，与"自然史"当然并不精确相同，甚至还有着相当大的区别，但是在"搜集自然界的物品"这种最原始的意义上，两者确实也大有相通之处，故以"博物学"对译 Natural History 一词，大体仍属可取，而且已被广泛接受。

　　已故科学史前辈刘祖慰教授尝言：古代中国人处理知识，如开中药铺，有数十上百小抽屉，将百药分门别类放入其中，即心安矣。刘教授言此，其辞若有憾焉——认为中国人不致力于寻求世界"所以然之理"，故不如西方之分析传统优越。然而古代中国人这种处理知识的风格，正与西方的博物学相通。

　　与此相对，西方的分析传统致力于探求各种现象和物体之间的相互关系，试图以此解释宇宙运行的原因。自古希腊开始，西方哲人即孜孜不倦建构各种几何模型，欲用以说明宇宙如何运行，其中最典型的代表，即为托勒密（Ptolemy）的宇宙体系。

　　比较两者，差别即在于：古代中国人主要关心外部世界"如何"运行，而以希腊为源头的西方知识传统（西方并非没有别的知识传统，只是未能光大而已）更关心世界"为何"如此运行。在线性发展无限进步的科学主义观念体系中，我们习惯于认为"为何"是在解决了"如何"之后的更高境界，故西方的分析传统比中国的传统更高明。

然而考之古代实际情形，如此简单的优劣结论未必能够成立。例如以天文学言之，古代东西方世界天文学的终极问题是共同的：给定任意地点和时刻，计算出太阳、月亮和五大行星（七政）的位置。古代中国人虽不致力于建立几何模型去解释七政"为何"如此运行，但他们用抽象的周期叠加（古代巴比伦也使用类似方法），同样能在足够高的精度上计算并预报任意给定地点和时刻的七政位置。而通过持续观察天象变化以统计、收集各种天象周期，同样可视之为富有博物学色彩的活动。

还有一点需要注意：虽然我们已经接受了用"博物学"来对译 Natural History，但中国的博物传统，确实和西方的博物学有一个重大差别——即中国的博物传统是可以容纳怪力乱神的，而西方的博物学基本上没有怪力乱神的位置。

古代中国人的博物传统不限于"多识于鸟兽草木之名"。体现此种传统的典型著作，首推晋代张华《博物志》一书。书名"博物"，其义尽显。此书从内容到分类，无不充分体现它作为中国博物传统的代表资格。

《博物志》中内容，大致可分为五类：一、山川地理知识；二、奇禽异兽描述；三、古代神话材料；四、历史人物传说；五、神仙方伎故事。这五大类，完全符合中国文化中的博物传统，深合中国古代博物传统之旨。第一类，其中涉及宇宙学说，甚至还有"地动"思想，故为科学史家所重视。第二类，其中甚至出现了中国古代长期流传的"守宫砂"传说的早期文献：相传守宫砂点在处女胳膊上，永不褪色，只有性交之后才会自动消失。第三类，古代神话传说，其中甚至包括可猜想为现代"连体人"的记载。第四类，各种著名历史人物，比如三位著名刺客的传说，此三名刺客及所刺对象，历史上皆实有其人。第五类，包括各种古代方术传说，比如中国古代房中养生学说，房中术史上的传说人物之一"青牛道士封君达"等等。前两类与西方的博物学较为接近，但每一类都会带怪力乱神色彩。

"所有的科学不是物理学就是集邮"

在许多人心目中，画画花草图案，做做昆虫标本，拍拍植物照片，这类博物学活动，和精密的数理科学，比如天文学、物理学等等，那是无法同日而语的。博物学显得那么的初级、简单，甚至幼稚。这种观念，实际上是将"数理程度"作为唯一的标尺，用来衡量一切知识。但凡能够使用数学工具来描述的，或能够进行物理实验的，那就是"硬"科学。使用的数学工具越高深越复杂，似乎就越"硬"；物理实验设备越庞大，花费的金钱越多，似乎就越"高端"、越"先进"……

这样的观念，当然带着浓厚的"物理学沙文主义"色彩，在很多情况下是不正确的。而实际上，即使我们暂且同意上述"物理学沙文主义"的观念，博物学的"科学地位"也仍然可以保住。作为一个学天体物理专业出身，因而经常徜徉在"物理学沙文主义"幻影之下的人，我很乐意指出这样一个事实：现代天文学家们的研究工作中，仍然有绘制星图，编制星表，以及为此进行的巡天观测等等活动，这些活动和博物学家"寻花问柳"，绘制植物或昆虫图谱，本质上是完全一致的。

这里我们不妨重温物理学家卢瑟福(Ernest Rutherford)的金句："所有的科学不是物理学就是集邮(All science is either physics or stamp collecting)。"卢瑟福的这个金句堪称"物理学沙文主义"的极致，连天文学也没被他放在眼里。不过，按照中国传统的"博物"理念，集邮毫无疑问应该是博物学的一部分——尽管古代并没有邮票。卢瑟福的金句也可以从另一个角度来解读：既然在卢瑟福眼里天文学和博物学都只是"集邮"，那岂不就可以将博物学和天文学相提并论了？

如果我们摆脱了科学主义的语境，则西方模式的优越性将进一步被消解。例如，按照霍金(Stephen Hawking)在《大设计》(*The Grand Design*)中的意见，他所认同的是一种"依赖模型的实在论(model-dependent realism)"，即"不存在与图像或理论无关的实在性概念(There is no picture- or theory-independent concept of reality)"。在这样的认识中，我们以前所坚信的外部世界的客观性，已经不复存在。既然几何模型只不过是对外部世界图像的人为建构，则古代中国人干脆放弃这种建构直奔应用（毕竟在实际应用中我们只需要知道七政"如何"运行），又有何不可？

传说中的"神农尝百草"故事，也可以在类似意义下得到新的解读："尝百草"当然是富有博物学色彩的活动，神农通过这一活动，得知哪些草能够治病，哪些不能，然而在这个传说中，神农显然没有致力于解释"为何"某些草能够治病而另一些则不能，更不会去建立"模型"以说明之。

"帝国科学"的原罪

今日学者有倡言"博物学复兴"者，用意可有多种，诸如缓解压力、亲近自然、保护环境、绿色生活、可持续发展、科学主义解毒剂等等，皆属美善。编印《西方博物学大系》也是意欲为"博物学复兴"添一助力。

然而，对于这些博物学著作，有一点似乎从未见学者指出过，而鄙意以为，当我们披阅把玩欣赏这些著作时，意识到这一点是必须的。

这百余种著作的时间跨度为15世纪至1919年，注意这个时间跨度，正是西方列强"帝国科学"大行其道的时代。遥想当年，帝国的科学家们乘上帝国的军舰——达尔文在皇家海军"小猎犬号"上就是这样的场景之一，前往那些已经成为帝国的殖民地或还未成为殖民地的"未开化"的遥远地方，通常都是踌躇满志、充满优越感的。

作为一个典型的例子，英国学者法拉在（Patricia Fara）《性、植物学与帝国：林奈与班克斯》（*Sex, Botany and Empire, The Story of Carl Linnaeus and Joseph Banks*）一书中讲述了英国植物学家班克斯（Joseph Banks）的故事。1768年8月15日，班克斯告别未婚妻，登上了澳大利亚军舰"奋进号"。此次"奋进号"的远航是受英国海军部和皇家学会资助，目的是前往南太平洋的塔希提岛（Tahiti，法属海外自治领，另一个常见的译名是"大溪地"）观测一次比较罕见的金星凌日。舰长库克（James Cook）是西方殖民史上最著名的舰长之一，多次远航探险，开拓海外殖民地。他还被认为是澳大利亚和夏威夷群岛的"发现"者，如今以他命名的群岛、海峡、山峰等不胜枚举。

当"奋进号"停靠塔希提岛时，班克斯一下就被当地美丽的土著女性迷昏了，他在她们的温柔乡里纵情狂欢，连库克舰长都看不下去了，"道德愤怒情绪偷偷溜进了他的日志当中，他发现自己根本不可能不去批评所见到的滥交行为"，而班克斯纵欲到了"连嫖妓都毫无激情"的地步——这是别人讽刺班克斯的说法，因为对于那时常年航行于茫茫大海上的男性来说，上岸嫖妓通常是一项能够唤起"激情"的活动。

而在"帝国科学"的宏大叙事中，科学家的私德是无关紧要的，人们关注的是科学家做出的科学发现。所以，尽管一面是班克斯在塔希提岛纵欲滥交，一面是他留在故乡的未婚妻正泪眼婆娑地"为远去的心上人绣织背心"，这样典型的"渣男"行径要是放在今天，非被互联网上的口水淹死不可，但是"班克斯很快从他们的分离之苦中走了出来，在外近三年，他活得倒十分滋润"。

法拉不无讽刺地指出了"帝国科学"的实质："班克斯接管了当地的女性和植物，而库克则保护了大英帝国在太平洋上的殖民地。"甚至对班克斯的植物学本身也调侃了一番："即使是植物学方面的科学术语也充满了性指涉。……这个体系主要依靠花朵之中雌雄生殖器官的数量来进行分类。"据说"要保护年轻妇女不受植物学教育的浸染，他们严令禁止各种各样的植物采集探险活动。"这简直就是将植物学看成一种"涉黄"的淫秽色情活动了。

在意识形态强烈影响着我们学术话语的时代，上面的故事通常是这样被描述的：库克舰长的"奋进号"军舰对殖民地和尚未成为殖民地的那些地方的所谓"访问"，其实是殖民者耀武扬威的侵略，搭载着达尔文的"小猎犬号"军舰也是同样行径；班克斯和当地女性的纵欲狂欢，当然是殖民者对土著妇女令人发指的蹂躏；即使是他采集当地植物标本的"科学考察"，也可以视为殖民者"窃取当地经济情报"的罪恶行为。

后来改革开放，上面那种意识形态话语被抛弃了，但似乎又走向了另一个极端，完全忘记或有意回避殖民者和帝国主义这个层面，只歌颂这些军舰上的科学家的伟大发现和成就，例如达尔文随着"小猎犬号"的航行，早已成为一曲祥和优美的科学颂歌。

其实达尔文也未能免俗，他在远航中也乐意与土著女性打打交道，当然他没有像班克斯那样滥情纵欲。在达尔文为"小猎犬号"远航写的《环球游记》中，我们读到："回程途中我们遇到一群黑人姑娘在聚会，……我们笑着看了很久，还给了她们一些钱，这着实令她们欣喜一番，拿着钱尖声大笑起来，很远还能听到那愉悦的笑声。"

有趣的是，在班克斯在塔希提岛纵欲六十多年后，达尔文随着"小猎犬号"也来到了塔希提岛，岛上的土著女性同样引起了达尔文的注意，在《环球游记》中他写道："我对这里妇女的外貌感到有些失望，然而她们却很爱美，把一朵白花或者红花戴在脑后的髮髻上……"接着他以居高临下的笔调描述了当地女性的几种发饰。

用今天的眼光来看，这些在别的民族土地上采集植物动物标本、测量地质水文数据等等的"科学考察"行为，有没有合法性问题？有没有侵犯主权的问题？这些行为得到当地人的同意了吗？当地人知道这些行为的性质和意义吗？他们有知情权吗？……这些问题，在今天的国际交往中，确实都是存在的。

也许有人会为这些帝国科学家辩解说：那时当地土著尚在未开化或半开化状态中，他们哪有"国家主权"的意识啊？他们也没有制止帝国科学家的考察活动啊？但是，这样的辩解是无法成立的。

姑不论当地土著当时究竟有没有试图制止帝国科学家的"科学考察"行为，现在早已不得而知，只要殖民者没有记录下来，我们通常就无法知道。况且殖民者有军舰有枪炮，土著就是想制止也无能为力。正如法拉所描述的："在几个塔希提人被杀之后，一套行之有效的易货贸易体制建立了起来。"

即使土著因为无知而没有制止帝国科学家的"科学考察"行为，这事也很像一个成年人闯进别人的家，难道因为那家只有不懂事的小孩子，闯入者就可以随便打探那家的隐私、拿走那家的东西、甚至将那家的房屋土地据为己有吗？事实上，很多情况下殖民者就是这样干的。所以，所谓的"帝国科学"，其实是有着原罪的。

如果沿用上述比喻，现在的局面是，家家户户都不会只有不懂事的孩子了，所以任何外来者要想进行"科学探索"，他也得和这家主人达成共识，得到这家主人的允许才能够进行。即使这种共识的达成依赖于利益的交换，至少也不能单方面强加于人。

博物学在今日中国

博物学在今日中国之复兴，北京大学刘华杰教授提倡之功殊不可没。自刘教授大力提倡之后，各界人士纷纷跟进，仿佛昔日蔡锷在云南起兵反袁之"滇黔首义，薄海同钦，一檄遥传，景从恐后"光景，这当然是和博物学本身特点密切相关的。

无论在西方还是在中国，无论在过去还是在当下，为何博物学在它繁荣时尚的阶段，就会应者云集？深究起来，恐怕和博物学本身的特点有关。博物学没有复杂的理论结构，它的专业训练也相对容易，至少没有天文学、物理学那样的数理"门槛"，所以和一些数理学科相比，博物学可以有更多的自学成才者。这次编印的《西方博物学大系》，卷帙浩繁，蔚为大观，同样说明了这一点。

最后，还有一点明显的差别必须在此处强调指出：用刘华杰教授喜欢的术语来说，《西方博物学大系》所收入的百余种著作，绝大部分属于"一阶"性质的工作，即直接对博物学作出了贡献的著作。事实上，这也是它们被收入《西方博物学大系》的主要理由之一。而在中国国内目前已经相当热的博物学时尚潮流中，绝大部分已经出版的书籍，不是属于"二阶"性质（比如介绍西方的博物学成就），就是文学性的吟风咏月野草闲花。

要寻找中国当代学者在博物学方面的"一阶"著作，如果有之，以笔者之孤陋寡闻，唯有刘华杰教授的《檀岛花事——夏威夷植物日记》三卷，可以当之。这是刘教授在夏威夷群岛实地考察当地植物的成果，不仅属于直接对博物学作出贡献之作，而且至少在形式上将昔日"帝国科学"的逻辑反其道而用之，岂不快哉！

2018年6月5日
于上海交通大学
科学史与科学文化研究院

《不列颠鱼类志》是英国学者威廉·亚雷尔（William Yarrell，1784-1856）的一部博物学著作。亚雷尔是 19 世纪初叶英国颇有名气的全才式学者，出生在伦敦威斯敏斯特的公爵街，是父母所生十二个孩子中唯一生存到成年的。尽管家里拥有一家书店，他幼时即在书香熏陶中长大，却因 10 岁丧父而不得不紧缩开支，失去了进入大学学习的机会。18 岁时他成为银行小职员，20 岁时和堂兄爱德华·琼斯一起进入叔父的报社。但亚雷尔并不习惯坐办公室，经常跑到乡间钓鱼射鸟，当然，他的渔猎爱好并非游手好闲，而是观察、研究自然生态的副产品。因此，他迅速自学成才，成为一名优秀的博物学家。1817 年，亚雷尔进入皇家科学研究所。在不惑之年，他通过《动物学期刊》出版了自己的第一部专著《英国珍稀鸟类谈》，而后也成为该杂志的编辑之一。此书刊行后，他就被选入了林奈学会。之后，他专注于撰写有关鸟类呼吸系统和雄鸟求偶时期羽毛变化模式的论文，也慷慨地为包括威廉·贾丁爵士、约翰·塞尔比和尼古拉斯·威格斯等博物学大家们提供鸟类和鱼类标本。1833 年，亚雷尔在伦敦参与设立了皇家昆虫学会。他也是伦敦动物学会的初创成员之一。除操持诸多学会事务外，亚雷尔也笔耕不辍，于 1836 年出版了重要著作《不列颠鱼类志》，在这部两卷本共 1100 余页的书中，他以三百多幅精致插画和详细的调查笔录，剖析了英国土生鱼类百态，并与前人的记载一一对照，纠错立新。本书即据原版影印。

BRITISH FISHES.

LONDON :
PRINTED BY SAMUEL BENTLEY,
Dorset Street, Fleet Street.

A

HISTORY

OF

BRITISH FISHES.

BY

WILLIAM YARRELL, V.P.Z.S. F.L.S.

ILLUSTRATED BY NEARLY 400 WOODCUTS.

IN TWO VOLUMES.

VOL. I.

ALL WORSHIP BE TO GOD ONLY

LONDON:

JOHN VAN VOORST, 3, PATERNOSTER ROW.

M.DCCC.XXXVI.

PREFACE.

THE geographical situation of the British Islands renders a knowledge of the productions of the numerous and valuable fisheries by which they are surrounded a subject worthy of inquiry to every one interested in the welfare of his country.

The large and constant supply of excellent food obtained from the seas all round the coast by moderate labour and expense, and the employment afforded to a numerous and valuable class of men, who become not only good seamen, but able pilots, since the successful exercise of their occupation depends on an intimate knowledge of the nature of the ground surface, the situation of banks and channels, with the particular direction and force of tides and currents, render the British fisheries also, in many points of view, a branch of political economy of great national importance.

It has long, however, been matter of general regret that the subjects of this particular branch of natural history, so valuable as articles of food and commerce, and so interesting from their organization, and the peculiarities and beauty of their varied forms and colours, should, with the exception of those inhabiting the fresh water, and the marine species most in request for the table, be almost wholly unknown.

Bewick's work, illustrated with engravings on wood of unrivalled excellence, may justly claim the distinguished merit of having done more towards rendering Ornithology popular in this country than any other book that could be named ; and it was hoped that this eminent artist would have devoted his great talents to the delineation of the subjects of

other classes in natural history, as well as to the Quadrupeds
and Birds. It is certain that he had once contemplated producing a work on BRITISH FISHES: but he had abandoned
the design before his lamented death in 1828. Some few
engravings of British Fishes were executed by him with his
usual success, impressions of which were occasionally to be
seen in the possession of his most intimate friends.

The time that has now elapsed since his death leaves but
little room to hope that this branch was ever so far prose-
cuted by him as to be made available in the present day, or
that an object so desirable as a work on BRITISH FISHES
executed to any extent by him can now be expected.
Could it have been ascertained that any such intention still
existed at Newcastle, the present work would never have
been attempted.

An extensive collection of BRITISH FISHES in the possession of the author, containing upwards of one hundred and
sixty species, to increase which no opportunity has been neglected during several past years, and the cordial assistance of
some of the best naturalists from Scotland to the Land's End,
are among the advantages the author of this work has enjoyed
to assist him in his undertaking.

To Mr. Couch of Polperro, the indefatigable ichthyologist
of Cornwall, the author is indebted for several examples of
the most rare species found on the Cornish coast, for the use
of a large and valuable collection of characteristic coloured
drawings, and the whole of his manuscript notes.

W. J. Broderip, Esq. Vice-President of the Zoological
and Geological Societies, having in his possession an inter-
leaved copy of Mr. Donovan's Natural History of British
Fishes, which formerly belonged to the late Colonel Montagu, the author of the Ornithological Dictionary and Tes-
tacea Britannica, containing voluminous notes in his own

writing, of observations on fishes and fishing, made during his long residence near the coast in Devonshire, has with the greatest kindness most liberally allowed the whole of these notes to be transcribed.

The author is also indebted to W. Walcott, Esq. of Bristol, for the use of a valuable manuscript, with a collection of more than one hundred drawings of British Fishes, executed by his father, the author of the Synopsis of British Birds, and other works on natural history, during his residence at Teignmouth.

To Sir William Jardine, Bart. the author is under obligations for many examples, and various communications on the species of the genus *Salmo*, from which materials were drawn for the elucidation of this difficult but important genus.

Dr. George Johnston of Berwick-upon-Tweed, whose name will be found to occur very frequently in this work, has very kindly transmitted rare specimens from that locality, with various notices of the natural history of some of the most remarkable species of the eastern coast.

The author is also desirous of recording his thanks to Dr. Edward Moore of Plymouth, for frequent communications on the fishes taken in that neighbourhood, and an extensive catalogue of local names.

To W. Thompson, Esq. of Belfast, Vice-President of the Natural History Society of that town, the author's thanks are due for many valuable notes of the fishes of the Irish lakes, and communications of the occurrence of many marine species at various localities on the Irish coast.

The author's acknowledgments are especially due to his friend E. T. Bennett, Esq. Secretary of the Zoological Society, for his valuable advice and assistance throughout the progress of the work.

But, without entering into a farther enumeration of the names of other liberal friends who have assisted, it may be sufficient to state, so great has been the success in obtaining species either entirely new, or new to our coast, and so extensive the resources available in the present instance, that this work contains a greater number of species by one-fourth than has yet appeared in any British catalogue, with an extensive list of well-authenticated localities and local names. Two hundred and twenty-six species are described and figured, several of them in different stages of growth. The number of representations of fishes amounts to two hundred and forty. The drawings in almost every instance have been made under the author's superintendence from the specimens. The best artists, both as draughtsmen and engravers on wood, have been employed, and the representations will be found characteristic of the species, and highly creditable as works of art. Besides the figures of the fishes, there are upwards of one hundred and forty illustrative vignettes subservient to the general subject, representing teeth, scales, gill-covers, swimming-bladders, and other viscera, occasionally, when interesting in structure, form, or function. The different boats, nets, and apparatus in use at our various fishing stations on the coast, are figured, and the modes of employing them described.

The systematic arrangement of Baron Cuvier, as detailed in the last edition of the *Règne Animal,* has been adhered to ; and the author hopes that the care bestowed on every part of the work will merit the approbation of all lovers of natural history.

Ryder-street, St. James's,
    June 1836.

INTRODUCTION.

THE external characters of fishes in general are too well known to require particular description. The form of the body, however, is subject to great variety. In some, it is short and rounded, almost spherical, as in the Globe Tetrodon, vol. ii. p. 347; in others, elongated, as in the Eel; it is remarkably compressed in the Dory and Opah, and depressed in the Flatfishes and Skate. The most common form is that of a cylinder, more or less pointed at each end, and slightly compressed at the sides: the Mackerel, at page 121, as a familiar instance, may perhaps be named as exhibiting the highest degree of elegance in shape, and, when very recently taken from the water, is so rich and so varied in its colour, as to be fairly entitled to be considered one of the most beautiful among British Fishes.

The surface of the body is in most instances covered by numerous scales, which vary considerably in size and substance in different species. The great importance of these productions of the skin, as the organs of protection and relation between the animal and the medium in which it resides, has been more particularly enforced by M. Agassiz in his most valuable researches on Fossil Fishes. The arrangement of the scales exhibits considerable uniformity: the almost vertical lines in which each series is placed, crossing each other at an acute angle, resembles the letter X, the scale on the lateral line forming the apex of both cones. This arrangement may be observed in the species of Carp at pages 305, 311, 314, and 315, although the illustrations in this work are from necessity of small size. Each scale is attached to the skin of the fish by its anterior edge; and the manner in which the scales overlap each other in different genera is variable, and gives an appearance of form to each scale which in reality it does not possess. By maceration in water, scales exhibit a series of laminæ, the smallest in size having been the first produced : they resemble a cone, the apex of which is outwards, the smallest being in the centre ; hence the appearance of numerous concentric lines, all of the same shape, which mark the growth. Scales from the same fish differ in size, depending on the part of the body from which they are taken : those above the lateral line are smaller than those immediately below it, but the scales near the vent are the smallest.

The row of scales along the side, forming the lateral line, in addition to the structure common to the scales of the other parts of the body, are pierced through near the centre by a tube which allows the escape of the mucous secretion produced by the glands beneath. Each of the various scales represented at pages 5, 33, 339, and 357, exhibit this tube, with the numerous concentric, and some radiating lines, by which the scales of particular species are distinguished.

The fins are important not only as organs of motion, but as affording by their structure, position, and number, materials for distinguishing orders, families, and genera. The membranes of the fins are thin, and more or less transparent, supported by slender elongated processes of bone, some of which consist of a single piece, which is pointed at the end : such fin-rays are called spinous rays. Others are formed of numerous portions of bone united by articulations, and frequently divided at the end into several filaments : these from their pliant nature are called soft or flexible rays, and two leading divisions in systematic arrangement are founded on this difference in structure. The number of fin-rays in each fin of different examples of the same species of fish is not always exactly alike.

The names given to the different fins are derived from the part of the body to which they are attached. The position of one pair, the ventral fins, attached as they are to the anterior and sometimes to the middle of the belly, affords a valuable character for distinction. These ventral fins are considered analogous to the hinder limbs in other animals; the pectoral fins, to those attached to the shoulder; and many points of resemblance exist in the structure. The principal organ of motion in fishes is the tail, assisted by the simultaneous action of the pectoral and ventral fins. The older writers on Ichthyology considered that the perpendicular position maintained by the fish was owing to the presence of the dorsal fin; but an experiment detailed in this volume, at page 230, appears to indicate that the power of sustaining a particular position in the water is due, in part at least, to other causes.

The economy of Nature is conspicuous in the habits of fishes. Some always swim at or near the surface, others about mid-water, and many close to the bottom: all parts of the water are alike occupied, and some peculiar qualities and powers being found to belong to fishes affecting by choice these different stations in the water, they will be occasionally referred to by the terms of surface, mid-water, and ground swimmers. To include the extremes to which their powers of motion are occasionally applicable, some, as the Flying-fish, page 398, are able to sustain themselves for a short time in the air, from the momentum obtained by their previous exertions before quitting the water; and others, by the strength of the serrated bony ray in each pectoral fin, are able to transport themselves overland from one pool in search of another.

Other important external characters are derived from the operculum or gill-cover : a knowledge of its various parts, and the names by which they are designated, may be learned by a reference to the figure ·in vol. ii. page 3 ; where figure 1 marks the posterior edge of the preoperculum ; 2, the operculum ; 3, the suboperculum ; 4, the interoperculum ; 5, the branchiostegous rays. When a line or division occurs anterior to the preoperculum, it marks the boundary of the cheek, as in the head represented in volume i. at page 8, in which the different portions are not referred to by figures.

The use of the operculum is to close the aperture behind the gills. The blood in fishes, while passing through the gills or branchiæ, receives the influence of oxygen from water which enters by the mouth and goes out by this aperture. In the fishes included in the first three orders, the gills are so formed, and so freely suspended, that the water bathes in its passage every part of their surface.

In the Sturgeon, while swimming, respiration is carried on in the same manner : but when the Sturgeon adheres to any substance by the mouth—which it has the power of doing by extending its lips—some other mode of respiration is required ; and it is found that by the act of extending the mouth the gill-covers are drawn up so as to leave a large channel between them and the gills, through which the water is brought into the mouth, and returned through the gills.

In the Sharks and Rays, the temporal orifices probably assist in the act of respiration by allowing entrance and egress to water while the mouth is closed : they also enable the fish to expel the water taken into the mouth with the prey previous to deglutition.

In the Lampreys and Myxine, the branchial cells which

admit water are lined by the delicate membrane through which the blood is aërated. In the Lampreys, the external apertures of the branchial cells are placed on the side of the neck; but in the Myxine, which feeds upon the internal parts of its prey, and buries its head and a part of its body in the flesh, the openings of the respiratory organs are removed sufficiently far from the head to admit of respiration going on while the animal's head is so inserted.

The branchiæ or gills in fishes possess complex powers, and are capable of receiving the influence of oxygen not only from that portion of atmospheric air which is mixed with the water, but also directly from the atmosphere itself. When fishes confined in a limited quantity of water are prevented by any mechanical contrivance from taking in atmospheric air at the surface, they die much sooner than others that are permitted to do so. The consumption of oxygen, however, is small; and the temperature of the body of fishes that swim near the bottom, and are known to possess but a low degree of respiration, is seldom more than two or three degrees higher than the temperature of the water at its surface. Dr. John Davy, however, in a paper read before the Royal Society of London in 1835, on the temperature of some fishes allied to the Mackerel, all of which are surface-swimmers with a high degree of respiration, observed that the Bonito had a temperature of 90 degrees of Fahr. when the surrounding medium was 80° 5′; and that it therefore constituted an exception to the generally received rule, that fishes are universally cold-blooded. Physiologists have shown that the quantity of respiration is inversely as the degree of muscular irritability. It may be considered as a law, that those fish which swim near the surface of the water have a high standard of respiration, a low degree of muscular irritability, great necessity for oxygen, die soon—almost immediately when taken out of water, and have flesh prone to rapid decomposition : Mackerel, Salmon, Trout, and Herrings are examples. On the contrary, those fish that live near the bottom of the water have a low standard of respiration, a high degree of muscular irritability, and less necessity for oxygen ; they sustain life long after they are taken out of the water, and their flesh remains good for several days : Carp, Tench, Eels, the different sorts of Skate, and all the Flatfish, may be quoted. But as this subject is occasionally referred to in the body of this work when describing the powers of particular species, farther details here will be unnecessary.

With tenacity of life is connected the extraordinary power observed in some fishes of sustaining extremes of high and low temperature. The Goldfish not only lives, but thrives and breeds to excess, in water the temperature of which is constantly kept as high as 80° Fahr. Fishes exist in the hot springs and baths of various countries the temperatures of which are found to range between 113 and 120 degrees of Fahr.; and Humboldt and Bonpland, when travelling in South America, perceived fishes thrown up alive, and apparently in health, from the bottom of a volcano, in the course of its explosions, along with water and heated vapour, that raised the thermometer to 210 degrees, being but 2 degrees below the boiling point.

On the other hand, in the Northern parts of Europe, Perch and Eels are advantageously transported from place to place while in a frozen state, without destroying life. Mr. Jesse, in the second series of his Gleanings in Natural History, page 277, says, a friend of his, who resided near London, had a single Goldfish with the water in a marble basin frozen into one solid body of ice. He broke the ice around it, took it out, and found it to all appearance lifeless, and looking perfectly crystallized. This was about noon. Leaving the fish with the ice in the basin, and a fire having been lighted, he after dinner, more from accident than any other cause, looked at the basin, and to his astonishment saw the ice in a great measure thawed, and the fish moving. At midnight, when he went to bed, it was as lively as usual. Dr. Richardson, in the third volume of his *Fauna Boreali-Americana*, devoted to Fishes, says of the Grey Sucking Carp, a common species in the fur-countries of North America, that, like its congeners, it is singularly tenacious of life, and may be frozen and thawed again without being killed. Other instances of both extremes are detailed in this volume, page 317.

The eyes in fishes are observed to occupy very different positions in different species. In some they are placed high up near the top of the head, but more frequently on the flattened side of the head, but always so situated as best to suit the exigencies of the particular fish. The external surface of the eye itself is but slightly rounded, but the lens is spherical—a structure that in a dense medium affords intense power of vision at short or moderate distances, rather than a long sight. When water is clear, smooth, and undisturbed, the sight of fishes is very acute : this is well known to anglers, who prefer a breeze that ruffles the surface, well knowing that they can then approach much nearer the objects of their pursuit, and carry on their various deceptions with a much better chance of success.

The sense of hearing has by some been denied to fishes—perhaps because they exhibit no external sign of ears : the internal structure, however, may be most successfully demonstrated in the various species of Skate, in which the firmer parts of the head being formed of soft and yielding cartilage, the necessary divisions may be effected with great ease.

The Chinese, who breed large quantities of the well-known Goldfish, call them with a whistle to receive their food. Sir Joseph Banks used to collect his fish by sounding a bell; and Carew, the historian of Cornwall, brought his Grey Mullet together to be fed by making a noise with two sticks.

From the rigid nature of the scaly covering in the generality of fishes, it is probable they possess but little external sense of touch; but they are not wholly unprovided with organs which in the selection of their food are of essential service. The lips in many species are soft and pulpy; the mouths of others are provided with barbules or cirri, largely supplied with nerves, which are doubtless to them delicate organs of touch, by which they obtain cognizance of the qualities of those substances with which they come in contact. The Gurnards may be said to be provided with elongated, flexible, and delicate fingers, to compensate for their bony lips. It is a rule, almost without an exception that I am aware of, that those fishes provided with barbules or cirri about the mouth obtain their food near the ground; and these feelers, as they are popularly called, appear also to be a valuable compensation to those species which, restricted by instinctive habits to feeding near the bottom of water that is often both turbid and deep, must experience more or less imperfect vision there from the deficiency of light.

The olfactory nerves in fishes are of very large size, and the extent of surface over which the filaments are disposed is very considerable. The nostrils are generally double on each side, but both openings lead to one common canal. Their sense of smell may be presumed to be acute from the selection they are known to make in their search after food; and the advantage said to be gained by the use of various scented oils with which some anglers impregnate their baits. A Pike in clear water has been seen to approach and afterwards turn

away from a stale Gudgeon, when at the distance of a foot
from his nose, as if perfectly aware at that distance of the
real condition of the intended prey.   Mr. Couch has observed in a Fifteen-spined Stickleback of large size, kept in
a glass vessel, that the opening and closing of the nostrils
was simultaneous with the action of the gill-covers, and he
felt convinced from his observations, that the fluid was received and rejected for the purpose of sensation.   Among
the ground-feeders in fishes, the various species of Skate are
remarkable for the extent of the surface over which the olfactory nerves are disposed, produced by numerous laminæ
radiating from a centre, which in appearance may be compared in form to the under surface of a mushroom, of which
the trunk of the nerve is the stem.   In the absence of cirri
or feelers in the various Skate, very considerable branches of
the fifth pair, the nerve of touch, are distributed over the
angular snout with which these fish turn over the sand in
search of proper food.   It will be recollected that the mouth
in this family of fishes is on the under surface.   They are
probably among the lowest of the ground-feeders.

Whether fishes possess any high degree of taste is a subject not easily proved.   Obliged unceasingly to open and
close the jaws for the purpose of respiration, they cannot long
retain food in the mouth when quite shut ; the substance if
of small size must be swallowed quickly, and without being
much altered by anything like mastication.   From the cartilaginous hardness of the tongue in many species, more or
less covered with recurved teeth, which assist in conveying
food to the back part of the mouth, the sense of taste may
pervade the surface of the soft and fleshy portions of the
pharynx.

The teeth in fishes are so constant as well as permanent in
their characters, as to be worthy particular attention.   In the

VOL. I.   b opinion of the best Ichthyologists, they are second only to the fins ; which, in their number, situation, size, and form, are admitted to be of first-rate importance. Some fishes have teeth attached to all the bones that assist in forming the cavity of the mouth and pharynx ; to the intermaxillary, maxillary, and palatine bones, the vomer, the tongue, the branchial arches supporting the gills, and the pharyngeal bones. Sometimes the teeth are uniform in shape on the various bones ; at others differing. One or more of these bones are sometimes without teeth of any sort ; and there are fishes that have no teeth whatever on any of them. The teeth are named according to the bone upon which they are placed ; and are referred to as intermaxillary, maxillary, palatine, vomerine, &c.—depending upon their position.

A reference to page 3 in the second volume will show the situation of the teeth in the Trout, with five rows on the upper surface of the mouth, and four rows below ; the particular bones upon which these rows are placed are also referred to. The form of the teeth in fishes is various ; in general it represents that of an elongated cone, slightly curved inwards to assist in holding a prey which is frequently alive. Sometimes the form is that of a short and rounded tubercle, adapted for crushing ; in some fishes the teeth are so small and numerous as to have the appearance of the hairs of a brush ; while in others they are thin and flat, with a cutting edge like the incisor teeth in the human subject. Some fishes that are without teeth in the mouth, have them in the throat ; this is particularly the case in the Carp, and the allied species in the family of the *Cyprinidæ* generally. In this family the pharynx is provided with five pair of branchial arches, the four most anterior of which support the four rows of gills ; the fifth pair, remarkable for the strength of the bone, support powerful teeth. The woodcut here introduced

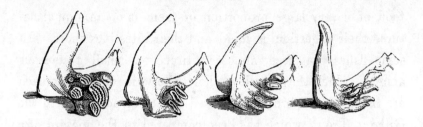

represents one half of this pharyngeal arch of bone, that of
the left side, looking from behind, with the teeth in the
Carp, Tench, the Roach, and the Barbel. In the Carp,
the first on the left hand, the crowns of the teeth are observed to be so worn down as to have the appearance of the
crowns of the molar teeth in the Hare. In the Tench, the
second figure from the left, the structure is less complicated.
In the Roach, the form and number again varies ; and in
the Barbel, on the extreme right, the teeth are crooked,
pointed, and disposed in three regular rows : other fishes
belonging to this family have the teeth in four rows, and
some in six rows. Considerable difference of form exists in
the four examples of teeth here shown ; and a reference to
the illustrations of other teeth at pages 98, 103, 106, 113,
116, and 248, will show the great variety of teeth that are
to be found among fishes, two or more distinct forms of
which are sometimes possessed by the same individual.
Some further particulars in reference to the teeth will be
found in other parts of this work.

　　Closely connected with this part of the subject is their
food, and the organs of digestion.

　　The food of fishes is very different at different periods of
the year, and this may be one of the causes, among others, of
the peculiar excellence of the flesh of some species of fishes
at particular seasons.

b 2

The parietes of the stomach in fishes are thin ; and as the food of a very large proportion of them is of an animal nature, their digestion is rapid and their intestines short. In some fishes that feed almost entirely on small testaceous animals, which they swallow whole, the walls of the stomach are thickened, affording increased muscular power, as in the Gillaroo Trout, which has been compared to the gizzard of a bird. The most gizzard-like stomach among fishes that I am acquainted with is that of the Grey Mullet. As in the higher classes of animals, those fishes which feed on vegetable substances have a long intestinal canal, with many convolutions : the most indurated vegetable matter requiring the greatest powers of assimilation. The longest intestines in the class of Birds will be found among the Grouse tribe, which feed principally on the more tender parts of heath ; and in Mammalia among the *Rodentia*, and in the Camel, the Dromedary, the Giraffe, and others that are known to browse.

Of the swimming-bladder, an important organ lodged in the abdomen of some fishes, a detailed account of the structure, its contents and use, will be found in this volume at page 36 ; and various representations of the form as it exists in different species are given at pages 36, 37, 43, and 94.

Some observations communicated to me by Mr. Couch in reference to the air-bladder and the partial inflation of some fishes deserve notice here.

Mr. Couch reports that some of the *Gadidæ* while under terror become distended with air, at least in the fins ; the Bib also in its eyes ; " and I have often seen," says that gentleman, " small fishes of this family caught and turned free again, when they have been unable to descend through the water, notwithstanding their utmost efforts, which have not been deficient in vigour. When in the early part of last summer I was preparing a bottle of fishes for your use, I pierced a Lesser Forked Beard with a pointed probe through the mouth into the air-bladder in order to render the fish small enough to enter the bottle ; but being obliged also to squeeze it with some force for that purpose, the dorsal fin became distended with air—a circumstance that would direct our attention to the air-bladder as the source of the air distending the fins and tunic of the eyes in the *Gadidæ*."

The analogy to the air-cells in birds, and the passage of air from thence into the bones of the limbs, is too obvious to be unobserved, and will give interest to further investigation.

Except in the cartilaginous Sharks and Rays, there are no very obvious external signs by which the sexes in fishes can be distinguished. As in the higher animals, however, the respiratory organs occupy more space in the males than in the females ; and on the other hand, the abdomen is larger in the females than in the males : the males may therefore be known from the females by their somewhat sharper or more pointed head, the greater length of the gill-cover, and the body from the dorsal fin downwards being not so deep compared with the whole length of the fish.

Among fishes generally a few are viviparous, bringing forth their young alive, which are able from the time of exclusion to shift for themselves. Of these some notice is taken in the body of the work when describing the particular species. The sexual parts are of a higher degree of organization in the Sharks and Rays, and more complicated in their structure than those of the bony fishes, resembling the sexual organs in reptiles ; and their mode of producing their young is described also at the commencement of the history of each, and need not therefore be repeated here.

The sexual organs in by far the greater number of fishes

are much more simple, consisting, as will be found towards the season of producing their young, of two elongated oval lobes of roe, one on each side of the body, placed between the ribs and the intestinal canal ; these lobes, in the female called hard roe, contain a very large number of roundish grains called ova or eggs, which are enclosed in a delicate membranous tunic or bag, reaching to the side of the anal aperture, where an elongated fissure permits egress at the proper time. In the males, the lobes of roe are smaller than in the females, and have the appearance of two elongated masses of fat, which are called soft roe ; they remain, however, firm till the actual season of spawning, when they become by degrees more and more fluid, and the whole is ultimately voided by small portions at a time under slight abdominal pressure.

A few exceptions to this rule appear to exist ; but which may perhaps rather be considered malformation than natural structure. According to Cavolini and Cuvier, some species of the genus *Serranus* have each lobe of roe made up of a portion of hard and of soft roe, and these fishes have been considered as hermaphrodites, each fish capable of producing fertile ova without the assistance of a second fish. Among other accidental malformations may be included the appearance of a hard or female roe on one side, and a soft or male roe on the other side of the same fish. This has been observed occasionally in the Perch, Mackerel, Carp, Cod, Whiting, and Sole ; and the probability is, that in these cases the fishes are prolific alone, since the two lobes of roe are observed to be of equal growth, advancing to maturity together. Pallas believed that in the genus *Syngnathus* there were no males ; but the singular anomaly of both sexes being found to carry ova, the females in the abdomen, and the males for a time in their caudal pouch, is now understood. The supposed hermaphroditism of the Lampreys has been disproved by various modern observers.

At the season for depositing the spawn, which varies with almost every genus, some species repair to the gravelly shallows of rivers, and others to the sandy bays of the sea. This movement is called by some fishermen, " going to hill, or roading ;" other species resort to bunches of weeds. In many instances, when ready to deposit her spawn, a female is accompanied by two males, one on each side,—a provision of nature which seems intended to secure the impregnation of the largest quantity of ova, and the range of the influence of the male fluid is enormously increased by diffusion in water. The adhesive nature of the surface of each egg supplies the means of attachment to any of the various substances near which it may happen to be left ; and the time required for the appearance of the young fish is very variable, depending upon the species, the season, and its temperature. The young fish is first apparent as a line wound round the central vitelline portion of the egg, and ultimately escapes by rupturing the external capsule with its tail.

Considerable attachment is often exhibited between the parent fish. Mr. Jesse relates that he once caught a female Pike during the spawning season, and nothing could drive the male away from the spot at which the female disappeared, whom he had followed to the very edge of the water. In some species this attachment is not confined to the season of spawning. A person who had kept two small fishes together in a glass vessel, gave one of them away ; the other refused to eat, and showed evident symptoms of unhappiness till his companion was restored to him. Some few species show also an attachment to their young, and even watch and defend their own spawn. I shall confine myself to a notice of two

British examples. Pennant says of the River Bullhead, " It deposits its spawn in a hole it forms in the gravel, and quits it with great reluctance ;" I have also been favoured by an excellent observer with the following notice on the same fish :—" It evinces a sort of parental affection for its ova, as a bird for its nest, returning quickly to the spot, and being unwilling to quit it when disturbed." It is believed also of the Lump Sucker, that the male fish keeps watch over the deposited ova, and guards it from every foe with the utmost courage. If driven from the spot by man, he does not go far, but is continually looking back, and in a short time returns.

A few observations on the impregnated roe may be worthy attention. Dr. Walker of Edinburgh, in an essay on the Natural History of the Salmon, published in the Transactions of the Highland Society, quoting the experiments of Jacobi of Berlin, says, he found that when the spawn of both sexes were extracted from dead fishes, the ova by mixture can be fecundated by the milt ; and when placed under water in a proper situation can be brought forth into life. He further discovered that this artificial fecundation can be accomplished with the roe and milt of fishes which have been dead two and even three days. This appears to point out the mode of obtaining the fishes of neighbouring countries by the transportation as far as possible of the living gravid fishes, afterwards for a time while dead, and finally by the mixture and further transportation of the mixed roes.

But there appear to be other, and still greater facilities. Mr. Jesse states, that he has been assured by persons who have lived many years in the East Indies, that ponds which become perfectly dry, and the mud hard, have after the rainy season been found with fish in them, although no stream communicated with them, or any passage or other means by which fish could be admitted. This curious fact has been confirmed to me by Colonel Sykes and other observers who have lived long in India, who state that the tanks and ditches near fortifications are alternately filled and empty on the occurrence of every rainy and dry season, but that a few days after the commencement of each rainy season these tanks and ditches are replenished not only with water, but also with small fish. The solution appears to me to be this.—The impregnated ova of the fish of one rainy season are left unhatched in the mud through the dry season, and from their low state of organization as ova, the vitality is preserved till the occurrence and contact of the rain and the oxygen of the next wet season, when vivification takes place from their joint influence. If this solution of the problem be the true one, it points at once to what perhaps may be effected after a few experiments,—namely, the artificial fecundation of the roe, the drying of that roe, (or of other roe naturally impregnated,) sufficiently to prevent decomposition, and its possible transportation to, and vivification in, distant countries.

The growth of young fish is rapid in proportion to the size of the parent fish, or the ultimate size attained by the species. They appear to be liable to occasional malformation, and two instances are figured, vol. i. page 110, and vol. ii. page 59, and a third of the same kind has been seen, where the upper jaw is deficient in the requisite length. Hervey is said to have been the first who observed that most irregularities in human structure were to be found in the lower animals, and modern physiologists have shown that various gradations of structure permanent in the lower animals are successively assumed by those of higher organization in their passage towards their ultimate developement. These usually transitory conditions sometimes become permanent, and constitute monstrosities. The most frequent

malformation in the human subject is that which is usually termed the hare-lip ; the divided lip, and imperfectly
closed palate, representing the state of these parts in some
species of mammalia of a lower grade of organization than
man.   In the case of the malformations in fishes here alluded
to, the deficiency appears to have arisen from an arrest of the
formative process at that point in which the shortened state
of the upper jaw resembles the rounded upper part of the
mouth in the Lampreys, a grade in structure preceding that
of the bony fishes.

The unclosed state of the bones of the head in the human
infant, which are not firmly united till some months after
birth, is a permanent condition of the cranium in some reptiles and fishes, as noticed and figured at page 380.

Wounds in fishes heal rapidly ; and they appear to have
but few diseases, probably owing to the uniformity of the
temperature in the medium in which they reside.

As previously stated, the food of a very large proportion
of fishes is of an animal nature, and they feed to a great
extent indiscriminately upon one another.   From their extraordinary voracity, their rapid digestion, and the war of
extermination they carry on among themselves, the greater
and more powerful fishes consuming the smaller and weaker,
from the largest to the most diminutive; add to this, the
constant and extensive destruction effected by the numerous sweeping nets of ruthless man, and it is even probable
that comparatively but few fishes die a natural death.

GENERAL INDEX.

The systematic names are printed in italics.

ERRATA.

Vol. ii. near the top of pages 317, 318, 320, 322, 324, for *Anguillidæ*, read *Murænidæ*.

BRITISH FISHES.

THE PERCH.

Perca fluviatilis, Linnæus. Bloch, pt. ii. pl. 52.
      ,,   ,,   Cuvier et Valenciennes, Hist. Nat. des Poiss. t. ii. p. 20.
            Perch. Pennant, Brit. Zool. edit. 1812, vol. iii. p. 345,
              pl. 59.
         ,,   Donovan, Brit. Fishes, plate 52.
         ,,   Fleming, Brit. Animals, page 213, species 142.

Generic Characters.—Two dorsal fins, distinct, separated ; the rays of the first spinous, those of the second flexible ; tongue smooth ; teeth in both jaws, in front of the vomer, and on the palatine bones ; preoperculum notched below, serrated on the posterior edge ; operculum bony, ending in a flattened point directed backwards ; branchiostegous rays 7 ; scales rough, hard, and not easily detached.

Baron Cuvier has chosen the Perch as representing the type of his first genus Perca, but has separated from that genus, as it was established by Linnæus, several species,

---

* Fishes with some of their fin-rays spinous, the others flexible.
† The family of the Perches.

VOL. I.   B

on account of certain variations which the generic characters and descriptions hereafter appended to such as are
British will sufficiently explain.   The Perch was well known
to the Greeks, and Aristotle has described its habits under
the name of Πέρχη.   It was the *Perca* of the Romans; and
is named *Pergesa* in Italy, *Perscke* in Prussia, *la Perche* in
France, and *Perch* in England.   As a species, it is common
to the whole of the temperate parts of Europe; and in this
country there is scarcely a river or lake of any extent where
this fish does not occur in abundance.   It is found in most
of the lakes of Scotland, in those of the North of England,
where it is sometimes called a Basse, and also in the lakes of
Wales.   In the various historical and statistical accounts of
the counties of Ireland, the Perch may be traced through
the southern, eastern, and northern districts from Cork to
Londonderry, and is probably to be found also in the rivers
and lakes of most, if not all the other counties.   In rivers,
the Perch prefers the sides of the stream rather than the
rapid parts of the current, and feeds indiscriminately upon
insects, worms, and small fishes.   So remarkable is the
Perch for its boldness and voracity, that in a few days
after some specimens had been placed in a vivarium, in
Bushy Park, Mr. Jesse tells us, they came freely and took
worms from his fingers; and the Perch is generally the first
prize of the juvenile angler.   They have been known to
breed in small vases; and Bloch mentions having watched
some while depositing their ova in long strings in a vessel
kept in his room.   A Perch of half a pound weight has been
found to contain 280,000 ova; and the spawning season is
at the end of April, or beginning of May.   Perch live for
some hours out of water, and bear a journey of forty or fifty
miles, if carried steadily, and watered occasionally.   They
are constantly exhibited in the markets of Catholic countries,

and, if not sold, are taken back to the ponds from which they were removed in the morning, to be reproduced another day. The flesh of this fish is firm, white, of good flavour, and easy of digestion.

A Perch of three pounds weight is considered a fish of large size; Perch, however, of four pounds have been taken from the Richmond Park ponds. Mr. Donovan, in his History of British Fishes, records one of five pounds taken in Bala Lake. Mr. Hunt, of the Brades, near Dudley, Stafford-shire, took a Perch of six pounds from the Birmingham Canal. Montagu once saw a Perch of eight pounds taken in the Avon, in Wiltshire, by a runner, or night-line, baited with a roach for a pike: and a Perch of eight pounds was caught in Dagenham Breach. Pennant records his having heard of one that was taken in the Serpentine River, Hyde Park, that weighed nine pounds; and it is stated by Bloch and others, that the head of a Perch is preserved in the church of Luehlah, in Lapland, which measures near twelve inches from the point of the nose to the end of the gill-cover.

The body of the Perch is compressed, and its height about one-third of its whole length. The length of the head is equal to the height of the body, and compared to the length of the body is as two to seven: the jaws are nearly equal, and the opening of the mouth is about one-fourth·of the whole head: the teeth are small, uniform in size, curving backwards, and the inside of the mouth is furnished with a transverse palatine membrane. There are two external openings to each nostril, surrounded by several orifices, which allow the escape of a mucous secretion. These apertures are larger and more nu-merous about the heads of fishes generally than over the other parts, the viscous secretion defending the skin from the action of the water. The distribution of the mucous orifices over the head is one of those beautiful and advantageous provi-

B 2

sions of nature which are so often to be observed and ad-
mired. Whether the fish inhabits the stream or the lake,
the current of the water in the one instance, or progression
through it in the other, carries this defensive secretion backwards, and spreads it over the whole surface of the body. In
fishes with small scales, this defensive secretion is in proportion more abundant ; and in those species which have the
bodies elongated, as the eels, the mucous orifices may be
observed along the whole length of the lateral line.

The formula of the number of fin-rays may be thus
stated :—

$$D. \; 15, \; 1 + 13 : P. \; 14 : V. \; 1 + 5 : A. \; 2 + 8 : C. \; 17.$$

And the mode of fin-ray notation employed is thus explained :—D. the dorsal fin, has, in the first fin, 15 rays, all spinous ; in the second fin, 1 spinous + plus 13 that are soft.
P. pectoral fin, 14 rays, all soft. V. the ventral fin, with
1 spinous ray + plus five that are soft. A. the anal fin, with
2 spinous rays + plus 8 that are soft. C. the tail or caudal
fin, 17 rays. In counting the rays of the caudal fin, those
only from the longest ray of the upper portion to the longest
ray of the lower portion, both inclusive, are enumerated.

The Perch, though very common, is one of the most
beautiful of our fresh-water fishes, and, when in good condition, its colours are brilliant and striking. The upper part
of the body is a rich greenish brown, passing into golden
yellowish white below ; the sides ornamented with from five
to seven dark transverse bands ; the irides golden yellow ;
the first dorsal fin brown, the membrane connecting two or
three of the first and last rays spotted with black ; the second dorsal and pectoral fins pale brown ; ventral, anal, and
caudal fins, bright vermilion. A deformed variety of Perch,
with the back greatly elevated and the tail distorted, has

been noticed by Linnæus as occurring at Fahlun, in Sweden, and in other lakes in the North of Europe. Similar Perch are also found in Llyn Raithlyn, in Merionethshire. A fish of this description is figured in the volume of Daniel's Rural Sports devoted to Fishing and Shooting, page 247. Specimens of the Perch, almost entirely white, have also been found in the waters of particular soils.

Two Continental naturalists have pointed out the necessity of attending to the scales of fishes, as affording the most valuable and constant characters; and these productions of the skin, important also as the organs of protection and relation between the animal and the medium in which it resides, will occasionally be figured and referred to as additional marks of specific distinction in several instances of closely allied species. It has already been observed that the lateral line in fishes marks the situation of an extended series of mucous orifices. The scales placed in a row immediately upon this lateral line mark its particular course along the side; and these scales, besides bearing the characters of those of the other parts, are perforated by a tube through which escapes this mucus, or slime, as it is more commonly called, to be spread over the surface of the body. The vignette below represents a scale from the lateral line of the Perch,[1] the Basse,[2] and the Ruffe.[3]

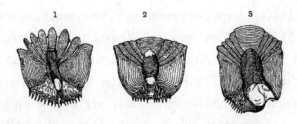

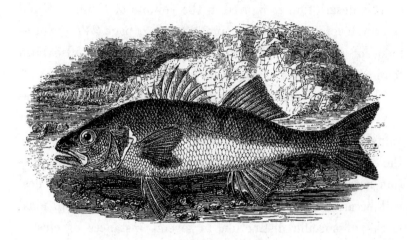

THE BASSE.

Labrax lupus,　Cuv. et Valenc. Hist. Nat. des Poiss. t. ii. p. 56, pl. 11.
Perca labrax,　Linnæus.　Bloch, pt. ix. pl. 301.
　　　　　　Basse, Penn. Brit. Zool. 1812, vol. iii. p. 348, pl. 60.
　　　　　　　,,　Don. Brit. Fish. pl. 43.
　　　　　　　,,　Flem. Brit. An. p. 213, sp. 143.

Generic Characters.—Two dorsal fins, distinct, separated; the rays of the first spinous, those of the second flexible; branchiostegous rays 7; tongue covered with small teeth; teeth on both jaws, on the vomer and palatine bones; cheeks, preoperculum, and operculum, covered with scales; suborbital bone and suboperculum without serrations; preoperculum notched below, serrated on its posterior edge; operculum ending in two points directed backwards.

The Basse, a marine perch, with two dorsal fins, abundant in the Mediterranean, was well known to the Greeks, who called it Λάϐραξ, and esteemed it highly.　Aristotle distinguished it from the fresh-water perch by the scales on the various parts of the gill-cover, the spines of the operculum, and the roughness of the tongue.　It was also well known to the Romans, who called it *Lupus,* on account of its voracity ; and these terms Cuvier has united for its modern distinction. This fish is found along the whole line of the southern coast of England, in the Bristol and St. George's Channel ; and, though less numerous farther north, on our eastern coast has been noticed by Dr. Johnston and Mr. Neill as occurring in Berwick Bay and the Frith of Forth, but is not included in Low's *Fauna Orcadensis*. On the Irish coast the Basse is taken along the line of the eastern shore from Waterford to Belfast Bay. It is stated by Willughby that this fish sometimes attains the weight of fifteen pounds ; but the more ordinary size is from twelve to eighteen inches in length, and the flesh is then excellent food. The Basse swim in shoals along the coast, depositing their spawn in summer, and generally near the mouths of rivers, up which they frequently pass to a considerable distance : they have been retained with success in Mr. Arnold's fresh-water lake in Guernsey, and Dr. M'Culloch has vouched for the superiority of the flavour obtained by the change. Their food consists generally of living prey. Mr. Neill took from the stomach of one, the fry of the sandlaunce and two young specimens of the father-lasher : they feed also on small crustaceous animals ; and Mr. Couch, of Cornwall, states, that " this fish is particularly fond of *onisci,* in pursuit of which it ventures among the rocks in the midst of a tempest, as at that time these insects are frequently washed from their hiding-places." They are captured at sea by various means : by the trawl-net, and by hooks attached either to hand-lines or deep sea-lines. They take a bait freely ; and many are caught by angling, during the flood-tide, with a long rod and strong line, from a projecting pier-head or jutting rock. " We have seen several taken in Bideford Bay," says Col. Montagu, " with a small Seine net, manageable by two men. The men wade a considerable way into the water on this gradually-inclining sandy shore, and when the water reaches above their middle, the net is strained by the men separating, and drawn on shore, each man holding by a cord at the ends."—*Montagu's MS*.

D. 9, 1+12 : P. 16 : V. 1+5 : A. 3+11 : C. 17 : Vertebræ, 25.

The position and form of the fins are shown in the woodcut, and the character of the parts of the head in the additional outline at the bottom of this page. The body of the fish is elongated as compared with that of the perch, and in shape resembles that of the salmon. The teeth uniform in size, short, and sharp ; those on the tongue assist in drawing the food back towards the throat. The nostrils are double ; the mucous pores numerous ; the irides silvery ; the back dusky blue, passing into silvery white on the belly ; the scales of moderate size, adhering firmly ; the fins pale brown.

At Ramsgate, and some other places along the line of the Kentish coast, the Basse is called a *sea-dace*.

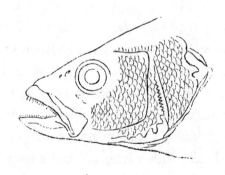

ACANTHOPTERYGII.   *PERCIDÆ.*

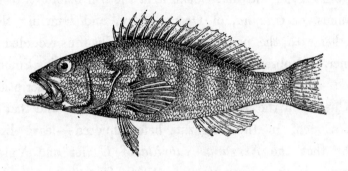

THE SMOOTH SERRANUS.

Serranus cabrilla, Cuv. et Valenc. Hist. des Poiss. t. ii. p. 223, pl. 29.
Perca cabrilla, Linnæus.
  ,, *channus,* Couch, Mag. Nat. Hist. vol. v. p. 19, fig. 6.

Generic Characters.—A single elongated dorsal fin, the rays of the anterior portion spinous, the others flexible; branchiostegous rays 7; small teeth in both jaws, on the palatine bones and the vomer; some elongated teeth among the smaller ones; cheeks and operculum covered with small scales; preoperculum serrated; operculum ending in two or three flattened points projecting backwards.

Cuvier's sub-genus Serranus, the term being derived from the serrated operculum, and the fishes belonging to the division distinguished from those of Perca and Labrax by the single elongated dorsal fin, is new to the History of British Fishes; and we are indebted to Mr. Couch, of Cornwall, for the only specimens known to have been taken on our coast, and which, it is believed, will be found to belong to three distinct species. The first is the Smooth Perch, *Perca channus,* a fish made known by Mr. Couch, as frequently occurring on the coast of Cornwall, in an article in the Magazine of Natural History, conducted by Mr. Loudon; which contained also a notice of a second species of the same genus, and also several other interesting species in other genera, some of which were likewise new.

Both Cuvier and Mr. Couch refer the fish before us to the Channus, or *Channa*, of Gesner, Ray, and Gmelin : this, together with the peculiar habit of the Channus recorded by Gesner, and observed by Mr. Couch to prevail in his Smooth Perch —and the close resemblance between the descriptions by Cuvier, in the *Hist. des Poiss.* t. ii. p. 223, and that by Mr. Couch, in the Magazine before quoted—leave little doubt that the *Serranus cabrilla* of Cuvier and Valenciennes, and the *Perca channus* of Mr. Couch, are in reality the same species. It has therefore been placed among the British fishes, under the name of *Smooth Serranus*, which the distinction of possessing but a single dorsal fin appears to render necessary, and which, it is hoped, Mr. Couch will not disapprove.

This species of Serranus is abundant in the Mediterranean, and passing in the ocean northward to a considerable distance, is, in the opposite direction, taken as far south as Teneriffe and Madeira. Mr. Couch considers it a common fish, well known to the Cornish fishermen ; "that it keeps in the neighbourhood of rocks not far from land ;" and adds, " it is singular that the spasm, which seizes this fish when taken, never passes off: hence it is found, long after death, in a state of rigidity and contortion, with the fins preternaturally erect."

D. 10 + 14 : P. 15 : V. 1 + 5 : A. 3 + 8 : C. 17.

The peculiarities of the teeth and gill-cover are expressed in the generic characters : " the irides are yellow ; the body about ten inches long, compressed, deep. Colour of the back brown, in some specimens having distinct bars running round to the belly ; sides yellow, reddish, or saffron-coloured, more faint below : two irregular whitish lines pass along the side from head to tail ; a third, more imperfect, on the belly. On the gill-plates are several faintish blue stripes, running obliquely downward. The fins are striped longitudinally with red and yellow ; pectorals wholly yellow." The description is from Mr. Couch ; the figure, from the work of Cuvier.

One peculiarity of the Serrani must not be passed over. Cavolini and Cuvier have, after repeated examinations, described the Smooth Serranus, and some other species of this genus, as true hermaphrodites, one portion of each lobe of roe consisting of true ova, the other part having all the appearance of a perfect milt, and both advancing to maturity simultaneously. A structure of a different kind, which must be considered as accidental, has been observed by others in the perch, mackerel, carp, cod, whiting, and sole. This occasional malformation, to speak in a popular phrase, consists of a lobe of hard female roe on one side, and of soft male roe on the other side, of the same fish. Observations are still wanting to prove whether such fishes have the power of impregnating their own ova.

Cavolini believed that the Serrani had this power ; and the probability is that in the other cases the fish are also prolific, since the two sides are observed to be of equal growth.

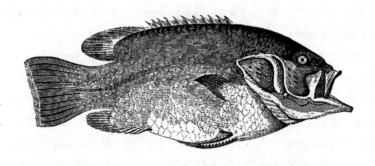

COUCH'S SERRANUS.

Serranus Couchii, Yarrell.
Stone-Basse, Couch, Linn. Trans. vol. xiv. p. 81.

Favoured by Mr. Couch with a drawing of the Stone-
Basse, included in his memoir on the Natural History
of Fishes found on the coast of Cornwall, published in
the fourteenth volume of the Transactions of the Linnean
Society, it appears to be a species of Serranus; and judg-
ing by a comparison with the detailed descriptions of all
the *Serrani* of Cuvier and Valenciennes, in their *Histoire Naturelle des Poissons*, it also appears to be an undescribed species; and the name therefore of its discoverer
has been accordingly appended to it, as a proper tribute
to a gentleman who has for many years devoted his un-
ceasing attention to the natural history of his particular
county.

This species, Mr. Couch states, in the memoir above referred to, " approaches the Cornish coast under peculiar circumstances. When a piece of timber, covered with barnacles, is brought by the currents from the more southern regions, which these fishes inhabit, considerable numbers of them sometimes accompany it. In the alacrity of their exertions, they pass over the wreck in pursuit of each other, and sometimes, for a short space, are left dry on the top, until a succeeding wave bears them off again. From the circumstance of their being usually found near floating wood covered with barnacles, it might be supposed that this shell-fish forms their food ; but this does not appear to be the case, since, in many that were opened, nothing was found but small fishes. Perhaps these young fishes follow the floating wood for the sake of the insects that accompany it, and thus draw the Stone-Basse after them."

It would be unsafe to venture on a statement of the number of fin-rays from a drawing ; but the woodcut at the head of the page is an exact copy, reduced in size, of the original representation. Mr. Couch uses the terms *totus argenteus* in reference to the colour of this fish, and it may therefore be concluded that its prevailing tint is silvery white, the ends of all the fins considerably darker. The attention of naturalists on our southern coast is respectfully invited to a close examination of such species of Serrani as come under their notice, in the hope of obtaining a more perfect knowledge of an interesting species apparently new. A figure of the Stone-Basse of Sloane is added at the foot of the next page, to show by comparison that the Stone-Basse of Mr. Couch is not, as has been supposed, the Stone-Basse of Sloane ; which latter fish is the *Gerres rhombeus* of Cuvier and Valenciennes, *Hist. Nat. des Poiss.* t. vi. p. 459. It is an additional reason in favour of the new name here proposed, that, if sanctioned by naturalists—which, it is hoped, it will be,—the term Stone-Basse will not then refer to two distinct fishes.

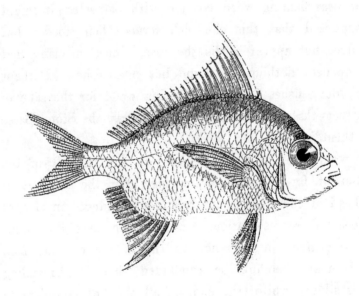

ACANTHOPTERYGII. *PERCIDÆ.*

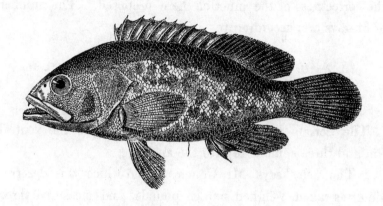

THE DUSKY SERRANUS.

Serranus gigas, Cuv. et Valenc. Hist. Nat. des Poiss. t. ii. p. 270, pl. 33.
Perca robusta, Couch, Mag. Nat. Hist. vol. v. p. 21, fig. 7.
  ,, *gigas,* Brunnich and Gmelin.

THE third species of Serranus to be added to the British catalogue, which, as before stated, was first made known as occurring on our shore by Mr. Couch, is his Dusky Perch, *Perca robusta,* which, from a careful comparison of descriptions, appears to be identical with the *Serranus gigas* of Cuvier and Valenciennes, above quoted, and the synonymes have been brought together accordingly.

This species inhabits the Mediterranean, and is also, but less frequently, taken in the ocean. Among the islands of its more congenial sea, this fish sometimes attains the weight of sixty pounds, and this circumstance originally suggested its specific name Gigas ; but specimens of ordinary occurrence weigh only from ten pounds to twenty pounds, and the flesh is in some estimation as food. The females deposit their spawn in shallow water during the months of April and May.

In the present instance, the figure and fin-ray formula of Cuvier are given; to which the description of Mr. Couch's fish is added, the better to prove, by their general accordance, the correctness of the junction here proposed. The number of fin-rays are, according to

Cuvier.

B. 7 : D. 11 + 16 : P. 17 : V. 1 + 5 : A. 3 + 8 : C. 15.

Couch.

„ 7 : „ 11 + 17 : „ 19 : „ 6 : „ 2 + 9 : „ 16.

The Serrani have usually one spinous ray to the ventral fin, and three spinous rays to the anal.

" The fish," says Mr. Couch, " from which this description was taken, weighed sixteen pounds, and measured three feet in length, and seven inches in depth, exclusive of the fins ; the body thick and solid. Under jaw longest ; both, as well as the palate, having numerous slender incurved teeth : in front of the under jaw was a bed of them. Lips like those of the cod-fish ; two large open nasal orifices, and a large hole under the projection of the nasal bone. First plate of the gill-cover serrate, the second with a broad flat spine projecting through the skin, and pointing backward ; the fleshy covering of the gill-covers elongated posteriorly ; seven rays in the gill membrane. Body and head covered with large scales ; lateral line gently curved. Dorsal fin single, long, expanding towards its termination, with eleven spinous rays, the first short, and seventeen soft rays, the two last from one origin. Pectoral fin round, nineteen rays ; ventrals fastened down by a membrane through part of their course, six rays. Vent an inch and a half from the origin of the anal fin, which fin has two spinous and nine soft rays, the last two from one origin. Tail roundish, sixteen rays. Colour of the back reddish brown, lighter on the belly : two slightly-marked lines on the gill-covers running obliquely downward, one on each plate. The gill-covers are not ridged. In its aspect this fish has some resemblance to the *Labri*, yet it has none of the generic characters by which these fishes are distinguished. That it should be placed among the Perches, I make no question; but my most industrious search has not been able to find that it has been either figured or described: until, therefore, some other naturalist shall be more fortunate, I venture to denominate it *Perca robusta*, from its great size and strength. I have never seen more than one specimen, which was taken with a line." In accordance with the remark made by Mr. Couch, Cuvier mentions that the Spanish name for this fish signifies a *Labrus*.

The term *Dusky Serranus* is suggested for it, instead of Dusky Perch, the better to identify it with the sub-genus to which it belongs.

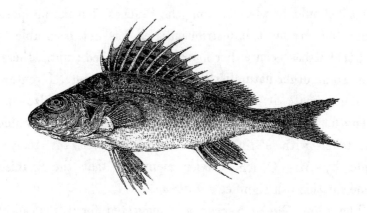

THE RUFFE, OR POPE.

Acerina vulgaris,   Cuv. et Valenc. Hist. Nat. des Poiss. t. iii. p. 4, pl. 41.
Perca cernua,   Linnæus. Bloch, pt. ii. pl. 53.
    ,,   ,,   *Ruffe,* Penn. Brit. Zool. 1812, vol. iii. p. 350.
    ,,   ,,   ,, Don. Brit. Fish. pl. 39.
Cernua fluviatilis,   ,, Flem. Brit. An. p. 212, sp. 141.

Generic Characters.—Dorsal fin single, elongated, the rays of the first portion spinous, the others flexible; branchiostegous rays 7; teeth very small, uniform, numerous; head without scales: suborbital bone and preoperculum indented; operculum ending in a single point.

The Ruffe, a fresh-water fish, closely allied to the Perch, but with a single dorsal fin, appears to have been unknown to the ancients, and Cuvier assigns the credit of its first discovery to an Englishman whose name was Caius.* He found it in the river Yare, near Norwich, and called it *Aspredo,* a translation of our name of Ruffe (rough), which is well applied to it on account of the harsh feel of its denticulated scales. Caius sent the first figure of this fish to Gesner, who published it.

The Ruffe is common to almost all the canals and rivers

* The learned Dr. Caius, well known for his various zoological writings.

of England, particularly the Thames, the Isis, and the Cam ; and, though said to be unknown in Spain, Italy, and Greece, is found over the colder portion of the European Continent, preferring slow, shaded streams, and a gravelly bottom. In its habits also the Ruffe resembles the Perch, and feeds, like that fish, on the fry of others and on aquatic insects. A small red worm used as a bait generally proves too tempting to be long resisted ; it seldom, however, when caught, exceeds six or seven inches in length, but its flesh is considered excellent. The spawning season is in April; and the ova, which are of a yellowish white colour, are deposited among the roots and stems of flags and rushes at the sides of the stream.

The generic characters, and the engraved outline at the bottom of the page, show the peculiarities of the various parts of the head : around the eyes are several oval depressions. Fin-rays :—

D. 14 + 12 : P. 13 : V. 1 + 5 : A. 2 + 5 : C. 17.

The prevailing colour of the upper part of the body and head is a light olive brown, passing into a yellowish brown on the sides, and becoming almost silvery white on the belly. The lateral line prominent and strongly marked. A tinge of greenish pearl pervades the gill-cover ; the irides are brown, the pupil blue. Small brown spots are disseminated over the back, dorsal fin, and tail, assuming on the latter from arrangement the appearance of bars ; pectoral, ventral, and anal fins, pale brown.

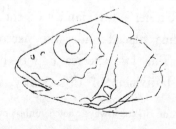

c 2

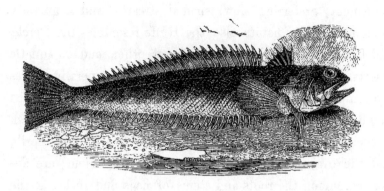

THE GREAT WEEVER, STING-BULL,

SEA CAT. *Sussex.* CHANTICLEER
AND GOWDIE. *Scotland.*

Trachinus draco, LINNÆUS.
    ,,   ,,   CUV. et VALENC. Hist. Nat. des Poiss. t. iii. p. 238.
    ,,   *major,* *Greater Weever,* PENN. Brit. Zool. edit. 1812,* vol. iii.
                             p. 229, pl. 33.
    ,,   ,,   ,,   ,,   DON. Brit. Fish. pl. 107.
    ,,   ,,   ,,   ,,   FLEM. Brit. An. p. 214, sp. 146.

Generic Characters.—Head and body compressed, eyes approximate ; branchiostegous rays 6 ; teeth in both jaws, on the front of the vomer and palatine
bones : two dorsal fins, the first very short, the rays spinous ; the second long,
the rays flexible : operculum with one long spine directed backwards ; anal fin
elongated.

THE six species of fishes already described belong to the
first division of Cuvier's first family of the Perches, which
have the ventral fins placed under the pectorals, and hence
called thoracic. The two species now to be mentioned belong to the second division of this family, distinguished by

* The octavo edition of 1812 is always quoted, unless otherwise expressed.

having the ventral fins situated before the pectorals, and called jugular.

Rondeletius believed the fish now called the Great Weever to be the Draco of the ancient naturalists; and their references to the injuries effected by the spines of the dorsal fin and operculum of this species, which they also called a sea-dragon, appear to confirm his opinion. The generic name *Trachinus* is derived from the Greek, and the fish is called in several languages by a term that signifies a spider, in reference to its supposed venom.

The English name of Weever, or Wiver, according to Merrett, is considered to be derived from the French term for this fish, *La Vive;* a name bestowed upon it from the circumstance of its living a long time after it has been taken out of the water; which latter power, with some other peculiarities in the habits of the Weevers, will be again adverted to.

The Great Weever generally measures about twelve inches in length, but has been known to attain seventeen inches: its food is the fry of other fishes, and its flesh is excellent. It swims very near the bottom, is sometimes taken in deep water by the trawl-net, and occasionally with a baited hook attached to deep-sea lines. When caught, it should be handled with great caution. " I have known," says Mr. Couch, " three men wounded successively in the hand by the same fish, and the consequences have been in a few minutes felt as high as the shoulder. Smart friction with oil soon restores the part to health;" but such is the degree of danger, or apprehension of it rather, arising from wounds inflicted by the spines of the Weevers, that our own fishermen almost invariably cut off the first dorsal fin, and both opercular spines, before they bring them on shore: the French have a police regulation by which their fishermen are directed to cut off the spines before they expose the fish for sale ; and in Spain there is a positive law by which fishermen incur a penalty if they bring to market any fish whose spines give a bad wound, without taking them off.

That the Great Weever prefers deep water, that it lives constantly near the bottom, that it is tenacious of life when caught, and that its flesh is excellent, are four points that have been already noticed ; but this subject, in reference to fishes generally, may be farther illustrated. It may be considered as a law, that those fish that swim near the surface of the water have a high standard of respiration, a low degree of muscular irritability, great necessity for oxygen, die soon— almost immediately, when taken out of water, and have flesh prone to rapid decomposition. On the contrary, those fish that live near the bottom of the water have a low standard of respiration, a high degree of muscular irritability, and less necessity for oxygen ; they sustain life long after they are taken out of the water, and their flesh remains good for several days. The carp, the tench, the various flat fish, and the eel, are seen gaping and writhing on the stalls of the fishmongers for hours in succession ; but no one sees any symptom of motion in the mackerel, the salmon, the trout, or the herring, unless present at the capture. These four last-named, and many others of the same habits, to be eaten in the greatest perfection, should be prepared for table the same day they are caught ;* but the turbot, delicate as it is, may be kept till the second day with advantage, and even longer, without injury ; and fishmongers generally are well aware of

---

* The chub swims near the top of the water, and is caught with a fly, a moth, or a grasshopper, upon the surface ; and Isaac Walton says, " But take this rule with you—that a chub newly taken and newly dressed is so much better than a chub of a day's keeping after he is dead, that I can compare him to nothing so fitly as to cherries newly gathered from a tree, and others that have been bruised and lain a day or two in water."

the circumstance, that fish from deep water have the muscle more dense in structure—in their language, more firm to the touch,—that they are of finer flavour, and will keep longer, than fish drawn from shallow water.

The law referred to has its origin in the principles of organization ; and though it would be difficult for the anatomist to demonstrate those deviations in structure between the trout and the tench which give rise to these distinctions and their effects, it is only necessary to make the points of comparison wider to be assured of the fact.

Between a fish with a true bony skeleton, the highest in organization among fishes, and the lamprey, the lowest, the differences are most obvious. If we for a moment consider the lamprey, which is the lowest in organization of the vertebrated animals, with only a rudimentary vertebral column, as the supposed centre of zoological structure, and look from thence up and down the scale of organization, we at the extreme on one side arrive at man, to whom division of his substance would be destruction ; but, on the other, we come to the polype, the division of which gives rise to new animals, each possessing attributes, not only equal to each other, but equal also to the animal of which they previously formed but a small part.

To return to the Great Weever : the number of fin-rays are,

D. 6—30 : P. 15 : V. 1 + 5 : A. 1 + 31 : C. 14.

Head and body compressed ; teeth small and numerous ; two small spines before each eye, irides golden yellow ; interoperculum and suboperculum smooth and without scales, cheeks and operculum with small scales ; gill-opening large ; vent in a line under the last spine of the first dorsal fin ; scales of the body arranged in oblique lines descending from

above backwards ; colour of the body reddish grey, browner
on the back, paler on the belly, marked with dark and dull
yellow lines in the same oblique direction as the scales ; head
brown with darker brown spots, gill-covers striped with yellow ; membrane of the first dorsal fin black to the fourth
spine, the remainder and the second dorsal fin pale brown,
almost white ; other fins light brown. The spawning season
is in June.

The following lines, referring to various qualities in the
Weever, may be quoted by way of conclusion.

> " The Weever, which although his prickles venom be,
>   By fishers cut away, which buyers seldom see,
>   Yet for the fish he bears, 'tis not accounted bad."
>
>   DRAYTON, POLY-OLBION, Song xxv.

ACANTHOPTERYGII. *PERCIDÆ.*

LESSER WEEVER, OTTER-PIKE, STING-FISH.

Trachinus vipera, Cuv. et Valenc. Hist. Nat. des Poiss. t. iii. p. 254.
 ,, *draco,* *Common Weever,* Penn. Brit. Zool. vol. iii. p. 226, pl. 32.
 ,, ,, ,, ,, Don. Brit. Fish. pl. 23.
 ,, ,, ,, ,, Flem. Brit. An. p. 213, sp. 145.

THE LESSER WEEVER is more frequently met with on different parts of our coast than the Greater Weever; it occurs also in the bays of Dublin and Belfast, and being much smaller and quicker in its motions, is even more difficult to handle with security. In its habits it is active and subtle, burying itself in the loose soil at the bottom of the water, the head only being exposed; it thus waits for its prey—aquatic insects, or minute crustaceous animals—which the ascending position of its mouth enables it to seize with certainty. If trod upon or only touched while thus on the watch, it strikes with force either upwards or sideways; and Pennant states, that he had seen it direct its blows with as

much judgment as a fighting-cock.   Montagu says, " Whether the supposed venomous quality of the sharp spines is
justly founded, is difficult to determine; but it appears to be
a fact, that the wounds inflicted by these offensive weapons
usually exhibit symptoms of great inflammation and pain,
and which has given rise to the vulgar name of Sting-fish.
It is caught sometimes in the shore-nets, or seine, about
Teignmouth and Torcross, but rarely exceeding five or six
inches in length."   This small species appears to have been
much less perfectly known than the Greater Weever : neither
Bloch nor Lacépède make any mention of it, and other
writers have included in their descriptions of a single species
some of the peculiarities of both.   Pennant, in the octavo
edition of his British Zoology, dated 1776, says this small
one " grows to the length of twelve inches ;" and this statement appears to have misled Dr. Turton, Mr. Donovan, and
Dr. Fleming, who have each assigned to it a length of ten or
twelve inches.   From the examination of many specimens, it
is more probable that it very seldom exceeds five inches.

D. 5 or 6 — 24 : P. 15 : V. 1 + 5 : A. 1 + 24 : C. 11.

Cheeks devoid of scales; mouth placed more vertical; teeth
stronger in proportion to its size, but less numerous ; and the
obliquity of the lines on the side less apparent,—are other
specific distinctions.   The back is reddish grey ; lower part
of the sides and the belly silvery white ; membrane of the
first dorsal fin black ; caudal fin tipped with black, the other
fins pale brown.   The Lesser Weever spawns in spring, the
Greater Weever spawns in summer : neither species possess
any swimming bladder.

ACANTHOPTERYGII. *PERCIDÆ.*

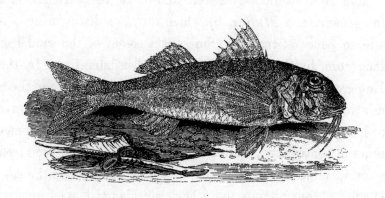

STRIPED RED MULLET.

Mullus surmuletus, Linnæus. Bloch, pt. ii. pl. 57.
,, ,, Cuv. et Valenc. Hist. Nat. des Poiss. t. iii. p. 433.
,, ,, *Striped Surmullet,* Penn. Brit. Zool. vol. iii. p. 368,
pl. 64.
,, ,, ,, ,, Don. Brit. Fish. pl. 12.
,, ,, ,, ,, Flem. Brit. An. p. 216, sp. 158.

Generic Characters.—Body thick, oblong; profile of the head approaching to a vertical line; scales large, deciduous; two dorsal fins widely separated, the rays of the first spinous, those of the second flexible; teeth on the lower jaw and palate only; two cirri at the symphysis of the lower jaw; branchiostegous rays 4.

The characters which distinguish the two species of Red Mullet common in the Mediterranean, both entitled to a place in the catalogue of British Fishes, have been long known, and figures of both are given in Willughby's *Historia Piscium,* plate S. 7, figs. 1 and 2. One species, the well-known Striped Red Mullet, is of frequent occurrence along the extended line of our southern coast from Cornwall to Sussex, but becomes more rare in proceeding from thence northward by the eastern coast.

The Red Mullets were well known to the ancients, and the generic term *Mullus*, by which they are distinguished, is said to have reference to the scarlet colour of the sandal or shoe worn by the Roman Consuls, and in later times by the Emperors, which was called *mulleus*. So much were these fish in estimation, that a Mullet of large size appears always to have been an object of particular admiration, and sometimes of contention. A fish of three pounds weight produced a considerable sum to the fortunate fisherman, while the cost of a fish of four pounds and a half, says Martial, was ruinous. A Mullet of six pounds is recorded to have produced a sum equal to 48*l.* ; one still larger, 64*l.* ; and even 240*l.* were given for three of very unusual size, procured on the same day for a repast of more than usual magnificence. The Striped Red Mullet is the species which occasionally only attains to so enviable a size in the Mediterranean ; the second species, which on our coast is very rare, is much smaller, but more beautiful in colour, and is the species which on that account the Romans exhibited in vases of glass to their friends and guests. They also kept Mullets in their numerous *vivaria ;* but, thus confined, the fish did not continue to increase in size. At the present time, the Mullets of Provence and Toulon are in high estimation. The flesh is white, firm, of good flavour, and being free from fat, is considered easy of digestion. The liver is the part of the fish in the greatest request. On our own coast the Striped Red Mullet seldom exceeds fourteen inches in length, and even this would be considered a fish of

large size. The largest for which I possess any authority occurred several years since. This Mullet weighed three
pounds six ounces, was in the highest perfection, and
beautiful in colour. It was sent from Weymouth as a
present to the late Thomas Palmer, Esq. of Berkeley-
square.

The Striped Red Mullet has been considered migratory ;
but it appears in the shops of the London fishmongers
throughout the year, though in much greater plenty during
May and June, at which time their colours are most vivid,
and the fish, as food, in the best condition. If closely examined, it will be observed that where the scales happen not
to have been removed, the natural colour is little more than a
pale pink, passing into white on the belly, the lower part of
the sides having three or four yellow longitudinal stripes ; but
that the mixture of purple and bright red which ornaments
various parts of the fish is the consequence of violence :
every scale removed by force—and but little is necessary—
increases this colour ; it is produced by extravasated blood
lying under the transparent cuticle, but above the true
skin.

These fish take a wide range through the water. Many
are caught in mackerel-nets near the surface during that
fishing season ; but the principal supply is derived from the
trawl-net, which traverses the bottom, and encloses these and
other fish in a manner that will be hereafter described. The
Mullets occur sometimes in profusion, at other times are
exceedingly scarce, owing to the fish shifting or changing
their ground, remaining unmolested till accident or perseverance betrays to the trawler their new locality, which on the
southern coast is sometimes several miles east or west of their
previous position.

The Striped Red Mullet spawns in the spring, and the young are five inches long by the end of October. The food appears to be selected from among the softer crustaceous and molluscous animals. In connexion with their food and the search made for it, the long *cirri* articulated to the under jaw require to be noticed. These *cirri* are generally placed near the mouth, and they are mostly found in those fishes that are known to feed very near the bottom. On dissecting these appendages in the Mullet, the common Cod, and others, I found them to consist of an elongated and slender flexible cartilage, invested by numerous longitudinal muscular and nervous fibres, and covered by an extension of the common skin. The muscular apparatus is most apparent in the Mullet, the nervous portion most conspicuous in the Cod. These appendages are to them, I have no doubt, delicate organs of touch, by which all the species provided with them are enabled to ascertain, to a certain extent, the qualities of the various substances with which they are brought in contact, and are analogous in function to the beak, with its distribution of nerves, among certain wading and swimming birds which probe for food beyond their sight ; and may be considered another instance, among the many beautiful provisions of Nature, by which, in the case of fishes feeding at great depths, where light is deficient, compensation is made for consequent imperfect vision.

D. 7 — 1 + 8 : P. 17 : V. 1 + 5 : A. 2 + 6 : C. 13.

The forehead, nape, cheek, and operculum are covered with scales ; irides pale yellow ; mucous pores abundant ; the teeth and the colours of the body have been already noticed ;

the membrane of the first dorsal fin is tinged with yellow, those of the other fins transparent ; the axilla of the ventral fin furnished with a pointed scale ; the vent placed under the commencement of the second dorsal fin.

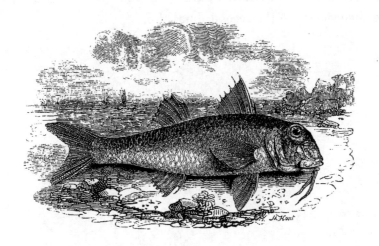

PLAIN RED MULLET.

Mullus barbatus, Linnæus.
,,   ,,   Cuv. et Valenc. Hist. Nat. des Poiss. t. iii. p. 442, pl. 70.
,,   ,,   Bloch, pt. x. pl. 348, fig. 2.
,,   ,,   *Surmullet,*   Penn. Brit. Zool. vol. iii. p. 365.
,,   ,,   *Red Mullet,* Couch, MS.

Pennant admitted this fish in his British Zoology on
account of one taken on the coast of Scotland, but which it
does not appear he had any opportunity of examining.   Mr.
Couch, according to the manuscript obligingly lent for this
work, has had the good fortune to obtain two specimens of
this very rare Mullet on the coast of Cornwall; which are
described as showing one yellow line a little below the lateral
line, the sides and part of the belly dark red, and the back
lighter in colour than the Striped Mullet.   A specimen of
this Plain Mullet in the collection at the British Museum,
and another in my own possession, have the colour of the
most delicate carmine on the back and sides, the belly silvery

white, but without any appearance of a yellow line, and very similar to the coloured figure in Bloch, plate 348, fig. 2, and the figure in the coloured copies of the work of Cuvier and Valenciennes before quoted, plate 70.

The habits of this species are stated to be the same as those of the Striped Red Mullet, and the number of fin-rays are as follows :—

$$\text{D. } 7 - 1 + 8 : \text{P. } 16. : \text{V. } 6 : \text{A. } 1 + 6 : \text{C. } 15.$$

The positions of the fins differ a little in the two species, as shown in the woodcuts on comparison, and the colour of the connecting membrane is a pale yellow; the irides also are yellow, the scales somewhat smaller in size than those of the Striped Mullet, and equally deciduous, but decidedly distinct in structure, as the vignettes exhibit. The trivial term *barbatus* applied to this species is objectionable, as the *cirri*, to which it is intended to refer, are common not only to our Striped Mullet, but also to several Indian and American species of Red Mullets, which were till lately included in the genus *Mullus :* the *cirri* are in reality a generic rather than a specific character.

The head is remarkable for its almost vertical profile, and the fish seldom exceeds six inches in length.

A scale from the lateral line of each fish is added in farther proof of the distinction of the species; that on the right hand is from the Plain Red Mullet, the other from the Striped Red Mullet.

VOL. I.   D

ACANTHOPTERYGII.   *WITH HARD CHEEKS.*

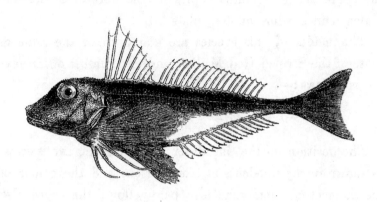

THE RED GURNARD, CUCKOO GURNARD.

Trigla cuculus, LINNÆUS.
    ,,   ,,   CUV. et VALENC. Hist. Nat. des Poiss. t. iv. p. 26.
    ,,   *pini,*   BLOCH, pt. xi. pl. 355.
    ,,   *lineata,* MONTAGU, Mem. Wern. Soc. vol. ii. pt. ii. p. 460.
    ,,   ,,   FLEM. Brit. An. p. 215, sp. 153.

Generic Characters.—Head nearly square, covered with bony plates ; gill-cover and shoulder-plate ending in a spine directed backwards ; body elongated, nearly round ; two dorsal fins, the rays of the first spinous, those of the second flexible ; teeth in both jaws and on the front of the vomer, pointed small and numerous ; branchiostegous rays 7 ; gill-opening large ; three detached rays at the base of each pectoral fin.

CUVIER's second family of the Acanthopterygii contains those genera, the species of which have their cheeks defended by indurated plates, which are sometimes spinous.   Of the first genus of this family, *Trigla,* the Gurnards, the British coast produces six species, three of which are common, the others are of rarer occurrence.   They are chiefly caught by the trawl-net used in deep water ; as the Gurnards mostly swim near the bottom, and are tenacious of life after they have been

taken from the sea. Excellent amusement is occasionally to be obtained by fishing for them with hand lines,
the hooks baited with a shining silvery piece of a sandlaunce.

The Red Gurnard is very common on the English coast,
and in Ireland is taken from Waterford on the south, up the
eastern shore to Londonderry in the north, but seldom found
larger than twelve or fourteen inches in length: it feeds on
crustaceous animals, spawns in May or June, and I have
found the characters well marked in young Gurnards only
an inch and a half long, taken in the small pools among
the rocks under Portland Island, by the end of August.
Their flesh is good food, and they are in greatest perfection
about October, and through the winter months. The number of fin-rays are as follows :—

D. 9—18 : P. 10—3 : V. 1+6 : A. 16 : C. 11.

Few fishes have the head so well defended as the Gurnard : its form is nearly square; the nose, in the Red Gurnard, with four projecting, but short tooth-like processes on
each side; the mouth small, a band of small teeth on both
jaws, and a small row on the vomer; the cheeks hard, gill-
openings large; operculum with one small spine directed
backwards, and one much larger on the scapular region above
the pectoral fin; three free rays at the base of this fin, which
are abundantly supplied with nerves, and assist the fish as an
organ of touch to find its food at the bottom; the eyes large,
the edge of the orbits with two or three small spines directed
upwards; both dorsal fins placed in a groove between two
rows of short triangular spines directed backwards; the body
is covered with small oval ciliated scales; the lateral line is
not armed, runs parallel to the line of the back of the fish,
and is crossed throughout its length with small short straight
elevated lines, which have the appearance of a series of pins,

D 2

Bloch compared them to the acicular leaves of the pine, a resemblance which suggested to him the trivial name of *pini* for this species.* The lateral line bifurcates at the caudal end. The colour of the body of this fish when quite fresh is a beautiful bright red, the sides and belly silvery white ; the first ray of the first dorsal fin slightly crenated ; the colour of the fins reddish white, becoming paler the second or third day after the fish has been caught.

As the Gurnards are remarkable for the various forms of the swimming-bladders in the different species, outlines of two of which are added as vignettes, an account of the structure, functions, and peculiarities of this singular and anomalous organ is here added.

Rondeletius was the first to notice that the swimming or air-bladder was more constantly found in fresh-water fishes than in those of the sea ; and Needham and Redi soon after pointed out the diversity of form in the swimming-bladder that prevailed in different species. Redi afterwards described the duct or tube by which this air-vessel communicates with the alimentary canal, and valuable additions to our knowledge on this subject have been since made by Monro, Lacépède, St. Hilaire, and Cuvier.

The swimming-bladder, as before stated, varies considerably in form in different species. In the Sapphirine Gurnard it is composed of three lobes, placed side by side, as shown

* Montagu called this species *lineata*, for the same reason.

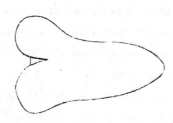

in the outline appended as a vignette to that fish ; in other Gurnards it is bilobed, but not very deeply cleft at the anterior part : the common Red Gurnard, the subject of the present notice, and the Grey Gurnard, are instances. In the Salmon, the Herring, and the Eel tribes, it is one elongated cylindrical tube, lying close to the under surface of the backbone. In the *Sciæna aquila*, the edges of the single-chambered swimming-bladder of that fish are fringed all round, of which a representation will be added ; but in the Carp, this organ is formed of two oblong cavities, the larger one lying behind the other, and communicating by an aperture in the neck or narrow portion connecting the two parts. From the anterior surface of the posterior lobe in the Carp, (a section of the whole subject reduced in size from Monro's Anatomy of Fishes being here added, with a probe introduced through the aperture in the neck, to show the communication between the two chambers,) a tube is given off, which, passing forwards, opens into the œsophagus, but is closed against the admission of any extraneous bodies by a delicate valve, which can only be passed in the outward direction.

The air-bladders are usually made up of two membranes. The inner one has a moist, smooth, and, apparently, a secreting surface ; the outer membrane is fibrous in its structure, and a portion of the bladder is in some species invested by a fold of the peritoneum : the three coats, when

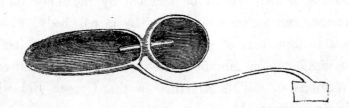

present, are nourished by blood-vessels, which are very apparent.

The air-bladder does not occur in all fishes: some fishes, and those principally that live near the bottom of the water, are without any. Among those species that have an air-bladder, many appear on the closest examination to have no canal or tube by which the air, with which the bladder is more or less distended, can escape. Muscles for compressing the air-bladder are obvious in some species, and wanting in others, yet the air-bladder apparently performs the same service in all.

The gas contained in these air-bladders has also been the subject of repeated investigations. Priestley and Fourcroy determined the gas in the Carp to be nearly pure nitrogen; other chemists found the air in different fishes to consist of nitrogen, oxygen, and carbonic acid; the nitrogen in greater proportion, and the oxygen in smaller, than in atmospheric air. In the air-bladder of marine fishes the oxygen is in excess, varying from forty to eighty-seven per cent., depending on the depth at which the different species usually remained. The Gurnards were frequently selected for these experiments, their air-bladders having no canal of communication admitted of being removed without losing their contents. It should be borne in mind that fresh-water contains more oxygen than that of the sea.

The air thus found in these bladders, however variable in its nature, is believed to be secreted by the inner lining membrane, and in some instances by a red body, which appears to form part of the walls of the air-bladder itself, and is made up of minute blood-vessels arranged between the membranes. This structure in the Conger Eel will amply repay the trouble of examination.

That the air found in this bladder is not taken in at the

mouth, is proved not only by the perfection of the valves of the canal, which only open outwards, but also by the want of uniformity in the quality of the air itself, and its existence in those swimming-bladders that have no canal of communication. That one use of these air-bladders to the fishes possessing them is to enable them to alter their specific gravity with reference to that of the fluid they inhabit, seems almost certain. We see the gold-fishes in our ornamental vases ascend and descend in the water without making any visible external muscular effort. In this respect their action is to be understood and explained by the well-known hydrostatic toy of the philosophical instrument makers, in which a small glass-balloon, or other figure, confined in a column of water, has its weight, by the introduction of a small quantity of air, so nicely balanced in reference to the specific gravity of the water, that it is made to ascend or descend according to the degree of pressure made by the finger on the elastic cover of the top.

In other respects, however, the function is quite as anomalous and uncertain as the quality of its contained gas. Our two Red Mullets have no swimming-bladder, yet they appear in the water to possess all the powers of the Indian or American species, which are well provided with them. The two British species of Mackerel, hereafter to be described, both swim near the surface of the water with the same apparent swiftness and ease : one has a swimming-bladder, the other none. Of our two species of *Orthagoriscus*, which, as far as the habits of such rare fishes are known, appear to possess the same powers, one has a swimming-bladder, the other not.

" The swimming-bladder of fishes," says Dr. Roget in his excellent Bridgewater Treatise, " is regarded by many of the German naturalists as having some relations with the respiratory function, and as being the rudiment of the pulmonary cavity of land animals ; the passage of communication with the œsophagus being conceived to represent the trachea."

Hervey long ago observed " that the air in birds passed into cells beyond the substance of the lungs ; thus showing a resemblance to the cellular lungs in reptiles, and the air-bladder in fishes." M. Agassiz, in dissecting a species of *Lepisosteus*, a fresh-water fish of the rivers of America, found the air-bladder composed of several cells, with a canal proceeding upwards into the pharynx, and ending in an elongated slit, with everted edges, resembling a glottis or tracheal aperture. However obvious may be these relations of structure, it is still difficult to believe there can be any analogy in function, when it is recollected that one-fourth of the fishes known are entirely without air-bladders, and that two-thirds of the other three-fourths have neither canal nor aperture for external communication, but that all are provided with gills.

The search for these relations of structure in animals of different classes is among the most interesting of the investigations of the comparative anatomist. The sexual organs of the Sharks and Rays very closely resemble those of some of the reptiles, and the young of both these families of cartilaginous fishes, as far as they have been examined, are now known to possess, for a short time, external branchial filaments. Linnæus called the cartilaginous fishes AMPHIBIA NANTES.

The trivial names of *cuculus* and Cuckoo Gurnard are said to have been appropriated to this species on account of the similarity in the sound which issues from this fish when taken out of the water to the note of the well-known bird.

ACANTHOPTERYGII.   *WITH HARD CHEEKS.*

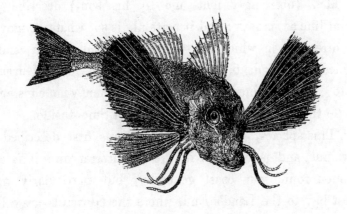

THE SAPPHIRINE GURNARD

Trigla hirundo,   Linnæus.   Bloch, pt. ii. pl. 60.
    ,,   ,,   Cuv. et Valenc. Hist. Nat. des Poiss. t. iv. p. 40.
    ,,   ,,   Penn. Brit. Zool. vol. iii. p. 376, pl. 68.
    ,,   ,,   Don. Brit. Fish. pl. 1.
    ,,   *lævis,*   Montagu, Mem. Wern. Soc. vol. ii. pt. 2, p. 455.
    ,,   ,,   Flem. Brit. An. p. 214, sp. 148.

The large size of the pectoral fins, and their fine blue colour on the inner surface, probably suggested both the specific names of this Gurnard, which is also the most valuable of the British species. In addition to its being equal to either of the others as food, it is not only much more abundant, but attains also a larger size, occasionally measuring two feet in length. That this species is the *Trigla lævis* of Montagu, there can be but little doubt; and the words of Linnæus, " *linea laterali aculeata*" are certainly incorrect in reference to this fish, and induced Montagu to

consider his *Trigla lævis* as distinct.   Pennant, Mr. Donovan, and Dr. Fleming, following Linnæus, have each described this species as having a rough lateral line.   Cuvier and
Valenciennes, in their fourth volume, and Mr. Walcott in
his MS. (obligingly lent me by his son,) describe the
lateral line as smooth ; and it certainly is so, whatever may be
the direction in which the finger is passed over it, and so
decidedly different in this respect from the other Gurnards
as to have obtained among the fishermen who constantly
handle them, the distinguishing name of Smoothside.

" This species," says Lacépède, " was first described by
Salvianus, and is common in the Mediterranean : it is also
common round our coast generally, but particularly from
West bay to the Land's End, where the Gurnards are called
Tubs, Tubfish, and, in reference to colour, Red Tubs.
Like the other species of Gurnards, they are taken by the
trawl-net chiefly, but many are also caught on the long lines
called bulters, with their baited hooks.   The flesh is of good
flavour, though rather dry, and requires sauce.   In the
North of Europe the flesh is salted for keeping.

D. 9 — 16 : P. 11 — 3 : V. 1 + 5 : A. 15 : C. 11.

This species bears some general resemblance to the Red
Gurnard in form, but the head is larger and more flattened ;
the eyes large, irides yellow, and pupil dark greenish blue ;
the prevailing colour of the head and body brownish red ;
the pectoral fins large and long, reaching beyond the vent,
blue on the inner surface, brownish red without, the fin-rays
white ; the spines of the operculum and scapula similar to
those of the Red Gurnard, but the supporting rays of the
first dorsal fin are not so strong as in that species.   The
scales are small, oval, and smooth ; those on the lateral line
slightly elevated, but perfectly smooth, and the line bifurcates
at the tail.   The air-bladder, as shown by the outline, has

three lobes, with strong lateral contracting muscles. Risso says this species deposits its spawn in spring, but Mr. Couch considers that this takes place in winter : and the ova are certainly very large towards the end of the year.

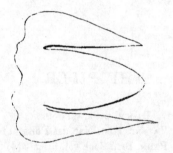

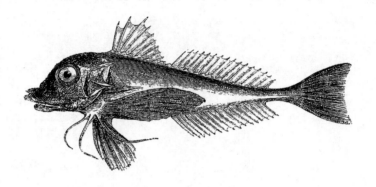

THE PIPER.

Trigla lyra,   Linnæus. Bloch, pt. x. pl. 350.
   ,,   ,,   Cuv. et Valenc. Hist. Nat. des Poiss. t. iv. p. 55.
   ,,   ,,   *Piper,* Penn. Brit. Zool. vol. iii. p. 374, pl. 67.
   ,,   ,,   ,,   Don. Brit. Fish. pl. 118.
   ,,   ,,   ,,   Flem. Brit. An. p. 215, sp. 154.

The Piper is at once distinguished from the other species
of British Gurnards, by the large size of the head, the greater
extent of the nasal projections, and the length and strength of
the opercular and scapulary spines ; the arming of the dorsal
crest is also more decided.   This fish was described by Belon
and figured by Rondeletius, and is a species well known in
the Mediterranean Sea.   On our own coast it is rare ; it was
however obtained by Pennant, and since his time by Mr.
Donovan and Mr. Couch.   Pennant says the Piper is frequently taken ; but this apparent contradiction to what is
stated above, is explained by an observation made by Mr.
Couch.   " The Piper wanders about more than the others,

at least, of the Cornish species ; consequently it is sometimes common, and at others somewhat rare." It is chiefly obtained on the western shores of Devonshire and Cornwall, occasionally off Anglesey, and is also said to have been taken in Belfast bay. It attains the length of two feet, weighing then three and a half pounds, and is supposed to have gained the name of Piper from the sound which escapes from it when taken in hand from the sea. All the species, however, emit a grunting noise at intervals for a considerable time ; which may probably have given origin to the name that distinguishes them by some corruption from the Latin *grunnio* or the French *gronder*. Perhaps a little assisted by its rarity, its flesh has been considered superior to that of the other Gurnards ; even Quin has borne testimony to the merits of a West-country Piper.

D. 9—16 : P. 11—3 : V. 1 + 5 : A. 16 : C 11.

The head is large, but the body declines rapidly to the tail ; eyes large, irides yellow, pupils dark blue ; one strong orbital spine in front, a smaller one behind ; anterior lateral portions of the muzzle very much produced on both sides, and notched, the central indentation deep in proportion ; under-jaw the shortest ; gill-openings large ; both opercular and the scapulary spines large and strong. In one of my own specimens, twenty inches long, the scapulary spine measures two inches and a quarter ; pectoral fins reaching beyond the vent ; the arming on the ridges of the back more conspicuous in this than in any other British species ; lateral line slightly elevated above the general surface, and rising gradually to the upper edge of the operculum : scales of the body small, oval, and ciliated ; the general colour a brilliant red ; belly white, fins red. Mr. Donovan's figure, otherwise very good, is much too pale in colour.

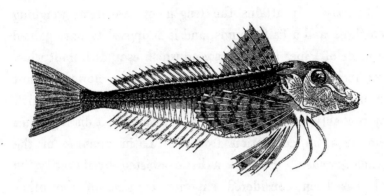

THE STREAKED GURNARD,

FRENCH GURNARD, AND ROCK GURNARD.

Trigla lineata,	Linnæus.	
„ *Adriatica,*	Gmelin.	
„ *lineata,*	Cuv. et Valenc. Hist. Nat. des Poiss. t. iv. p. 34.	
„ „	Bloch, pt. x. pl. 354.	
„ „	*Streaked Gurnard,* Penn. Brit. Zool. vol. iii. p. 377, pl. 66.	
„ „	„ „ Don. Brit. Fish. pl. iv.	
„ *Adriatica,*	„ „ Flem. Brit. An. p. 215, sp. 150.	

The Streaked Gurnard is the fourth species of the British Gurnards having large pectoral fins reaching beyond the vent, and which four species have here been placed in succession, the more readily to distinguish them from the two species hereafter to be described, which have short pectoral fins.

In the Gurnard now to be noticed, the head is much shorter, the profile more vertical, and the arming spines of the head but little produced, while the whole surface of the body exhibits transverse lines reaching from the dorsal ridge or crest on each side to the belly.

This species seldom exceeds twelve or thirteen inches in length, and was first described by Brunnich under the name of Adriatica. It is found at the Canaries, Teneriffe, the Mediterranean, on our southern and occasionally on our eastern coast, but in the last two places not in great numbers. Like the Gurnards generally, this species feeds principally on crustaceous animals, and is usually taken with the trawling-net. The formula of the fin-rays is as follows :—

D. 10 — 16 : P. 10 — 3 : V. 1 + 5 : A. 13 : C. 11.

The head is short, the upper jaw but little produced; the occipital, opercular, and humeral spines short and broad; eye rather small compared with those of other species, irides yellow, pupils dark blue; orbital spines two or three; the scales forming the lateral line elevated, carinated, and notched; the body crossed by as many lines as there are scales on the lateral line, with two rows of ordinary, square, ciliated scales to each line; the general colour of the body and fins a fine rich red; the fins spotted and sometimes edged with a darker colour; the belly white; the pectorals long, tipped with blue, and with four rows of large darkish blue spots, so arranged as to appear like continuous bands when the fins are closed. The swimming-bladder is a single oval chamber, with strong lateral muscles of contraction.

ACANTHOPTERYGII.   *WITH HARD CHEEKS.*

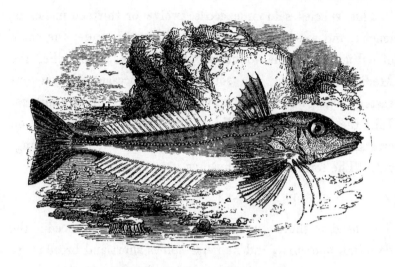

THE GREY GURNARD.
KNOUD OR NOWD. *Ireland.*

Trigla gurnardus,   Linnæus.   Bloch, pt. ii. pl. 58.
    ,,   ,,   Cuv. et Valenc. Hist. Nat. des Poiss. t. iv. p. 62.
    ,,   ,,   *Grey Gurnard,* Penn. Brit. Zool. vol. iii. p. 371, pl. 65.
    ,,   ,,   ,,   ,,   Don. Brit Fish. pl. 30.
    ,,   ,,   ,,   ,,   Flem. Brit. An. p. 215, sp. 152.

THE GREY GURNARD is much more common than either the Piper or the Streaked Gurnard, and is easily distinguished by its shorter pectoral fins, and by its elongated and slender body, generally of a greenish brown colour, spotted with white above the lateral line. This species was first described by Belon; there is also a good description in Willughby's *Historia Piscium,* and an excellent figure in Klein. The Grey Gurnard is taken along the line of our southern coast generally, up the eastern coast going northwards, on the coast of Scotland, and at the Orkney Islands; it is found also in the Baltic and on the west coast of Norway. In Ireland the Grey Gurnard occurs in all the localities which produce the Red Gurnard, *T. cuculus;*—namely, from Waterford in the south, up the eastern coast to Londonderry in the north. This species spawns in May or June; its swimming-bladder in shape resembles that of the *Trigla cuculus* of Linnæus, but it is not considered so good a fish to eat. The fin formula is—

D. 8 — 20 : P. 10 — 3 : V. 1 + 5 : A. 20 : C. 11.

The head is less elevated than in the other Gurnards, and the profile of the face is concave; the anterior prominences of the upper jaw armed with two or three denticulations; eyes large, irides silvery white, pupils black, each orbit with one small spine on its edge; opercular and humeral spine slender and sharp: the form of the body of the fish long and attenuated; the general colour brownish grey or greenish grey, with a few irregularly placed white spots on the back; the belly silvery white; the lateral line strongly marked with a sharp crest formed by scales of a white colour; the scales of the body small, oval, and smooth: first dorsal fin brown, sometimes spotted with black; the three or four first rays granulated, and rough to the touch: second dorsal fin and tail light brown: pectoral fins short, not reaching the vent; dusky grey in colour, but liable to some variation: ventral and anal fins nearly white. Occasional varieties in colour occur among the Gurnards, but these variations are mostly confined to the species *cuculus* and *gurnardus* of Linnæus. The varieties of the latter are frequently red, resembling *cuculus*, but are distinguished by the short pectoral fins, the three or four granulated spines of the first dorsal fin, and the long and slender body. The varieties of *cuculus* are mostly brown, resembling in this respect the general appearance of the Grey Gurnard, but are distinguished by their long pectoral fins reaching beyond the vent, as well as their shorter and thicker body. However different in colour varieties of the Gurnards may appear, the other specific characters remain unchanged.

ACANTHOPTERYGII.   *WITH HARD CHEEKS.*

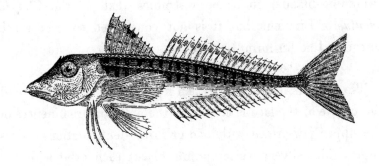

BLOCH'S GURNARD.

Trigla Blochii, YARRELL.
  ,,   *cuculus,* BLOCH, pt. ii. pl. 59.
  ,,   ,,   CUV. et VALENC. Hist. Nat. des Poiss. t. iv. p. 67.
  ,,   ,,   *Red Gurnard,* PENN. Brit. Zool. vol. iii. p. 373, pl. 66.
  ,,   ,,   ,,   ,,   MONTAGU, Mem. Wern. Soc. vol. ii. pt. ii.
                p. 457.
  ,,   ,,   ,,   ,,   FLEM. Brit. An. p. 215, sp. 153.

Two species of Red Gurnards having received the trivial name of *cuculus,* the first given by Linnæus, the second by Bloch, and both species entitled to a place in this work, I have followed the practice usually adopted in such cases, and propose for the second the name of its describer, as a tribute due to the author of the most valuable work on Ichthyology that has yet been completed.

This second species of Red Gurnard, the *T. cuculus* of Bloch, not the *T. cuculus* of Linnæus,—and which, for the reason assigned, as well as for distinction, is here called Bloch's Gurnard,—is not common as a British species on some parts of the coast. Pennant, by whom it is shortly described, and who has added a figure as quoted above, and

Bloch also, plate 59, considered this fish as the *T. cuculus* of Linnæus : Klein, however, appears to have been of a different opinion ; and Cuvier and M. Valenciennes have given it as a distinct species. Risso has also described it as a distinct species, among his Fishes of the Mediterranean, under the name of *granaou, T. cuculus,* and says the first spinous ray of the first dorsal fin is the longest ; which is not the case in the common *T. cuculus.* Compared with the common cuculus, (the true *T. cuculus* of Linnæus, the first of the Gurnards described and figured in the present work,) Bloch's Gurnard will be found to have the body longer and narrower, the head smaller but more powerfully armed, the pectoral fins short, not reaching to the anal fin, and the first dorsal fin having a conspicuous black spot on the margin of the membrane connecting the fourth, fifth, and sixth rays. The spot on the first dorsal fin, however, must not be considered as sufficient alone to identify this species ; as two specimens under comparison, both having this black spot, are in reality only varieties of the Grey Gurnard.

Montagu considered the Red Gurnard, described in the Memoirs of the Wernerian Society already quoted, as distinct from the Grey Gurnard ; but has certainly described the common Red Gurnard under the term *lineata,* considering this word as applicable to the linear elevations along the side which cross the lateral line, and which induced Bloch to call the species *T. pini.* This character is shown in the woodcut of the Red Gurnard, but is scarcely perceptible, from its diminished size, without the assistance of a lens.

Not possessing a specimen, the description of Colonel Montagu is adopted. " The forehead is more sloping than that of the Grey Gurnard ; the nose armed with three spines on each side ; the spine on the operculum of the gills, and

E 2

that behind it, are long and rough; lateral line and ridge of the back on each side serrated; a large black spot on the first dorsal fin at the margin, extending between the third and fifth rays. The whole body is rough : the spine on the gill-covers extends nearly as far as the spine behind it ; the lateral line and ridges on the back more strongly serrated than on the Grey Gurnard."—"Many of these," according to Colonel Montagu, "are taken in the summer months on the coast of Devon by the shore-nets; their size inferior to the other Gurnards, rarely exceeding a foot in length, and seldom above nine or ten inches."

The fin-ray formula, as given by Cuvier, is as follows :—

D. 8 — 19 : P. 11 — 3 : V. 1 + 5 : A. 17 : C. 11.

This species occurs also in the Channel, at Boulogne, and, as before mentioned, in the Mediterranean. Cuvier and M. Valenciennes, in their voluminous work now in progress, have described the internal anatomical distinctions.

Having stated that the various species of Gurnards are chiefly obtained by a particular mode of fishing in the sea called trawling, and representations being introduced at the foot of this and the next page of a trawl-net, and the sort of fishing-boat most common on the Sussex and Hamp-

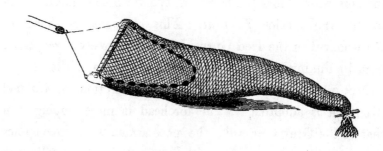

shire coasts, it remains to describe both, and the mode of using them. The boat is about twenty-five feet long, and ten feet in the beam, or breadth. The average burthen about ten tons; and they carry three tons of ballast—generally shingle, with some loose pigs of iron, which are shifted from side to side as occasion may require. The boat is fitted with two masts, with a square sail to each; sometimes a third mast and sail are set up when the wind is very light, and thus rigged they are called lugsail-boats. The trawl-net for a boat of this power has a beam of eighteen or twenty feet in length—the extent of the beam being the breadth of the mouth of the net; and the length of the net is from sixty to seventy-five feet. In the representation of this net, the rope on the extreme left that runs through the block is called the trawl-warp, and is the only connexion between the boat and the net when the net is overboard. The ropes passing obliquely from the block to the two sides are called the bridle, and serve effectually to keep the open mouth of the net square to the front, when the net is drawn along over the ground by the boat. The trawl-beam is four inches diameter, and is supported at the height of twenty or twenty-four inches above the ground by a heavy frame of iron of a particular form at each end

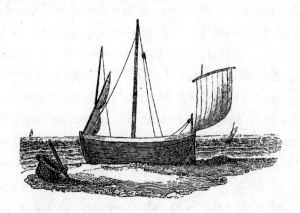

of the beam, called the trawl-heads, which assist by their weight to sink the net and keep it on the ground. The upper edge of the netting is attached along the whole length of the beam ; the lower edge is fastened along a heavy rope called the ground-rope, and follows considerably behind the advanced straight line of the beam, forming the portion of the circle seen through the upper surface of the net in the representation. This sort of net is only adapted for taking those fish that live upon or very near the bottom. When drawn along, the first part of the net that touches the fish is the ground-rope, from the contact of which the fish darts upward ; but that part of the net hanging from the beam is not only over, but also in advance of him, while the onward draft of the net by the progress of the boat brings the fish against the closed end of the tail, and if he then shoots forward towards the mouth of the net, he is stopped and entangled in pockets that only open backwards. As the fish in the tideway lie with their heads against the stream, the fishermen trawl with the tide ; that is, draw the net down the stream, carrying only so much sail on their boat as will give the net the proper draft along the ground—generally at the rate of two and a half or three miles an hour. When it is desirable to examine the contents of the net, the beam is hauled up to the side of the vessel by the trawl-warp, the tail of the net is handed in, untied, and the contents shaken out. The produce, depending somewhat on the nature of the ground, generally consists of Red Mullet, different species of Gurnards, flat fish, and Skate, with abundance of *asteria, crustacea,* and *echini.* The saleable fish being selected, the tail of the netting is retied, and the net again lowered to the ground ; and while the vessel continues its course, the refuse of one haul of the net is swept overboard to make room for the produce of the next. On some parts of the Dorsetshire and Devonshire coast, the trawling-boats and their apparatus are much larger than those here described; the former being cutter-rigged vessels of seventy or eighty tons burden, and their nets of thirty-six feet beam. Such vessels are constantly employed trawling in West Bay, and the Brixham and Torbay ground; even as near London as Barking Creek, boats and nets of this size are common; but the fishing-grounds for these vessels and their crews are in various parts of the North Sea, where a large and stout boat is absolutely necessary. The principal trawling off the Sussex and Hampshire coast is in the Channel, from twelve to thirty miles from the shore, and the men are seldom absent more than one night at a time.

Where the water is deep, this mode of fishing is successfully practised either in the day or night; but if the water is shallow and clear, but little success is to be obtained in the day.

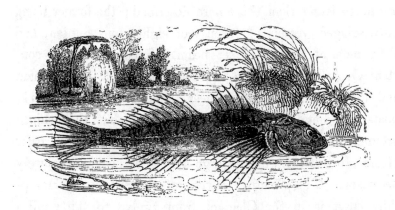

THE RIVER BULLHEAD, MILLER'S THUMB,

TOMMY LOGGE.

Cottus gobio, Linnæus. Bloch, pt. ii. pl. 39.
    ,,   ,,   Cuv. et Valenc. Hist. Nat. des Poiss. t. iv. p. 145.
    ,,   ,,   *River Bullhead,* Penn. Brit. Zool. vol. iii. p. 291, pl. 43.
    ,,   ,,   ,,   ,,   Don. Brit. Fish. pl. 80.
    ,,   ,,   ,,   ,,   Flem. Brit. An. p. 216, sp. 157.

Generic Characters.—Head large, depressed ; teeth in both jaws and in front of the vomer, small, sharp, none on the palatine bones ; preoperculum or operculum armed with spines, sometimes both ; branchiostegous rays 6 ; gill-openings large ; body attenuated, naked, without scales ; two dorsal fins, distinct or very slightly connected ; ventral fins small.

The River Bullhead is an inhabitant of almost all the fresh-water streams of the whole of Europe, from Italy to Sweden ; and most of the streams in this country that in their course run over sand or gravel produce this fish. It occurs also in the north of Ireland, in Belfast and Londonderry. Its length seldom exceeds four or five inches, and it is generally found among loose stones, under which, from the peculiarly flattened form of its head, it is enabled to thrust itself, and thus to find a hiding-place. When disturbed, it swims rapidly. The term Bullhead has been attached to all the species of the genus *Cottus*, on account of the large size of the head ; as we also use the words Bullfinch, Bullfrog, Bulltrout, and Bullrush, to indicate species of large comparative size.

As the term Bullhead is thus considered to refer to the large size of the head, so the name of Miller's Thumb given to this species, it has been said, is suggested by, and intended to have reference to, the particular form of the same part.

The head of the fish, it will be observed by the accompanying vignette, is smooth, broad and rounded, and is said to resemble exactly the form of the thumb of a miller, as produced by a peculiar and constant action of the muscles in the exercise of a particular and most important part of his occupation.

It is well known that all the science and tact of a miller is directed so to regulate the machinery of his mill, that the meal produced shall be of the most valuable description that the operation of grinding will permit when performed under the most advantageous circumstances. His profit or his loss, even his fortune or his ruin, depend upon the exact adjustment of all the various parts of the machinery in operation. The miller's ear is constantly directed to the note made by the running-stone in its circular course over the bed-stone, the exact parallelism of their two surfaces, indicated by a particular sound, being a matter of the first consequence : and his hand is as constantly placed under the meal-spout, to ascertain by actual contact the character and qualities of the meal produced. The thumb by a particular movement spreads

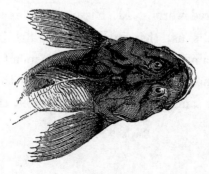

the sample over the fingers ; the thumb is the gauge of the value of the produce, and hence has arisen the sayings of, " Worth a miller's thumb;" and, " An honest miller hath a golden thumb;"* in reference to the amount of the profit that is the reward of his skill.   By this incessant action of the miller's thumb, a peculiarity in its form is produced which is said to resemble exactly the shape of the head of the fish constantly found in the mill-stream, and has obtained for it the name of the Miller's Thumb which occurs in the comedy of " Wit at several Weapons," by Beaumont and Fletcher, act v. scene i. ; and also in Merrett's " Pinax."

Although the improved machinery of the present time has diminished the necessity for the miller's skill in the mechanical department, the thumb is still constantly resorted to as the best test for the quality of flour.

This version of the cause of the application of the term Miller's Thumb to our River Bullhead, was communicated to me by John Constable, Esq. R.A.; whose father, being one of those considerable millers with which the counties of Essex and Suffolk abound, was early initiated in all the mysteries of that peculiar business.   He also very kindly lent me a view of an undershot water-mill at Gillingham, worked by a branch of the stream from Stourhead, which is represented in the vignette.

The larvæ of water-insects, ova, and fry, are the food of the Bullhead : it is voracious, and readily caught with a small portion of a red worm.   M. Risso says it is eaten in Italy ; and Pallas tells us, that in Russia this fish is used by some as a charm against fever, while others suspend it horizontally, carefully balanced by a single thread — and thus poised, but allowed at the same time freedom of motion, they believe this

* Ray's " Proverbs."

fish possesses the property of indicating, by the direction of the head, the point of the compass from which the wind blows. In Switzerland the children spear them in shallow water as they move from the stones under which they hide. Cuvier recommends this fish as a favourite bait for an eel.

D 6 to 9 — 17 or 18 : P. 15 : V. 3 : A. 13 : C 11.

The size and form of the head has been already noticed : the mouth is wide, jaws nearly equal, numerous small sharp teeth in both jaws and on the anterior part of the vomer ; no spines on the head ; irides yellow, pupils dark blue ; preoperculum with one spine curved upwards ; the operculum ending in a flattened point ; the dorsal fins united by a membrane ; rays of all the fins prettily spotted ; general colour of the body above dark brownish black, sides lighter, with small black spots ; under surface of the head and belly white ; the vent in a vertical line under the commencement of the second dorsal fin. This species spawns in summer.

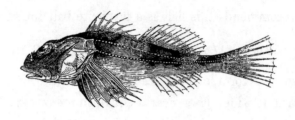

SEA SCORPION, SHORT-SPINED COTTUS.

Cottus scorpius,   Bloch, pt. ii. pl. 40.
,,   ,,   Klein, Miss. iv. pl. 13, fig. 2.
,,   ,,   Cuv. et Valenc. Hist. Nat. des Poiss. t. iv. p. 160.

The marine species of the genus *Cottus* appear to belong almost exclusively to the Northern Seas ; and although plentiful on most parts of our coast, M. Risso has not included them in his History of the Natural Productions of the Environs of Nice, which contains most of the fishes of the Mediterranean. Very various have been the names bestowed upon the species of the genus *Cottus* generally ; and under the term Father-Lasher two species have been constantly confounded in this country, and the habits and peculiarities of both included in one history. The Sea Scorpion, or short-spined *Cottus,* is common all round our coast, and, besides being less powerfully armed than the *Cottus bubalis,* or Father-Lasher, neither does it associate with that species. The Sea Scorpion is frequently found in estuaries, and measures from four or five to eight inches in length ; but it is said to acquire a much larger size in the North.

Like the other species of this genus, it is voracious in its appetite, and swims rapidly. There is reason to believe that this fish does not deposit its spawn at the same period of the year as the *Cottus bubalis;* some specimens of the former, examined in the month of November, exhibited little or no appearance of roe, while female specimens of the latter, if examined at the same time of the year, would be found to contain ova of large size, which are deposited in January, and are of a fine orange yellow colour. It has even been stated of *C. scorpius,* that it spawns in the spring, and that the ova are as black as ink.

In its habits this species resembles the Father-Lasher, and is found under stones and among *fuci* in the pools above low-water mark on our shores. They are very common, and every haul of a net of almost any description is nearly certain to produce examples of one species or the other, but seldom of both in the same immediate locality : no use, however, is made of them, and, on account of their numerous spines, they are handled with caution, only to be thrown overboard ; but if allowed to remain on the deck of the vessel, they are observed to be very tenacious of life. Their food is small crustaceous animals and the fry of other fish generally, which their wide mouths enable them to overcome without making any nicety in the selection necessary. Fin-rays :—

D 8 or 9—14 : P. 17 : V. 1 + 3 : A. 11 : C 12.

The head large, more elevated than that of the River Bullhead ; upper jaw rather the longer ; teeth small and sharp : eyes large, situated about half-way between the point of the nose and the occiput ; irides yellow, pupils bluish black : one pair of spines above the nostrils, with an elevated ridge between them ; the inner edges of the orbits elevated with a hollow depression above, but no occipital spines : preoperculum with three spines, the upper one the longest ; operculum with two spines, the upper one also the longest, the lower one pointing downwards ; there is besides a scapular and a clavicular spine on each side : gill-openings large ; the body tapers off rapidly, and is mottled over with dark purple brown occasionally varied with rich red brown ; the belly white ; the first dorsal fin slightly connected with the second by an extension of the membrane ; lateral line smooth ; the ventral fins attached posteriorly by a membrane to the belly.

ACANTHOPTERYGII. *WITH HARD CHEEKS.*

FATHER-LASHER, LONG-SPINED COTTUS.
LUCKY PROACH. *Scotland.*

Cottus bubalis, Euphrasen.
,, ,, Cuv. et Valenc. Hist. Nat. des Poiss. t. iv. p. 165, pl. 78.
,, *scorpius, Father-Lasher,* Penn. Brit. Zool. vol. iii. p. 294, pl. 44.
,, ,, ,, ,, Don. Brit. Fish. pl. 35.
,, ,, ,, ,, Flem. Brit. An. p. 216, sp. 156.

THE FATHER-LASHER is immediately recognised by its well-armed head and long spines, but seldom measures more than from six to ten inches in length on our shores. The general appearance of this fish is forbidding; yet in Greenland, besides attaining a much larger size, it is in such great request, that Pallas tells us it forms the principal food of the natives, and the soup made of it is said to be agreeable as well as wholesome. During the greater part of the year it is to be found on our coast from Cornwall to the Orkneys, and is frequently left by the receding tide in small pools among rocks. When touched, it distends its gill-covers, and sets out its numerous spines, assuming a most threatening appearance. This species spawns in January, and the ova at that time are very large, and of a fine orange yellow colour. These are deposited near the sea-shore, frequently in the estuaries, and sometimes even in rivers; the fish having prepared itself for this change by its previous residence in the brackish water, after which it appears to be able to bear either extreme. Its food is small crustaceous animals, and it is said to be particularly partial to feeding on the fry of the Blennies.

D. 8 — 12 : P. 16 : V. 1 + 3 : A. 9 : C. 10.

In *Cottus bubalis* the space between the eyes is much narrower than in the *C. scorpius*; the eyes in position more vertical, the crest above the eyes on each side more elevated, nearly straight, and ending at the nape in a spine directed backwards, forming a pair of occipital spines; irides yellow, pupils black : preoperculum with four spines, the upper one the longest; operculum with three spines, besides the scapular, clavicular, and nasal spines, similar to those of *C. scorpius :* gill-openings large ; in general colour very similar to that last described, and both species exhibit occasional variations in the intensity of the red and brown tints ; lateral line rough : the ventral fins in this species are devoid of the connecting membrane observable in *C. scorpius.*

Some circumstances observable in the economy of this species lead to the introduction here of a few observations on the respiration of fishes, in reference to their power of sustaining life when taken out of the water, and its supposed connexion with the size of the gill-aperture.

Most writers on Icthyology, even up to the present time, have stated that fishes with large gill-apertures, like the Herring, die soon when taken out of the water; and that, on the contrary, those with small gill-openings, like the Eel, have the power of sustaining life for a considerable time under the same circumstances. I will not say that the authors who have taken this view of the subject are in error; but I will venture to state the facts that appear to justify the belief that the duration of life in fishes is not altogether dependant on the size of the gill-opening.

That the Herring, the Mackerel, and many other fishes that swim near the surface, have large gill-apertures, and die almost immediately they are taken out of water, is most true; and that the Eel, with its small gill-aperture, does live for hours after it is taken out of water, is also true: but it will not be difficult to find many examples the very reverse of the instances supporting the rules, and also to show that in those fishes with large gill-apertures that do die quickly, the real cause of death has not been truly assigned.

The Carp, Tench, Barbel, Perch, and most of the various flat fish, have large gill-apertures, and yet they are all proverbially known to be able to sustain life long after they are removed from water. Cuvier, when writing on the genus *Trachinus*, says, in the *Histoire Naturelle des Poissons*, tome iii. p. 235, " Le nom François de Vive, que ces poissons portent sur nos côtes de l'océan, et celui de *Weever*, qu'on leur donne en Angleterre, viennent, dit-on, de ce qu'ils ont la vie dure et subsistent long-tems hors de l'eau." Yet, when describing *La Vive* and its gill-apertures, the words are (at p. 239) : " et l'on voit même que la fente des branchies est très-ample et s'ouvre jusque vis-à-vis la commissure des mâchoires." The two marine species of the genus *Cottus* just described have large heads and wide gill-pertures ; yet of them it is said (tome iv. p. 159), " Ces chaboisseaux vivent très long-tems hors de l'eau."

Of fishes with large gill-apertures it is said, in the same work (tome i. p. 519), that they die, " non pas faute d'oxigène, mais parce que leurs branchies se dessèchent ;" and of the Herring, that they die the instant they are taken out of the water. But may it not be objected to this view, that desiccation of the gills could not take place in so short an interval of time, and therefore could not be the cause of death ? Dr. Monro calculated that the surface of the gills in a large Skate was equal in extent to the whole surface of the body of a man ; yet, with this extent of surface exposed to the effects of desiccation, the different species of Skate are remarkable for the length of time they are able to sustain life after they are removed from water. Of fishes with small gill-apertures, our common Loche *Cobites fluviatilis,* and our most common species of the genus *Callionymus,* both die quickly. The Father-Lasher, with its large gill-aperture, will live a long time out of water, as has been already noticed ; yet, when taken out of the sea, if put into fresh water, it dies instantly.* The reverse of desiccation takes place in this instance : the gills are bathed with a fluid containing more oxygen than sea-water, and which also yields that oxygen much easier, yet death happens immediately. In this last instance it may be inferred that the fish, unable suddenly to accommodate its respiratory organs to fluids of such different densities as those of pure sea and fresh water, the blood is imperfectly aërated, the brain is affected, convulsions ensue, and, if not released, it soon dies ; and, from the previous examples, may we not

* Loudon's Magazine of Natural History, vol. ii. p. 217 and 218.

conclude that the power of fishes to sustain life for a time, when taken out of water, must be referred to a principle of internal organization, and is independent of the size of the gill-aperture.

M. Fleurens, a French physiologist, has explained what appears to be the true cause of death in a fish kept out of water.

If its motions be attentively watched, it will be seen that, although the mouth be opened and shut continually, and the gill-cover raised alternately, the arches supporting the branchiæ, or gills, are not separated, nor are the branchial filaments expanded—all remain in a state of collapse: the intervention of a fluid is absolutely necessary to effect their separation and extension; without it these delicate fibres adhere together in a mass, and cannot in that state receive the vivifying influence of oxygen. The situation of the fish is similar to that of an air-breathing animal enclosed in a vacuum, and death by suffocation is the consequence. To this may be added, that the duration of life in each species, when out of water, is in an inverse ratio to the necessity for oxygen.

F 2

ACANTHOPTERYGII.   *WITH HARD CHEEKS.*

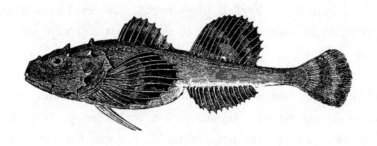

THE FOUR-HORNED COTTUS.

Cottus quadricornis, LINNÆUS.
„   „   BLOCH, pt. iii. pl. 108.
„   „   CUV. et VALENC. Hist. Nat. des Poiss. t. iv. p. 168.

I AM indebted to the communication of my friend Mr.
J. E. Gray, of the British Museum, for the knowledge of
the occurrence of the Four-horned Cottus on our shore;
and the figure at the head of the page was drawn from a
specimen in the National collection.

This species, first made known by Artedi, is common
in the Baltic, on the west coast of Norway, and in all the
Northern Seas even as far as Kamtschatka.   It has also been
taken on the north-east coast of England by our fishermen
in winter, when working nets with small meshes for sprats;
and in sorting for sale the many bushels of this common
fish brought to the London market, the Four-horned Cottus
has been occasionally found.

As a species, it is distinguished by four rough tubercles
on the top of the head, from which character its name has
been chosen : but Pallas observed occasional variations in

the number and size of these warty excrescences, and believed that the young had for a time but two of these tubercles, and were only provided with four by the time they had attained the length of seven or eight inches.

The Four-horned Cottus swims rapidly, but is generally observed lying in ambush, near stones or among sea-weed, ready to seize its food, and is known, by examination of the contents of the stomach, to feed more frequently on the young of the two species of Goby, that are there common, than upon any other small species. But little use is made of this fish, except as a bait for others. They spawn in winter, and the ova are white.

D. 8 — 14 : P. 17 : V. 1 + 3 : A. 15 : C. 11.

The head is large and flat ; mouth wide, jaws equal, teeth as described in the generic characters ; irides yellow, pupil black ; preoperculum with three spines, operculum with only one ; four horn-like tubercles on the top of the head, two of which are near the eyes, and two on the nape ; body elongated, compressed ; colour of the head brown, tinged with red on the gill-covers ; back brown, the sides yellow, the belly greyish white ; the lateral line nearly straight, and marked with rough points ; the body also freckled with scabrous points ; the fins prettily mottled with brown.

Two specimens of the Four-horned Cottus were the only fish taken in the sea with a net at Melville Island.—*Parry's First Voyage.*

ACANTHOPTERYGII.   *WITH HARD CHEEKS.*

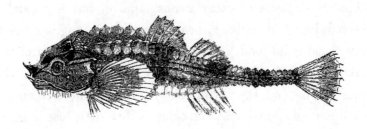

THE ARMED BULLHEAD, POGGE.
LYRIE, SEA-POACHER, PLUCK, NOBLE.   *Scotland.*

Aspidophorus Europæus,	Cuv. et Valenc. Hist. Nat. des Poiss. t. iv. p. 201.
Cottus cataphractus,	Linnæus. Bloch, pt. ii. pl. 39.
,, ,,	*Armed Bullhead,* Penn. Brit. Zool. vol. iii. p. 293, pl. 43.
,, ,,	*Pogge,* Don. Brit. Fish. pl. 16.
Cataphractus Schoneveldii, Pogge,	Flem. p. 216, pl. 155.

Generic Characters.—Body octagonal, covered with scaly plates; head thicker than the body, with points and depressions above, flattened below; teeth in both jaws only, none on the vomer; snout with recurved spines; branchiostegous rays 6 ; body tapering to the tail ; two dorsal fins, distinct.

THIS very marked species was first described by Schonevelde, a physician of Hamburgh, who published in 1624 a catalogue of the aquatic animals of Silesia and Holstein.   It is now known to exist not only in the Baltic, but on the coast of Norway, and in all the Northern Seas as far as Greenland and Iceland.   Mr. Couch says that it is not very common in Cornwall; and that, when found, it is most frequently near the mouths of rivers, but occasionally taken far out at sea.

Montagu considered this species as more common on the eastern parts of the kingdom than on the shores of the west, one or two instances only having occurred to him on the south coast of Devon ; and Mr. Neill has recorded its capture in the Forth.　It is not, however, uncommon along the line of our southern coast, where it is well known ; and the young of small size are frequently taken by the shrimpers in most of the sandy bays in the mouth of the Thames, and of other rivers : on the eastern coast it is very plentiful.　It seldom exceeds six inches in length ; its food is aquatic insects, and small crustaceous animals : it spawns in May, depositing the ova among stones, and its flesh is said to be firm and good.

D. 5 — 7 : P. 15 : V. 1 + 2 : A. 7 : C. 11.

The head is depressed, and wider than the body ; from the edge of each operculum the body tapers gradually to the tail ; the nose has three recurved spines ; the chin furnished with several minute *cirri ;* the eyes placed nearly vertical, irides yellow, pupils black : the mouth small ; teeth also small, but numerous : the suborbital bone and preoperculum each ending in a spine ; operculum surmounted by a spine, and an occipital tubercle on each side ;　a scapulary tubercle over the origin of each pectoral fin.　The body divided longitudinally by eight scaly ridges, of which those on the upper part of the body are the most produced.　The whole body defended by eight rows of strong scaly plates, of which the elevated ridges form the central lines ; the lateral line straight, lying parallel between the two ridges on the side.　Two dorsal fins slightly connected by a membrane, of a light brown colour mottled with dark brown ; pectoral fins large, with a broad bar of brown across the centre ; the general colour of the upper surface of the body brown, with four broad dark brown bands; tail brown; under surface of the body flattened; ventral and anal fins, and all the under parts of the head and body, very light brown, almost white. The vent placed very forward, on a line with the middle of the pectoral fin.

ACANTHOPTERYGII.   *WITH HARD CHEEKS.*

THE BERGYLT, AND NORWAY HADDOCK.

Sebastes Norvegicus,   Cuv. et Valenc. Hist. Nat. des Poiss. t. iv. p. 327
           pl. 87.
Perca marina,   Linnæus.
   ,, *Norvegica,*   Muller.
   ,, *marina,*   Sea Perch, Penn. Brit. Zool. vol. iii. p. 349, pl. 59.
Serranus Norvegicus,   Flem. Brit. An. p. 212, sp. 140.

Generic Characters.—Body oblong, compressed, covered with scales; all the parts of the head also covered with scales; eyes large; preoperculum and operculum ending in three or more spines; branchiostegous rays 7; teeth small, numerous, equal in size, placed on both jaws, the vomer, and palatine bones; a single dorsal fin, part spinous, part flexible; inferior rays of the pectoral fin simple.

According to Cuvier, the species of *Sebastes,* as separated by himself and M. Valenciennes from the genus *Scorpæna* of authors, so closely resemble some species of the genus *Serranus,* as to have deceived naturalists of the first order. The subject of the present article was arranged by Linnæus in his genus *Perca,* both in the *Systema Naturæ* and in the *Fauna Suecica,* and was confounded with the *Perca marina,* which, according to Cuvier, can be no other than the *Serranus scriba* of the Mediterranean: the words of Linnæus, " *Habitat in Norvegia, Italia,*" attached to his *Perca marina,* have induced authors to suppose that the Northern fish was also an inhabitant of the Southern Seas.

Pennant has engraved his *Perca marina,* and the figure has supplied the means of identifying his fish as the *Sebastes Norvegicus* of Cuvier.

This species inhabits all the Northern Seas, and is found in the deep bays on the southern coast of Greenland, where it is caught with baited hooks attached to very long lines: its general food is a small species of flat fish, *Pleuronectes cynoglossum,* which is there abundant. According to Fabricius, the flesh of *Sebastes,* though lean, is agreeable to the taste, and is eaten either cooked or dried; he states also, that the Greenlanders use the spines for needles.

Dr. Fleming obtained this fish in Zetland, where it is called Bergylt, and Norway Haddock; in several more Northern languages it is called by names that have reference to its prevailing red colour. " The late Dr. Skene," says Dr. Fleming, " observed this fish on the Aberdeenshire coast. Dr. George Johnston, of Berwick, has also obtained it on the shore of his own county; and I saw a well-preserved specimen of this fish, about twelve inches long, in the collection of Mr. John Hancock, of Newcastle-upon-Tyne: but this last example, if I recollect rightly, was obtained of the master of a Norwegian vessel.

D. 15 + 15 : P. 19 : V. 1 + 5 : A. 3 + 8 : C. 14.

The figure here given is taken from the plate of this fish in the *Histoire Naturelle des Poissons.* The peculiarities of the head are included in the generic characters. The mouth is large, the lower jaw the longest, the numerous teeth equal in size and small ; the eyes large, irides yellow, the pupils dark, the head depressed : the prevailing colour on the top of the head and back dark red, becoming lighter on the sides, and passing into a flesh-coloured silvery white on the under part of the head and body ; all the fins are red ; the flexible rays of the dorsal fin elongated.

ACANTHOPTERYGII.   *WITH HARD CHEEKS.*

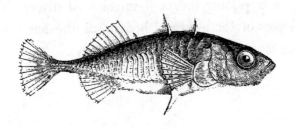

THE.ROUGH-TAILED STICKLEBACK.

BANSTICKLE, SHARPLIN.   *Scotland.*

Gasterosteus trachurus, Cuv. et Valenc. Hist. Nat. des Poiss. t. iv. p. 481,
            pl. 98, fig. 1.
    ,,   *aculeatus,* Bloch, pt. ii. pl. 53. fig. 3.
    ,,   ,,   Donovan, pl. 11.

Generic Characters.—Body without scales, more or less plated on the sides ;
one dorsal fin, with free spines before it ; ventral fin with one strong spine, and
no other rays ; bones of the pelvis forming a shield, pointed behind ; branchiostegous rays 3.

The Rough-tailed Three-spined Stickleback is
one of the smallest as well as one of the most common
of our fishes, and is found both in the salt and in the
fresh water : not only does almost every river, brook, and
lake produce this well-known species, but it is also
common all round the coast from the Land's End to the
Orkneys.

Cuvier and Valenciennes first noticed that three species

of Three-spined Sticklebacks had been constantly included under the term *G. aculeatus* of Linnæus; and the distinguishing characteristics being very obvious, all three species were shortly afterwards made known as inhabiting the waters of this country, and a figure of each given, with a short memoir, in the Magazine of Natural History, vol. iii. page 521.

The Three-spined Stickleback was first described by Belon, and figured by Rondeletius; and the history, habits, and peculiarities of the three species before referred to, have been constantly included in that of one only—the *aculeatus* of authors. Willughby and Pennant have figured the species now called *G. leiurus,* or the Smooth-tailed Stickleback; while Bloch and Mr. Donovan have given coloured representations of *G. trachurus,* the subject of the present article. It is probable that in their habits the species do not differ materially, and what is known on this subject will be added here.

They are active in their movements, and pugnacious in the extreme. A writer in the Magazine of Natural History, vol. iii. page 329, who appears to pay particular attention to the habits of fishes, has described their behaviour under confinement in wooden vessels of considerable size. " When a few are first turned in, they swim about in a shoal, apparently exploring their new habitation. Suddenly one will take possession of a particular corner of the tub, or, as it will sometimes happen, of the bottom, and will instantly commence an attack upon his companions; and if any one of them ventures to oppose his sway, a regular and most furious battle ensues: the two combatants swim round and round each other with the greatest rapidity, biting and endeavouring to pierce each other with their spines, which on these occasions are projected. I have witnessed a battle of this sort which lasted several minutes before either would give way; and when one does submit, imagination can hardly conceive the vindictive fury of the conqueror; who, in the most persevering and unrelenting way, chases his rival from one part of the tub to another, until fairly exhausted with fatigue. They also use their spines with such fatal effect, that, incredible as it may appear, I have seen one during a battle absolutely rip his opponent quite open, so that he sank to the bottom and died. I have occasionally known three or four parts of the tub taken possession of by as many other little tyrants, who guard their territories with the strictest vigilance; and the slightest invasion invariably brings on a battle. These are the habits of the male fish alone : the females are quite pacific ; appear fat, as if full of roe ; never assume the brilliant colours of the male, by whom, as far as I have observed, they are unmolested."

The woodcut represents this species of the natural size. Their appetite is voracious ; their food consists of worms and insects, and the minute fry and roe of other fishes. They spawn in summer ; the females, generally paler in colour than the males, depositing their ova of large size, but few in number, on aquatic plants. Although but few are thus produced by each female fish, their numbers are very great. Pennant states that they are occasionally so numerous at Spalding in Lincolnshire, that a man employed by a farmer to take them has earned four shillings a day for a considerable time by selling them at a halfpenny a bushel. Attempts have been made to obtain oil from them ; but they are more frequently strewed over the land for the purpose of manure.

This species seldom exceeds two and a half or three inches in length ; the body compressed ; the nostrils are pierced in a small depression rather nearer the eye than the end of the upper jaw : the mouth capable of slight projection ; teeth small, forming a narrow band in each jaw, but none on the vomer, palatine bones, or tongue : the gill-opening large; the fin-rays as follows .—

$$\text{D. III } 9 : \text{P. } 10 : \text{V. } 1 : \text{A. } 1 + 8 : \text{C. } 12.$$

The principal dorsal spine long and blunt, its lateral serrations small and few in number ; a membrane attached to the spine, by which it is depressed ; the ventral spine triangular at the base, the serrations on its upper edge large and not thickly set, those on the under edge small and numerous : the sides defended throughout their whole length by a series of elongated bony plates, arranged vertically ; a small fold of skin forms a horizontal crest on each side of the tail.

The Sticklebacks are said to live but two, or at most but three years ; and the males are generally to be distinguished by the pink colour of their under surface, but both sexes exhibit more than usual brilliancy at the season of spawning. The colour of the back is green ; the cheeks, sides, and belly, silvery white. The different species are of little value.

ACANTHOPTERYGII.   *WITH HARD CHEEKS.*

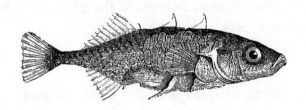

THE HALF-ARMED STICKLEBACK.

Gasterosteus semiarmatus, Cuv. et Valenc. Hist. Nat. des Poiss. t. iv. p. 493.

THIS species is distinguished from the preceding by the want of the arming by defensive plates along the sides of the tail, and in having rather larger teeth ; in other respects it does not differ much, and may be considered by some as only a variety or the young of *Gasterosteus trachurus,* that had yet by increased age to acquire the requisite number of lateral plates. I have, however, taken specimens of all sizes, which were uniform in the number of lateral plates, and close examination by a friend, who has paid particular attention to this subject, has shown that no point of ossification or induration is to be found posterior to the last perfect lateral plate, which seldom passes beyond the line of the vent. The figure makes farther description unnecessary. The number of fin-rays are

D. III + 10 : P. 10 : A. 1 + 9 : C. 12.

It occurs in similar situations to the other Sticklebacks, but not always in company with them.

ACANTHOPTERYGII.　　　　　　　　*WITH HARD CHEEKS.*

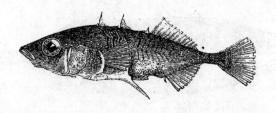

THE SMOOTH-TAILED STICKLEBACK.

Gasterosteus leiurus,　Cuv. et Valenc. Hist. Nat. des Poiss. t. iv. p. 481, pl. 98, fig. 4.
Pisciculus aculeatus,　Rond.　Willughby, X. 14, fig. 1.
Gasterosteus　　,,　　Penn. vol. iii. pl. 61.

The third species is the Smooth-tailed Stickleback, in which the lateral plates extend no farther than the ends of the rays of the pectoral fin; the whole length of the side beyond this being smooth and soft, without scale or fold, and only marked with the linear depressions produced on the surface by the divisions of the lateral muscles. The general colours of the three species are green above, passing into silvery white below. Some exhibit various shades of crimson and purple; but these colours are more frequent in males than females. Fin-rays:—

D. III + 10 : P. 11 : A. 1 + 8 : C. 12.

VOL. I.　　　　　　　　　　　　　　　　　　G

ACANTHOPTERYGII.   *WITH HARD CHEEKS.*

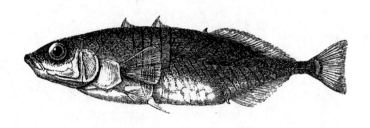

THE SHORT-SPINED STICKLEBACK.

Gasterosteus brachycentrus, Cuv. et Valenc. Hist. Nat. des Poiss. t. iv. p. 499, pl. 98, fig. 2.

Specimens of a large species of Three-spined Stickleback, with very short spines, taken in the North of Ireland, have been supplied me by William Thompson, Esq. Vice-President of the Belfast Natural History Society, who believes it to be identical with Cuvier's species as quoted above. In the number of lateral plates, this species agrees with *G. leiurus;* but the fish is of much larger size, while the spines, as may be seen by comparison, are very considerably shorter. The lateral plates do not extend beyond the limits of the pectoral fin, from whence the lateral line is a mere linear depression ; and whether the examples of this fish be taken from mountain streams, those of the lower grounds, or from the sea, the water of the lowest temperature produced specimens of the largest size. According to Mr. Thompson, the vertebræ in this species are more numerous than in *G. leiurus.* The plate represents this fish of the natural size. Fin-rays :—

D. III + 13 : P. 10 : A. 1 + 9 : C. 12.

ACANTHOPTERYGII.   *WITH HARD CHEEKS.*

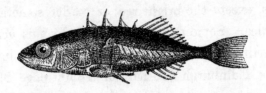

THE FOUR-SPINED STICKLEBACK.

Gasterosteus spinulosus, JENYNS and YARRELL.

I AM indebted to the kindness of Dr. James Stark for specimens of a Stickleback with four spines, taken in the pond of a meadow near Edinburgh in September 1830. This peculiarity in the number of spines has not that I am aware been made known, as occurring in this country, before the exhibition of these specimens by Dr. James Stark at a meeting of the Wernerian Natural History Society in 1831. These examples were of small size, measuring only one inch and one quarter in length, and were taken with the common Three-spined Stickleback; but other examples of this Four-spined Stickleback were afterwards found by Dr. Stark in other localities, where no species but those with four spines could be taken.

Dr. Stark succeeded in keeping these diminutive four-spined fishes in tumblers, and fed them with small leeches and aquatic insects, and found them quite as voracious, and even more pugnacious, than the more common ones with three spines.

In the MS. of John Walcott, Esq. which was written during a residence at Teignmouth, and which MS. has been most obligingly lent me by his son, I find a notice also of a Four-spined Stickleback; but no description is given, nor is

G 2

there any mention made of the locality from which it had been derived. Dr. Stark observed that his Four-spined Stickleback had all the varied colours of the other species of the genus, except the bright red or scarlet sometimes found in the males. Some experiments made by this gentleman— an interesting account of which was published in Professor Jameson's Edinburgh Journal for 1830, page 327—shows that the colour of these and some other small fishes is influenced, not only by the colour of the earthenware or other vessel in which they were kept, but also modified by the quantity of light to which they were exposed ; becoming pale when placed in a white vessel in darkness even for a comparatively short time, and regaining their natural colour when placed in the sun. From these circumstances, observed also in some species of other genera, Dr. Stark is led to infer that fishes possess, to a certain extent, the power of accommodating their colour to the ground or bottom of the waters in which they are found. The final reason for this may be traced to the protection such a power affords to secure them from the attacks of their enemies, and exhibits another beautiful instance of the care displayed by Nature in the preservation of all her species. Dr. Stark often observed that on a flat sandy coast the flounders were coloured so very much like the sand, that, unless they moved, it was impossible to distinguish them from the bottom on which they lay.

The specimens sent me have four spines, placed at equal distances from each, on the dorsal line, with one broad lateral plate nearly hid by the pectoral fin, and forming an ascending portion on each of the ventral plates. The fin-rays :—

D. IV $+$ 8 : P. 9 : V. 1 : A. 1 $+$ 8 : C. 12.

The colour has been already noticed. The figure is double the natural size.

ACANTHOPTERYGII. *WITH HARD CHEEKS.*

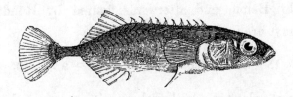

THE TEN-SPINED STICKLEBACK.

Gasterosteus pungitius, LINNÆUS.
     ,,   ,,   Cuv. et VALENC. Hist. Nat. des Poiss. t. iv. p. 506.
     ,,   ,,   BLOCH, pt. ii. pl. 53, fig. 4.
     ,,   ,,   *Ten-spined Stickleback,* PENN. Brit. Zool. vol. iii. p.
                                       335, pl. 61.
     ,,   ,,   ,,   ,,   DON. Brit. Fish. pl. 32.
     ,,   ,,   ,,   ,,   FLEM. Brit. An. p. 219, sp. 167.

THE TEN-SPINED STICKLEBACK is one of the smallest of the fishes that occur on our coast, and appears to be generally distributed, though by no means so numerous as those species with only three spines. It is found, however, in most of the creeks near the coast, as well as in many of our rivers, up which they are said to migrate in shoals in the spring. In size, it varies from one inch and a half to two inches and a quarter; and is distinguished from all the other Sticklebacks by the nine or ten spines on the back, all anterior to the dorsal fin, and by its sides being perfectly smooth, without any lateral plates,—which, with the number of dorsal spines before mentioned, forms its best specific character. Cuvier, in the last edition of the *Règne Animal,* tom. ii. p. 170, hints at the existence of a second species of Ten-spined Stickleback—the one having on the sides of the tail some carinated scales, the other (*G. lævis,* Cuvier) wanting this lateral arming. In the *Hist. Nat. des Poissons,* however, only the *G. pungitius* is retained, and a smooth tail forms part of its character. This species, like the former, was first described by Belon, and afterwards figured by Rondeletius. The fin-rays are :—

D. IX + 10 : P. 11 : V. 1 + 5 : A. 1 + 9 : C. 12.

The general colour is a yellowish or olive green on the back ; sides and belly silvery white, with minute specks of black ; fins pale yellowish white.

ACANTHOPTERYGII.   *WITH HARD CHEEKS.*

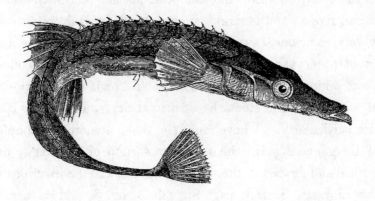

THE FIFTEEN-SPINED STICKLEBACK.

GREAT SEA ADDER, *Cornwall.*   BISMORE, *Orkney.*

Gasterosteus spinachia, LINNÆUS.
    ,,   ,,   BLOCH, pt. ii. pl. 53, fig. 1.
    ,,   ,,   CUV. et VALENC. Hist. Nat. des Poiss. t. iv. p. 509.
    ,,   ,,   *Fifteen-spined Stickleback,* PENN. Brit. Zool. vol. iii.
                                       p. 356, pl. 61.
    ,,   ,,   ,,   ,,   DON. Brit. Fish. pl. 45.
Spinachia vulgaris,   FLEM. Brit. An. p. 219, sp. 165.

THIS Stickleback, much more elongated in its form than any other of the British species, was first described and figured by Schonevelde, whose name as a naturalist has been mentioned before. It appears to be even more numerous northward than around the British Islands; and is found on the coast of Norway, as well as in the Baltic. Mr. Low includes it in his *Fauna Orcadensis,* and says it is found very frequent; and it has its Orkney name, quoted above, from the kind of balance there made use of, called bismores. Mr. Neill and Dr. George Johnston have taken it in the

Forth and Berwick bay ; from whence, southward and west-ward, it may be found all round our coast to the Land's End.

The Fifteen-spined Stickleback, however, though common on the coast, does not, like the other species of Sticklebacks, ascend rivers ; and is rarely, if ever, taken in fresh water. It is very voracious, swallowing indiscriminately the eggs and fry of other fishes, worms, and marine insects. The collec-tor of minute crustaceous animals should omit no opportunity of examining the stomachs of littoral fishes, and of this spe-cies particularly. I have found in them numerous examples of the genus *Mysis ;* the oppossum shrimp of Montagu, de-scribed and figured in the ninth volume of the Transactions of the Linnæan Society, page 90, tab. 5, fig. 3, and so named from the females having a pouch on the abdomen, formed by four concave scales turned upwards, in which she carries the ova, and afterwards the young. The species of this genus form the subject of the second memoir of the Zoological Researches of Mr. J. V. Thompson, of Cork.

For the following account of the habits of the Fifteen-spined Stickleback I am indebted to Mr. Couch : — " It keeps near rocks and stones clothed with sea-weeds, among which it takes refuge upon any alarm. Though less active than its brethren of the fresh water, it is scarcely less rapa-cious. On one occasion, I noticed a specimen, six inches in length, engaged in taking its prey from a clump of oreweed ; in doing which, it assumed every posture between the horizon-tal and perpendicular, with the head downward or upward, thrusting its projecting snout into the crevices of the stems, and seizing its prey with a spring. Having taken this fish with a net, and transferred it to a vessel of water, in com-pany with an eel of three inches in length, it was not long before the latter was attacked and devoured head foremost,— not, indeed, altogether, for the eel was too large a morsel, so that the tail remained hanging out of the mouth ; and it was obliged at last to disgorge the eel partly digested. It also seized from the surface a moth that fell on the water, but threw up the wings. The effect of the passions on the colour of the skin in the species of the genus *Gasterosteus* is remarkable ;* and the specimen now spoken of, under the influence of terror, from a dark olive with golden sides, changed to pale for eighteen hours, when it as suddenly regained its former tints. It spawns in spring ; and the young, not half an inch in length, are seen in considerable numbers at the margin of the sea in summer."—*Couch's MS*.

The whole length of this species is from five to seven inches. The jaws are elongated, the under one the most ; the mouth small ; the eye placed half-way between the point of the nose and the end of the gill-cover ; the irides silvery, the pupil black ; the head flat : the form of the body pentangular, the tail depressed ; the lateral line marked by a series of carinated scales throughout its whole length. The fin-rays are :—

$$\text{D. XV} + 6 : \text{P. 10} : \text{V. 2} : \text{A. 1} + 7 : \text{C. 12.}$$

The fifteen dorsal spines, curved backwards, are each furnished with its little membrane, and the last spine is the longest and most curved ; the belly, with two elongated bony plates, having, about midway on their inner edges, two unequally-sized ventral spines : the colour of the upper part of the head, body, and tail, is greenish brown, the sides inclining to yellow ; silvery white on the cheeks, gill-covers, under part of the head, and belly ; the dorsal and anal fins have each a black spot on the anterior part.

* See Magazine of Natural History, vol. iii. p. 329.

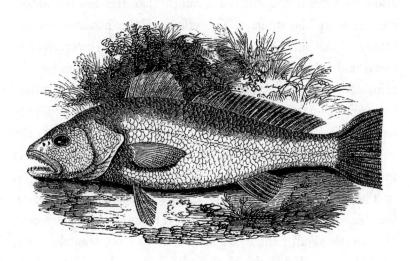

THE MAIGRE.

Sciæna aquila,   Cuv. et Valenc. Hist. Nat. des Poiss. t. v. p. 28, pl. 100.
Umbra Rondeletii, Willughby, p. 299, tab. S. 19.
Cheilodiptere aigle, Lacepede.
Sciæna aquila,   Flem. Brit. An. p. 213, sp. 144.

Generic Characters.—Body covered with scales : two dorsal fins ; spines of
the anal fin slender : a single row of strong teeth in each jaw ; a narrow line of
small ones in the upper jaw only, none on the vomer or palatine bones : preoperculum serrated, when young ; operculum ending with one or more spines :
branchiostegous rays 7.

THE limited space to be devoted to each species in this
work, will not allow an opportunity of following Cuvier
and M. Valenciennes through the long chain of historical
research by which they have succeeded in clearing the European *Sciænidæ* from the obscurity in which they were
involved by the older writers.   This important branch
of Ichthyological history, for which Baron Cuvier was so

* The family of the Maigres.

eminently qualified by his great talents and acquirements, his excellent memory, and the extensive materials by which he was surrounded, forms one of the most valuable features of all that part of the work on fishes he was spared to accomplish. It may be sufficient here to state, that, in the *Histoire Naturelle des Poissons,* the three best known species of the Mediterranean Sea have been considered the types of three genera, two of which will belong to British Fishes.

The name of *Sciæna,* as a generic term, has been given to those species which exhibit the peculiarities included in the generic characters, of which *Sciæna aquila,* the *Maigre* of the French, forms the type, or most characteristic example.

This fish, the largest and the most remarkable, is also the most common in certain localities ; and is celebrated for the goodness of its flesh. Salvianus has correctly described it under the name of *Umbrina,* but considered it the *Maigre* of the French. Rondeletius calls it *Peis Rei* (Royal Fish). It appears always to have been in great request with epicures ; and, as on account of its large size it was always sold in pieces, the fishermen of Rome were in the habit of presenting the head, which was considered the finest part, as a sort of tribute to the three local magistrates who acted for the time as conservators of the city.

Paulus Jovius relates a curious history of a head of one of these fishes, presented, as usual, to the conservators in the reign of Pope Sextus X. ; given by them to the Pope's nephew ; by him to one of the Cardinals ; from whom it passed as a noble donation to his banker, to whom he was deeply indebted ; and from the banker to his courtesan. It was followed through all its migrations by a parasite, whose industry was rewarded by at length partaking in the feast. This story forms much of the underplot of Beaumont and Fletcher's " Woman-Hater ;" where, as the condition of his becoming a sharer in the exquisite morsel, the parasite is made to marry the courtesan, with whom the head finally rested.

The Maigre, however, seems almost to have become forgotten at Paris; and Duhamel has afforded a clue that explains it. The fish has shifted its ground; and had, at the time the observation was made, taken up a new locality, near a hundred leagues distant from its previous position.

The southern side of the Mediterranean appears to be the situation in which the young of the Maigre are produced in the greatest numbers; and examples of small size have been brought from Egypt. The specimens that are taken on the northern shore are usually of large size. At Genoa, this fish is called *fegaro;* and at Nice, according to M. Risso, *figou*, and *vanloo.*

The Maigre is occasionally taken off the coast of Spain; and Duhamel considered it a fish that wandered continually, generally swimming in small shoals, and seldom remaining long in a place. In 1803, the fishermen of Dieppe caught nine or ten of these fishes, which were unknown to them before, and to which they gave the name of *aigle.* Specimens have also been taken occasionally since; and it has been observed, that, when these fishes are swimming in shoals, they utter a grunting or purring noise, that may be heard from a depth of twenty fathoms; and, taking advantage of this circumstance, three fishermen once took twenty Maigres by a single sweep of their net. They are described as possessing great strength, frequently upsetting the men in their struggles; and they are accordingly knocked on the head as soon as they are got into the boat.

As we advance northward, the Maigre becomes more rare. One specimen, five feet four inches in length, was taken in Zetland, in November 1819, as recorded by Mr. Neill.

It was first observed by some fishermen, as it was endeavouring to escape from a seal ; and when taken into the boat, made its usual purring sort of noise. A second specimen was taken in a seine-net, at Start bay, on the south coast of Devon, in August 1823, as communicated to the Zoological Society by the Rev. Robert Holdsworth. In September 1834, I saw a fine specimen, five feet two inches long, in the collection of Mr. John Hancock, of Newcastle-upon-Tyne, which had been taken on the coast of Northumberland, and preserved by himself; and in the month of November in the same year, a specimen, five feet four inches long, was taken by some herring-fishers off the Kentish coast, and brought to the London market. This fish was bought by Mr. Groves, of Bond-street, who very liberally gave the skin to the Zoological Society for preservation. Part of the flesh of this specimen was eaten by several persons, and by all reported to be good, particularly by those who prepared their portions by stewing. When plain boiled only, it was rather dry and tasteless. The two hard bones usually found just within the sides of the head in fishes are larger in proportion in the Maigre than in any other fish, and were supposed, the older writers say, to possess medicinal virtues. According to Belon, they were called colick-stones, and were worn on the neck, mounted in gold, to secure the possessor against this painful malady : to be quite effectual, it was pretended that the wearer must have received them as a gift; if they had been purchased, they had neither preventive nor curative power. These ear-bones are well represented by Klein, tab. 4, D. D.

The Maigre is seldom taken less than three feet, and sometimes as much as six feet, in length. It possesses many of the internal characters of some of the Percidæ, and has very much the general external appearance of a large Basse.

It differs, however, in having the tongue and the whole of the roof of the mouth quite smooth. The head is also shorter, and more rounded in form than that of the Basse. The mouth is furnished with one row of distinctly separated teeth in each jaw, pointed and curved, with a few smaller ones among those of the lower jaw, and a row of smaller ones behind those of the upper jaw; the eye placed high up on the head, distant about two of its own diameters from the end of the nose; and the nostrils pierced in a line between these two points, but nearer the eye. In both the specimens I had opportunities of examining, the serrations of the pre-operculum were nearly obliterated, probably by age: the fin-rays were in number—

$$D. \; 9-1+27 \; : \; P. \; 16 \; : \; V. \; 1+5 \; : \; A. \; 1+8 \; : \; C. \; 17 \; ;$$

but the membranes of the fins and the tail were very much worn: the lateral line is parallel to the line of the back throughout its length. When quite fresh, the colour of the body is a uniform greyish silver, slightly inclining to brown on the back, and lightest on the belly; but after keeping some days, the whole body became much darker. All the fins were reddish brown; the first dorsal, the pectoral, and ventral fins, rather more red than the others. The swimming-bladder in this species is peculiar, being fringed all round its edge. The figure of it here added is from the work of Cuvier and Valenciennes, before referred to.

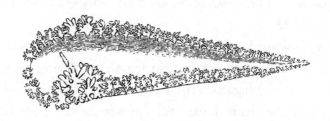

ACANTHOPTERYGII. *SCIÆNIDÆ.*

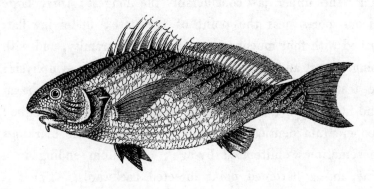

THE BEARDED UMBRINA.

Umbrina vulgaris, Cuv. et Valenc. Hist. Nat. des Poiss. t. v. p. 171.
Sciæna cirrosa, Linnæus.
   ,,   ,,   Bloch, pt. ix. pl. 300.

Generic Characters.—The Umbrina, besides the characters common to Sciæna, has a barbule, or cirrus, at the angle of the lower jaw ; the spines of the anal fin strong and sharp ; the teeth smaller and more numerous.

The Bearded Umbrina is a beautiful and excellent fish, which, though not attaining the size of the Maigre, is frequently taken two feet in length, and has been known to weigh forty pounds. It is very common on the coasts of Italy, France, and Spain. The flesh is white and of good flavour, and in considerable request, even at the best tables. Its food is small fishes, mollusca, and a particular sort of sea-weed, which have been found in its stomach.

On the British coast it appears to be a very rare visiter. In 1827, a fish, unknown to the oldest fisherman, was taken in the river Exe, which proved to be identical with that known at Gibraltar by the Spanish name of Umbrina, the

Sciæna cirrosa of Linnæus.—(Minute-book of the Linnæan Society.)

The head is short and blunt; the irides silvery, the pupil black; the upper jaw considerably the longest; three large mucous pores near the point of the nose; under jaw flat, marked with four mucous pores near its extremity, and with a single short and thick cirrus, or barbule, at the symphysis: the teeth very small, numerous, and arranged in a broad band in each jaw; none on the palate nor on the tongue: preoperculum denticulated while young, but these markings are sometimes obliterated by age; operculum ending in a spine, and a flattened point directed backwards. The fin-rays are :—

D. 10 — 22 : P. 17 : V. 1 + 6 : A. 2 + 7 : C. 17.

The lateral line parallel with, but much nearer, the dorsal line than in the Maigre; the scales large and rhomboidal; the ground colour of the body yellowish, traversed obliquely from the back downwards and forwards with bands of silver and blue; the belly white; dorsal fins brown, the second fin marked with two bars; pectoral and ventral fins nearly black; anal fin red.

The figure is taken from the work of Bloch, already quoted.

ACANTHOPTERYGII. *SPARIDÆ.**

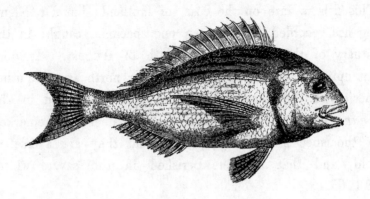

THE GILT-HEAD.

Chrysophrys aurata,	Cuv. et Valenc. Hist. Nat. des Poiss. t. vi. p. 85, pl. 145.
Aurata Rondeletii,	Willughby, p. 307, tab. V. 5.
Sparus aurata,	Linnæus.
,, ,,	Bloch, pt. viii. pl. 266.
,, ,,	Gilt-head, Penn. Brit. Zool. vol. iii. p. 327, but not plate 66.
,, ,,	,, Flem. Brit. An. p. 211, sp. 136 ?

Generic Characters.—Body deep, compressed ; dorsal fin single, the rays partly spinous, the posterior rays flexible ; teeth of two kinds, six incisors in each jaw, conical, with rounded and oval molar teeth, in four rows above and three rows below ; cheeks and operculum covered with scales ; branchiostegous rays 6.

The Gilt-head is one of the fishes most abundant in the Mediterranean : from Gibraltar it is found as far south as the Cape of Good Hope, and northward along the coast of Spain and France ; thence to the bold shore

* The family of Marine Bream.

VOL. I. H

of parts of our southern coast,—Colonel Montagu having examined two specimens taken at Torcross in 1802 ; and it is probably found occasionally at the Channel Islands. This fish is rare on the coast of Holland ; but Dr. Fleming has recorded having seen one specimen caught in the estuary of the Tay in the month of August. It does not appear, however, to proceed so far north as some other species of the same family, and is not included in the *Fauna* of Fabricius or Muller. Duhamel has remarked of the species of *Chrysophrys*, that they are averse to cold, and that numbers perished in the severe winter of 1766.

The species of *Chrysophrys*,—so called by the Greeks on account of their golden-coloured eyebrows, from whence also the names of *aurata, dorade*, and Gilt-head, have arisen, —like most of the *Sparidæ*, frequent deep water on bold rocky shores, from whence they are occasionally drawn by lines or nets. They are said to spawn in summer ; and their food consists of molluscous and testaceous animals, which their rounded teeth and strong jaws enable them to

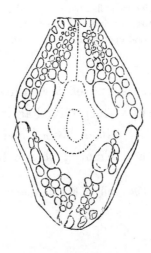

break down even in such thick and hard shells as those of the genera *Turbo* and *Trochus.*

The most ordinary form of teeth in fishes is that of an elongated cone, but varying greatly in size, and sometimes curving inwards : such has been the general form of those possessed by the different species already described. In the fish of the genus now under consideration, the teeth vary in shape, as the vignette will show : the varieties in the forms of the teeth in British fishes generally, the mode of growth and change, and the various bones to which they are attached, require to be noticed.

The forms of the teeth are not less varied than their position, and require various names. The most common form is that of an elongated cone, either straight or curved. When these conical teeth are small and numerous, they are compared to the points of the cards used for carding wool or cotton ; and they are sometimes so slender, yet so dense from their numbers, as to resemble the pile of velvet or plush ; and often, from their very minute size, their presence is more readily ascertained by the finger than by the eye. Some fishes have in the front of the jaws flat teeth with a cutting edge, like a true incisor : others have them rounded or oval ; they are then most frequently planted in rows, and adapted to bruise or crush the various substances with which they are brought in contact.

All the teeth of fishes are simple, each originating in its own simple pulpy germ.

Whatever the form of the tooth, it is produced by successive layers, as in the mammalia; but the growth is not directed downwards to form a root : there is no alveolar cavity ; the tooth consists only of that part which is usually called the crown, and it seems rather to be a production of the surface of the bone than of the interior.

H 2

The renewal of the teeth in fishes seems to take place at uncertain periods, apparently with some reference to the accidental wants of the animal; the new tooth sometimes grows beneath, sometimes at the side, or behind or before the old teeth, which are loosened at their attachment, not worn down, and thus thrown off.

Fishes may have teeth attached to all the bones that assist in forming the cavity of the mouth and pharynx; to the intermaxillary, maxillary, and palatine bones, the vomer, the tongue, the branchial arches supporting the gills, and the pharyngeal bones: there are genera, the species of which have teeth attached to all these various bones: sometimes these teeth are uniform in shape, at others differing. One or more of these bones are sometimes without teeth of any sort; and there are fishes that have no teeth whatever on any of them. The teeth are named with reference to the bone upon which they are placed, and are referred to as intermaxillary, maxillary, palatine, vomerine, &c., depending upon their position.

To return to the Gilt-head:—The body is deepest at the commencement of the dorsal fin: the head short and elevated; the irides golden yellow, the pupils black; the semilunar spot over the eye of a brilliant golden colour; and there is a violet-coloured patch at the upper part of the edge of the operculum: the scales of the cheeks smaller than those of the body: the teeth in an adult fish are as shown by the vignette, but in young fishes of this species the teeth are fewer in number. The fin-rays are:—

D. 11 + 13 : P. 16 : V. 1 + 5 : A. 3 + 11 : C. 17.

The back is silvery grey shaded with blue; the belly like polished steel, with longitudinal golden-coloured bands on

the sides, that give them a yellow appearance : the fins are
greyish blue ; the tail darker : the dorsal and anal fins
appear as if placed in grooves, from the rising edges of the
scales on each side. This fish seldom exceeds twelve inches
in length. The figures of the fish and teeth are derived
from the work of Cuvier and M. Valenciennes.

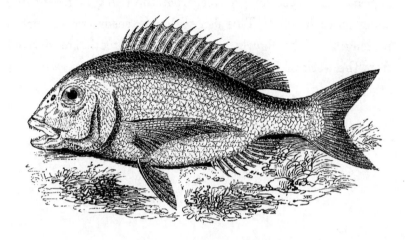

THE BRAIZE OR BECKER.

Pagrus vulgaris,	Cuv. et Valenc. Hist. Nat. des Poiss. t. vi. p. 142, pl. 148.
Sparus pagrus,	Linnæus.
,, ,,	*Becker,* Couch, Trans. Linn. Soc. vol. xiv. p. 79.
Pagrus vulgaris,	*Braize,* Flem. Brit. An. p. 211, sp. 137.

Generic Characters.—Four or six strong conical teeth in front, supported by smaller conical teeth behind them, with two rows of rounded molar teeth on each side of both jaws : the other characters as in the last genus *Chrysophrys.*

THERE is considerable similarity in outward form between the true *Pagrus,* the subject of the present article, and *Chrysophrys,* the fish last described ; but the red colour of the Braize, and the circumstance of its possessing but two rows of molar teeth, are sufficient to distinguish it. This fish was originally well figured by Rondeletius, lib. v. c. 15 ; but the number of the *Pagri* in the Mediterranean of a red colour, has led to some confusion in the accounts of many of the different authors since : neither Willughby nor Bloch can be quoted with certainty, and Pennant refers in his synonymes to both these authors, though they appear to have been considering two distinct fishes, neither of which accord with the true *Pagrus.* The name of this fish is said to be derived from *phagrus, e, phago,* 'to eat,' from its voracity; and its food is partly sea-weed, with shrimps and testaceous animals. Mr. Couch says that it appears on the Cornish coast in moderately deep water throughout the summer and autumn, but retires in winter and spring. The young are but rarely seen. In the North of Ireland, at Belfast bay, Antrim, and Londonderry, a fish belonging to the *Sparidæ* is taken, called the Brazier, which is said to be the *Pagrus,* but may, perhaps, prove to be the Sea Bream, *Pagellus centrodontus :*—farther observation is required to decide this species correctly. M. Risso says that in the Mediterranean this fish frequents deep water near rocks ; and the females are full of roe in summer. The number of fin-rays are :—

D. 12 + 10 : P. 15 : V. 1 + 5 : A. 3 + 8 : C. 17.

The muzzle is blunt, as in *Chrysophrys,* but the body is a little more elongated ; the eye large, the irides golden yellow; the mouth large ; the teeth are as described in the characters of the genus, and a representation of their arrangement forms the subject of the vignette. Part of the dorsal and anal fins

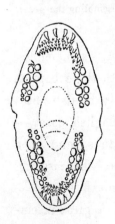

are hid in a groove formed by the elevation of the edges of the scales at the bases of the fins. The prevailing colour of the body is silvery tinged with red, without any metallic markings on the head, or any dark spot at the origin of the lateral line : fin-membranes white, tinged with rose colour ; the dorsal fin darkest at the posterior part ; the pectorals red, with occasionally a violet-coloured spot at their origin.

THE SPANISH BREAM.

Pagellus erythrinus,	Cuv. et Valenc. Hist. Nat. des Poiss. t. vi. p. 169, pl. 150.	
Sparus	,,	Linnæus.
Erythrinus Rondeletii,	Willughby, p. 311, tab. V. 6.	
,,	,,	*Spanish Bream,* Couch, Mag. Nat. Hist. vol. v. p. 17, fig. 3.
,,	,,	*Red Sea Bream,* Walcott's MS.

Generic Characters.—The teeth in front conical, slender, numerous; the molars rounded, smaller in size than in the preceding genera of the *Sparidæ,* those of the outer rank the most powerful : one dorsal fin, the rays of the anterior part spinous, the remainder flexible : in other respects resembling the genera *Chrysophrys* and *Pagrus.*

The *Pagellus erythrinus* of Cuvier and M. Valenciennes, the Spanish Bream of Mr. Couch,—who, with the exception of Mr. Walcott, seems to have been the only British naturalist acquainted with its appearance on the English coast,— was well known to Rondeletius and Salvianus, and is a common fish in the Mediterranean Sea, and when issuing thence, appears to pursue a course north and north-west.

" This species," says Mr. Couch, " bears a great resemblance to the Sea Bream, and will best be described in comparison with it. It is the size of a small Sea Bream; the body rather more slender; head flatter on the top; eyes smaller, inclining to oval: before the eye the head is more protruded; the mouth with a wider gape; front teeth as in the Sea Bream, grinders more broad and blunt; scarcely a depression before the eyes to receive the nasal orifices, though in the Sea Bream they are conspicuous: fins as in the Sea Bream, even to the numbers of the rays, except the pectoral, which in the Sea Bream reaches opposite to the third ray of the anal fin—in this fish opposite only to the vent: there is no lateral spot. This description agrees so well with what Ray delivers of the *Erythrinus, Syn. Pisc.* p. 132, that I suppose it to be the same fish. The name given above is that by which it is known to our fishermen. It is rare, as I have never seen above two or three specimens, which were taken with Sea Bream, and with the same kind of baits. Its habits seem to be like those of the Sea Bream."

To this may be added, that the food of this species consists of small fishes and testaceous animals. They swim in small shoals; visiting the shore in spring, and remaining till autumn. Neither Pennant nor Mr. Donovan have included the Spanish Bream in their accounts of British Fishes; but Mr. Walcott, whose MS. and drawings have been already mentioned, and will frequently be referred to, appears to have met with it at Teignmouth; and his drawings contain a most accurate representation of this fish.

The number of fin-rays are :—

D. 12 + 10 : P. 15 : V. 1 + 5 : A. 3 + 8 : C. 17.

The figure of this fish at the head of the page is from the work of Cuvier and M. Valenciennes. I should have availed myself of the drawing by Mr. Walcott, taken from an English specimen, but the wood-block had been engraved when the MS. and its illustrations came into my hands.

The teeth of this genus are represented in the vignette at the bottom of the page, but the teeth there figured belong to the next species of this genus,—the Sea Bream.

The colour of this fish when alive is a fine carmine red on the back, passing into rose colour on the sides, and becoming almost silvery white on the belly ; the membranes of the fins are rose colour, the anal and ventral fins being paler than the others.

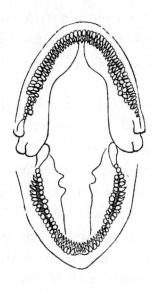

ACANTHOPTERYGII. *SPARIDÆ.*

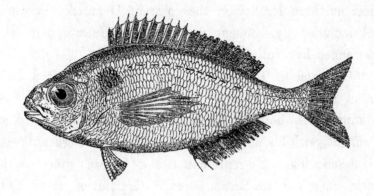

THE SEA BREAM.

Pagellus centrodontus,	Cuv. et Valenc. Hist. Nat. des Poiss. t. vi. p. 180.		
,,	,,	*Red Gilt-head,*	Penn. Brit. Zool. vol. iii. p. 329, and pl. 66, under the name of *Lunulated Gilt-head.*
,,	,,	*Lunulated Gilt-head,*	Don. Brit. Fish. pl. 89.
,,	,,	*Gilt-head,*	Flem. Brit. An. p. 211, sp. 136 ?
,,	,,	*Sea-Bream* of Couch and Montagu.	

The Sea Bream is a common fish in the Mediterranean, and in the ocean is taken frequently at Brest, Dieppe, and Boulogne: it is by no means an uncommon fish on the line of the southern shore of England, particularly on the coast of Sussex, and is constantly to be seen during summer and autumn in the fish-market at Hastings. Colonel Montagu obtained it in Devonshire; and Mr. Couch, whose account of this species I shall quote at some length, says it is abundant in Cornwall.

On the Irish coast, this fish may be traced from Waterford up the east coast to Dublin bay, and thence to Belfast bay and the north coast of Antrim, where it is called Murranroe and Barwin. On the east coast of England it is also not uncommon: Dr. Johnston has met with it in Berwick bay.

It is included by S. Nilsson in his *Prodomus Ichthyologiæ Scadinavicæ;* and Professor Reinhardt has ascertained its most northern locality on the coast of Denmark: but it is not included by Linnæus in his *Fauna Suecica,* nor is it mentioned by Muller or Fabricius.

"Common as this fish is," says Mr. Couch, "I have found a difficulty in assigning to it its proper synonymes. I suppose it, however, to be the Lunulated Gilt-head of Pennant, with his figure of which it agrees, though not with his description. He represents it as of a dusky green on the back, where our Sea Bream is red, with a tint of yellow. On the upper part of the gills, according to Pennant, is a black spot, and a purplish one beneath:" but our fish has only a broad dark brown spot at the origin of the lateral line. Dr. Fleming's description agrees with Pennant's; and Ray says it weighs ten pounds; but our fish would be thought enormous if of half that size.

The young fish, which are commonly known by the name of Chads, are without the lateral spot until their first autumn, when they are about half-grown.

The Sea Bream is found on the west coast of England throughout the year, but is most abundant in summer and autumn; and it retreats altogether in severely cold weather. The spawn is shed in the beginning of winter in deep water; and in January the Chads, about an inch in length, are found in the stomachs of large fishes, taken at two or three leagues from land: in summer, when from four to six inches long, they abound in innumerable multitudes, and are taken by anglers in harbours, and from the rocks; for they bite with great eagerness at any bait, even of the flesh of their own species. The food, both of the young and adult fish, is not, however, confined to animal substance; for they devour the green species of sea-weeds, which they bite from the rocks, and for bruising which their molar teeth are well suited, as are their long and capacious intestines for digesting it." In the stomach of one that was examined by Colonel Montagu, were several small sandlaunce, limbs of crustaceous animals, and fragments of shells. "In its general habits," Mr. Couch says, "the Sea Bream might be considered a solitary fish ; as when they most abound, the assemblage is formed commonly for no other purpose than the pursuit of food. Yet there are exceptions to this; and fishermen inform me of instances in which multitudes are seen congregated at the surface, moving slowly along as if engaged in some important expedition. This happens most frequently over rocky ground in deep water.

"The Sea Bream is not highly esteemed for the table, and is not at all in request when salted : hence, when abundant, I have known it sold at so low a rate as two shillings and sixpence the hundred weight !"

When at the sea-coast on fishing excursions, it has been one of my customs to eat of the various fishes I could either catch or purchase that are not in general use for the table. With the example of Isaac Walton before me, I will venture to suggest a mode of preparing a Sea Bream which materially improves its more ordinary flavour. When thoroughly cleaned, the fish should be wiped dry, but none of the scales should be taken off. In this state it should be broiled, turning it often, and if the skin cracks, flour it a little to keep the outer case entire. When on table, the whole skin and scales turn off without difficulty; and the muscle beneath, saturated with its own natural juices, which the outside covering has retained, will be found of good flavour.

The jaws are short, and equal in length ; the teeth as shown in the vignette page 106 : the eye very large, irides golden yellow : the head short, the line of the profile descends rapidly :

cheeks, operculum, and interoperculum covered with scales; preoperculum, part of the space before and under the orbit, of a metallic tinfoil appearance : two narrow stripes on each side behind the head, which meet on the central line at the top; at the origin of the lateral line, behind the edge of the operculum, a conspicuous dark patch made up of small spots : the colour of the body is reddish, tinged with grey; lighter on the sides, which are golden grey, and marked with faint longitudinal bands the whole length of the body : the belly nearly white; dorsal and anal fins brown, each appearing as if lodged in a groove, from the rising edges of the skin and scales along the base; pectorals and tail red; ventrals grey.

The number of fin-rays are :—

D. 12 + 13 : P. 17 : V. 1 + 5 : A. 3 + 12 : C. 17.

Among the drawings obligingly lent me by Mr. Couch, there is one of this fish in which a malformation has occurred in the want of intermaxillary bones. It is copied as a vignette. The fish is adult, as shown by its having acquired the dark spot; and it was taken on a line by a baited hook.

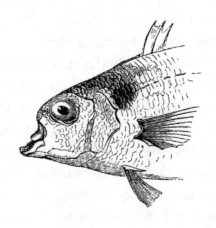

ACANTHOPTERYGII. *SPARIDÆ.*

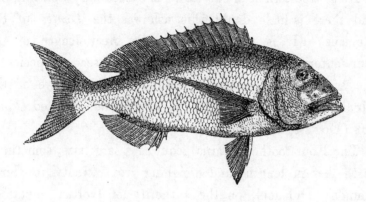

THE FOUR-TOOTHED SPARUS.

Dentex vulgaris, Cuv. et Valenc. Hist Nat. des Poiss. t. vi. p. 220, pl. 153.
   ,,   *Bellonii,* Willughby, p. 312, tab. V. 3.
Sparus dentex, Linnæus. Bloch, pt. viii. pl. 268.
  ,,   ,,   *Toothed Gilt-head,* Penn. Brit. Zool. vol. 331, but not the plate
                        bearing that name, which represents
                        Ray's Bream.
  ,,   ,,   *Four-toothed Sparus,* Don. Brit. Fish. pl. 73.
Dentex vulgaris, Flem. Brit. An. p. 212, sp. 139.

Generic Characters.—Body deep, compressed; dorsal fin single; head large; teeth conical, placed in a single row, four in the front, above and below, elongated, and curved inwards, forming hooks; teeth on the branchial arches, but none on the vomer or palatine bones : nose and suborbital space without scales ; branchiostegous rays 6.

The Four-toothed Sparus is here inserted as a British species on the authority of Mr. Donovan, the only English naturalist I am aware of who has recorded its capture on the British coast. In April 1805, a specimen of this fish, two feet six inches in length, which had been caught off Hastings, was brought to the London market, and, fortunately, fell into the hands of the author of the Natural History of British Fishes, who has given a good representation of it in his work, and whose English name for it is here adopted.

As a Mediterranean species it is exceedingly well known; and there is little doubt this fish was the *Dentex* of the Romans. It is remarkable for the great length of the four anterior teeth in each jaw; and a second species of the same genus, as now restricted—also a native of the Mediterranean—was from this peculiar character called *Cynodon* (Dog's-teeth).

The Four-toothed Sparus acquires a large size, sometimes three feet in length, and weighing from twenty to thirty pounds: Duhamel, on the authority of Gortier, mentions one instance of a *Dentex* that weighed no less than seventy pounds. They appear to be much more rare in the ocean, as well as smaller in size. The fish recorded by Mr. Donovan weighed sixteen pounds.

"A more voracious fish," says the same writer, "is scarcely known; and when we consider its ferocious inclination, and the strength of its formidable canine teeth, we must be fully sensible of the great ability it possesses in attacking other fishes, even of superior size, with advantage. It is asserted, that when taken in the fishermen's nets, it will seize upon the other fishes taken with it, and mangle them dreadfully. Being a swift swimmer, it finds abundant prey, and soon attains to a considerable size. Willughby observes, that small fishes of this species are rarely taken; and the same circumstance has been mentioned by later writers. During the winter it prefers deep waters; but in the spring, or about May, it quits this retreat, and approaches the entrance of great rivers, where it deposits its spawn between the crevices of stones and rocks.

"The fisheries for this kind of Sparus are carried on upon an extensive scale in the warmer parts of Europe. In the estuaries of Dalmatia and the Levant, the capture of this fish is an object of material consideration, both to the inhabitants generally as a wholesome and palatable food when fresh, and to the mercantile interests of those countries as an article of commerce. They prepare the fish, according to ancient custom, by cutting it in pieces, and packing it in barrels with vinegar and spices, in which state it will keep perfectly well for twelve months."

The fin-rays, according to Cuvier, are as follows :—

$$D. \ 11+11 : P. \ 14 : V. \ 1+5 : A. \ 3+7 : C. \ 17.$$

The form of the head is obtuse ; the character of the teeth is shown in the vignette, which is taken from Bloch's figure ; the eyes are rather small, the irides yellow ; the back is of a brownish red, slightly mottled with some darker spots ; the sides paler, and inclining to yellow ; the belly almost white. This fish is said to become of a greenish purple tint by age, and to be paler in colour during winter. The lateral line takes the curve of the back at an equal distance throughout its whole length, and at about one-fourth of the depth of the fish. All the fins pale reddish brown ; the rising edges of the skin and scales on each side the base of the dorsal and anal fins form grooves from which these fins appear to issue.

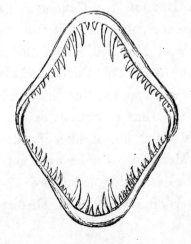

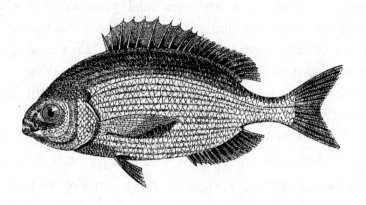

THE BLACK BREAM.

Cantharus griseus, Cuv. et Valenc. Hist. Nat. des Poiss. t. vi. p. 333.
Sparus lineatus,   *Black Bream,* Montagu, Mem. Wern. Soc. vol. ii. p. 451,
                 pl. 23. 1815.
Sparus vetula,   *Old Wife,*   Couch, Trans. Linn. Soc. vol. xiv. p. 79.
                 1822.
Pagrus lineatus, Flem. Brit. An. p. 211, sp. 138.

   Generic Characters.—Body deep, compressed ; a single elongated dorsal fin ; teeth of rather small size, numerous, conical, placed in several rows, those of the outer row rather larger and more curved than those forming the inner rows; mouth rather small ; branchiostegous rays 6.

THE BLACK BREAM, the *Cantharus griseus* of Cuvier and M. Valenciennes, was made known as a British fish in 1815, by Colonel Montagu, under the name of *Sparus lineatus ;* and in 1822, Mr. Couch included in his paper printed in the Transactions of the Linnean Society, a notice of a fish under the name of *Sparus vetula,* which that gentleman has since stated he considers identical with the *Sparus lineatus* of Montagu. Cuvier does not appear to have been aware of the description and figure of this fish in the Memoirs of the Wernerian Natural History Society, since, in 1830, in the sixth volume of the *Histoire Naturelle des Poissons*, he states, at page 319, that his fourth species, *C. griseus*, then appears for the first time ; but it had been also figured by Duhamel, under the name of *Sarde grise.*

Of the genus *Cantharus*, but one species, as far as I am acquainted, appears on our coast ; but some attention is necessary to the teeth of the different genera forming the *Sparidæ* of Cuvier.

The Black Bream,—for by this name is this species known along the Kentish and Sussex coasts, as well as in Devonshire,—though more rare than the Sea-Bream, *Pagellus centrodontus*, is not an uncommon species. The Zoological Society has received specimens from Madeira, sent by the Rev. R. T. Lowe. It is taken at Dieppe, Boulogne, and Calais : I have seen it at Dover and Hastings. Colonel Montagu saw it in considerable abundance on the coast in Devonshire, and Mr. Couch in Cornwall. They are taken by the hook, and also by the net ; are most abundant in July and August, but are not observed to grow so large as the Sea Bream. Mr. Couch says, " it takes the common baits which fishermen employ for other fish, but feeds much on marine vegetables, upon which it becomes exceedingly fat." It enters harbours, and is frequently taken by anglers from rocks and pier-heads ; but he has never known it assemble in shoals, and it is very rare to take the young of small size. Of three examples obtained by myself in the London market, the largest measures seventeen inches in length, and five inches and a half in depth, exclusive of the dorsal fin. The largest specimen recorded measured twenty inches in length.

The fin-rays are :—

D. 11 + 12 : P. 16 : V. 1 + 5 : A. 3 + 10 : C. 17.

From the upper and back part of the head two dark lines

I 2

descend to the upper edge of the operculum, enclosing between them a space covered with scales ; preoperculum, suborbital ring, nose, and the part over the eye, smooth ; cheeks, operculum, and interoperculum, covered with scales ; irides reddish orange ; lips and region of the mouth pale reddish brown : the prevailing colour of the body is bluish grey, marked with alternate dark and light narrow longitudinal bands, the centres of the scales being darker than the edges ; the lateral line darkest of all, and receding from the dorsal line as it approaches the top of the operculum : dorsal fin pale brown, and lodged in a groove throughout its whole length : the pectoral fins in colour resemble the body, as do also the rays of the ventral, anal, and caudal fins ; but the membranes of these fins are much darker, approaching to dusky lead colour : the upper division of the tail the largest.

The vignette of the teeth was drawn from the large specimen of seventeen inches before mentioned as obtained in the London market.

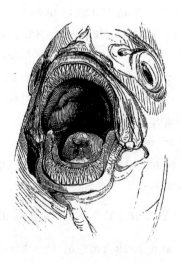

ACANTHOPTERYGII.   *SQUAMMIPENNES.**

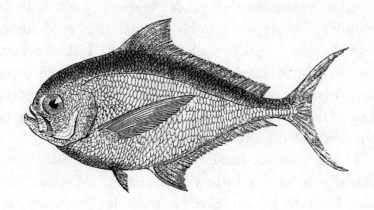

RAY'S BREAM.

Brama Raii, Cuv. et Valenc. Hist. Nat. des Poiss. t. vii. p. 281, pl. 190.

    „   „   Willughby, Appendix, p. 17, tab. V. 12.

Sparus Raii, Bloch, pt. viii. pl. 273.

    „   „   *Rayan Gilt-head,*   Penn. Brit. Zool. vol. iii. p. 330, pl. 54,
                           under the name of *Toothed Gilt-head.*

    „   „   *Ray's Toothed Gilt-head,* Don. Brit. Fish. pl. 131.

    „   „   „   „   „   MS. of Couch and Montagu.

Generic Characters.—Body compressed, deep ; profile of the head almost vertical ; a single elongated dorsal fin ; dorsal and anal fins with scales attached to the membranes ; teeth slender, incurved, placed on the jaws and palatine bones, sometimes with two in the front more elongated than the others ; branchiostegous rays 7.

The very peculiarly formed Marine Bream to which Ichthyologists have assigned the name of our celebrated countryman and naturalist John Ray, appears to have been less perfectly known to the older writers than might have been expected from its singular shape and prevailing numbers. It is figured by Duhamel, and also by Willughby and Bloch. Duhamel obtained his specimen from Provence:

With scales on the membranes of some of the fins.

the species is said to be common in the Mediterranean. Willughby has given a figure of this fish, tab. V. 12, which he calls *Brama marina cauda forcipata*; and it is described in the Appendix to his Natural History of Fishes, page 17, from a specimen obtained, on the 18th of September 1681, in Middlesburgh Marsh, near the mouth of the Tees, having been left there on the sands by the retiring tide. Bloch has figured and described it, as quoted in the synonymes at the head of this subject.

This fish cannot certainly be so rare or so little known generally as various authors have related. Colonel Montagu has recorded one example taken in Devonshire, and another at Swansea: Mr. Couch has obtained one or two, if not more, in Cornwall. It has been taken at Belfast, where it is called Henfish; and a correspondent in Mr. Loudon's Magazine of Natural History, vol. vi. p. 529, says this fish is not uncommon on the west coast of Scotland: he had himself seen several individuals from the Frith of Clyde and from the Argyleshire coast.

I may farther state, that there are two specimens in the British Museum, one in the collection of the Zoological Society, and probably others in London. In 1828, a specimen was taken on the coast of Normandy; another at Stockton-upon-Tees—the spot of its first recorded occurrence in England—in 1821; it has been taken in Berwick bay, and Mr. Neill has recorded that several have been taken in the Frith of Forth; it has also been taken at St. Andrew's.

In the autumn of 1834, I saw no less than nine examples of *Brama Raii* in the museums of Edinburgh, Newcastle-upon-Tyne, and York; including, besides, but two private collections.

Ray's Bream is mentioned in Nilsson's *Prodromus*, which

has been quoted before, as occurring on the coast of Norway ; and Professor Reinhardt, in a paper read before the Royal Society of Natural History and Mathematics of Denmark, has ascertained the northern limits of this species on that coast.

From this enumeration of specimens and localities, it will be evident that Cuvier, in his history of this fish, was deceived in supposing it exclusively peculiar to the Mediterranean, and that only a straggler occasionally wandered into the ocean ; and, on the contrary, that Bloch and Lacépède were perfectly justified in considering this fish a native of the Northern Seas, as well as of the Mediterranean.

The following description of a recent fish is from the MS. of Mr. Couch :—" The specimen was twenty-three inches in length, and eight and a half inches in depth before the dorsal fin ; the figure much compressed ; head small, sloping in front ; snout short ; angle of the mouth depressed ; under jaw longest ; teeth slender, numerous, sharp, incurved, the inner row of the lower jaw longest ; tongue fleshy ; eye large, rather oval, not far from the mouth ; iris dark, pupil light ; nostril single ; gill-cover with two plates, the membrane concealed, seven rays. Measuring along the curve, the dorsal fin begins seven and a half inches from the snout, having three shorter rays like blunt spines, each longer than that before it, the fourth ray longest ; the fin then becomes narrower, and continues slender to within an inch of the root of the tail ; anal fin shaped like the dorsal, beginning farther back, and ending opposite the former ; pectoral fin six inches long, rather narrow, pointing obliquely upwards ; ventrals triangular, with a long pointed scale in the axilla ; tail deeply forked ; lateral line near the back obscure ; head, body, and fins, except the pectorals and ventrals, covered with firmly fixed scales, but a band across the forehead is without them, the colour of which, and also of the back, is a very dark blue ; copper-coloured brown over and before the eye ; somewhat silvery on the sides and belly ; the anal and dorsal fins, and a stripe along the base of the latter, sparkle like silver ; tinted with green before the dorsal fin ; coppery and lake along the upper part of the sides ; some dusky irregular stripes along the other parts of the sides.

" The scales on the fins of this fish are arranged on the membranes in lines, so as to admit a slight degree of motion ; the points of the rays were also free.

" My fish was caught with a line near Polperro, October 26th, 1828, and was immediately brought to me : no elongated teeth were to be seen in this specimen."

The number of fin-rays are as follows :—

D. 34 : P. 19 : V. 1 + 5 : A. 2 + 28 : C 17.

The flesh of this fish is said to be of exquisite flavour : specimens have been taken that measured two feet six inches in length ; but of twelve or fourteen examples that I have seen, the largest did not exceed sixteen inches.

Rudolphi pointed out six species of worms which infested either the flesh or the intestines of this fish.

Cuvier and M. Valenciennes consider the *Chætodon* of Mr. Couch to be an example of Ray's Bream.

ACANTHOPTERYGII. *SCOMBERIDÆ.**

THE MACKEREL.

Scomber scomber, LINNÆUS. BLOCH, pt. ii. pl. 54.
,,   scombrus, CUV. et VALENC. Hist. Nat. des Poiss. t. viii. p. 6.
,,   ,,   Common Mackrel, PENN. Brit. Zool. vol. iii. p. 357, pl. 62.
,,   ,,   Mackarel,   DON. Brit Fish. pl. 122.
,,   vulgaris, Mackerel, FLEM. Brit. An. p. 217, sp. 161.

Generic Characters.—Scales on the body small and smooth ; vertical fins not bearing scales ; two dorsal fins widely separated ; some of the posterior rays of the second dorsal and the anal fin free, forming finlets ; sides of the tail slightly carinated ; one row of small conical teeth in each jaw ; the parts of the gill-cover without denticulations or spines ; branchiostegous rays 7.

THE MACKEREL is so well known for the beauty and brilliancy of its colours, the elegance of its form, its intrinsic value to man as an article of food, both in ˙reference to quantity as well as quality, that farther observation on these points will be considered unnecessary.

The Mackerel was supposed by Anderson, Duhamel, and others, to be a fish of passage ; performing, like some birds, certain periodical migrations, and making long voyages from north to south at one season of the year, and the reverse

* The family of the Mackerel.

at another.   It does not appear to have been sufficiently considered, that, inhabiting a medium which varied but little either in its temperature or productions, locally, fishes are removed beyond the influence of the two principal causes which make a temporary change of situation necessary.   Independently of the difficulty of tracing the course pursued through so vast an expanse of water, the order of the appearance of the fish at different places on the shores of the temperate and southern parts of Europe is the reverse of that which, according to their theory, ought to have happened.   It is known that this fish is now taken, even on some parts of our own coast, in every month of the year.   It is probable that the Mackerel inhabits almost the whole of the European seas ; and the law of nature, which obliges them and many others to visit the shallower water of the shores at a particular season, appears to be one of those wise and bountiful provisions of the Creator, by which not only is the species perpetuated with the greatest certainty, but a large portion of the parent animals are thus brought within the reach of man ; who, but for the action of this law, would be deprived of many of those species most valuable to him as food.   For the Mackerel dispersed over the immense surface of the deep, no effective fishery could be carried on ; but, approaching the shore as they do from all directions, and roving along the coast collected in immense shoals, millions are caught, which yet form but a very small portion compared with the myriads that escape.

This subject receives farther illustration from a freshwater fish, as stated in the Magazine of Natural History, vol. vii. p. 637.   " When the char spawn, they are seen in the shallow parts of the rocky lakes (in which only they are found), and some of the streams that run into them : they are then taken in abundance ; but so soon as the spawning is over, they retire into the deepest parts of the lake, and are but rarely caught."

It may be observed farther, that as there is scarcely a month throughout the year in which the fishes of some one or more species are not brought within the reach of man by the operation of the imperative law of nature referred to, a constant succession of wholesome food is thus spread before him, which, in the first instance, costs him little beyond the exercise of his ingenuity and labour to obtain.

On the coast of Ireland, the Mackerel is taken from the county of Kerry in the west, along the southern shore, eastward to Cork and Waterford; from thence northward to Antrim, and north-west to Londonderry and Donegal. Dr. M'Culloch says it visits some of the lochs of the Western Islands, but is not considered very abundant. On the Cornish coast, this fish in some seasons occurs as early as the month of March, and appears to be pursuing a course from west to east. They are plentiful on the Devonshire coast, and swarm in West Bay about June. On the Hampshire and Sussex coast, particularly the latter, they arrive as early as March; and sometimes, as will be shown, even in February: and the earlier in the year the fishermen go to look for them, the farther from the shore do they seek for and find them. Duhamel says the Mackerel are caught earlier at Dunkirk than at Dieppe or Havre: up our own eastern coast, however, the fishing is later. The fishermen of Lowestoffe and Yarmouth gain their great harvest from the Mackerel in May and June. Mr. Neill says they occur in the Forth at the end of summer; and Mr. Low, in his *Fauna Orcadensis*, states that they do not make their appearance there till the last week in July or the first week in August.

The Mackerel spawns in June; and, according to Bloch,

five hundred and forty thousand ova have been counted in one female. I have observed, by the Mackerel sent to the London market from the shallow shores of Worthing and its vicinity, that these fish mature and deposit their roe earlier on that flat sandy shore than those caught in the deep water off Brighton. The young Mackerel, which are called Shiners, are from four to six inches long by the end of August. They are half grown by November; when they retire, says Mr. Couch, "to deep water, and are seen no more that winter: but the adult fishes never wholly quit the Cornish coast; and it is common to see some taken with lines in every month of the year." Their principal food is probably the fry of other fish; and at Hastings the Mackerel follow towards the shore a small species of *Clupea*, which is there called in consequence the Mackerel mint. I have been unable hitherto to obtain any specimens of this small fish; but, from various descriptions, I think it is probably the young of the sprat. It is described as being about one inch long in July.

The Mackerel as feeders are voracious, and their growth is rapid. The ordinary length varies from fourteen to sixteen inches, and their weight is about two pounds each: but they are said to attain the length of twenty inches, with a proportionate increase in weight. The largest fish are not, however, considered the best for the table.

As an article of food, they are in great request; and those taken in the months of May and June are generally considered to be superior in flavour to those taken either earlier in spring or in autumn. To be eaten in perfection, this fish should be very fresh: as it soon becomes unfit for food, some facilities in the way of sale have been afforded to the dealers in a commodity so perishable. Mackerel were first allowed to be cried through the streets of London on a

Sunday in 1698, and the practice prevails to the present time.

At our various fishing-towns on the coast, the Mackerel season is one of great bustle and activity. The frequent departures and arrivals of boats at this time form a lively contrast to the more ordinary routine of other periods; the high price obtained for the early cargoes, and the large return gained generally from the enormous numbers of this fish sometimes captured in a single night, being the inducement to great exertions. A few particulars from various sources may not be uninteresting.

In May 1807, the first Brighton boat-load of Mackerel sold at Billingsgate for forty guineas per hundred,—seven shillings each, reckoning six score to a hundred; the highest price ever known at that market. The next boat-load produced but thirteen guineas per hundred. Mackerel were so plentiful at Dover in 1808, that they were sold sixty for a shilling. At Brighton, in June of the same year, the shoal of Mackerel was so great, that one of the boats had the meshes of her nets so completely occupied by them, that it was impossible to drag them in; the fish and nets therefore, in the end, sunk together; the fishermen thereby sustaining a loss of nearly 60*l.*, exclusive of what the cargo, could it have been got into the boat, would have produced. The success of the fishery in 1821 was beyond all precedent. The value of the catch of sixteen boats from Lowestoffe, on the 30th of June, amounted to 5252*l.*; and it is supposed that there was no less an amount than 14,000*l.* altogether realized by the owners and men concerned in the fishery of the Suffolk coast.* In March 1833, on a Sunday, four Hastings' boats

* In an interesting and useful sketch of the Natural History of Yarmouth and its neighbourhood, by C. and J. Paget, it is stated at page 16, that, in 1823, one hundred and forty-two lasts of Mackerel were taken there. A last is ten thousand.

brought on shore ten thousand eight hundred Mackerel; and the next day, two boats brought seven thousand fish. Early in the month of February 1834, one boat's crew from Hastings cleared 100*l.* by the fish caught in one night; and a large quantity of very fine Mackerel appeared in the London market in the second week of the same month. They were cried through the streets of London three for a shilling on the 14th and 22nd of March 1834, and had then been plentiful for a month. The boats engaged in fishing are usually attended by other fast-sailing vessels, which are sent away with the fish taken. From some situations, these vessels sail away direct for the London market; at others, they make for the nearest point from which they can obtain land-carriage for their fish. From Hastings and other fishing-towns on the Sussex coast the fish are brought to London by vans, which travel up during the night.

The most common mode of fishing for Mackerel, and the way in which the greatest numbers are taken, is by drift-nets. The drift-net is twenty feet deep, by one hundred and twenty feet long; well corked at the top, but without lead at the bottom. They are made of small fine twine, which is tanned of a reddish brown colour, to preserve it from the action of the sea-water; and it is thereby rendered much more durable. The size of the mesh about two and a half inches, or rather larger. Twelve, fifteen, and sometimes eighteen of these nets are attached lengthways, by tying along a thick rope, called the drift-rope, and at the ends of each net, to each other. When arranged for depositing in the sea, a large buoy attached to the end of the drift rope is thrown overboard, the vessel is put before the wind, and, as she sails along, the rope with the nets thus attached is passed over the stern into the water till the whole of the nets are run out. The net thus deposited hangs suspended in the water perpendicularly twenty feet deep from the drift-rope, and extending from three quarters of a mile to a mile, or even a mile and a half, depending on the number of nets belonging to the party or company engaged in fishing together. When the whole of the nets are thus handed out, the drift-rope is shifted from the stern to the bow of the vessel, and she rides by it as if at anchor. The benefit gained by the boat's hanging at the end of the drift-rope is, that the net is kept strained in a straight line, which, without this pull upon it, would not be the case. The nets are shot in the evening, and sometimes hauled once during the night, at others allowed to remain in the water all night. The fish roving in the dark through the water, hang in the meshes of the net, which are large enough to admit them beyond the gill-covers and pectoral fins, but not large enough to allow the thickest part of the body to pass through. In the morning early, preparations are made for hauling the nets. A capstan on the deck is manned, about which two turns of the drift-rope are taken. One man stands forward to untie the upper edge of each net from the drift-rope, which is called, casting off the lashings ; others hand in the net with the fish caught, to which one side of the vessel is devoted; the other side is occupied by the drift rope, which is wound in by the men at the capstan. The whole of the net in, and the fish secured, the vessel runs back into harbour with her fish ; or, depositing them on board some other boat in company, that carries for the party to the nearest market, the fishing-vessel remains at sea for the next night's operation.

Near to land, another mode of fishing is adopted, which is thus described by Mr. Couch in his MS :—" A long deep net is employed, of which, unlike the former, the

meshes are too small to admit any of the fish. Two boats
are necessary ; one of which is rowed round the schull,*
while the net is thrown overboard by two men to enclose
it ; the other boat is employed in keeping steady the end
of the net, and warping it, the sooner and more surely to
prevent the escape of the fish. When this is effected, the
seine stands like a circle enclosing the captives, and the men
proceed to draw it together at the ends and bottom ; at the
same time throwing pebbles at that place where the circle
closes, to prevent the approach of the fish to the only place
where escape is possible. When at last the enclosure is
perfect, and the net raised from the ground, the fish thus
brought to the surface are taken on board in flaskets. Such
is the mode of proceeding with the seine in deep water, or
at a distance from shore ; but in some places it is hauled on
the beach in the manner of a ground-net, with less trouble
and expense."

" A third mode of fishing is with the line, and is called
railing (trailing). The Mackerel will bite at any bait that
is used to take the smaller kinds of fish ; but preference is
given to what resembles a living and active prey, which is
imitated by what is termed a lask,—a long slice cut from
the side of one of its own kind, near the tail. It is found,
also, that a slip of red leather, or a piece of scarlet cloth,
will commonly succeed ; and a scarlet coat has therefore been
called a Mackerel bait for a lady. The boat is placed under
sail, and a smart breeze is considered favourable; hence
termed a Mackerel breeze. The line is short, but is weighed
down by a heavy plummet ; and in this manner, when these

* Shoal.

——— ——— In sculls that oft
Bank the mid sea.
                                        MILTON.
This word is in Cornwall, I have been told, pronounced like school.

fish abound, two men will take from five hundred to a thousand in a day. It is singular that the greatest number of Mackerel are caught when the boat moves most rapidly, and that even then the hook is commonly gorged. It seems that the Mackerel takes its food by striking across the course of what it supposes to be its flying prey. A gloomy atmosphere materially aids this kind of fishing for Mackerel."

Mr. Couch adds, that " French fishing-boats from the eastern ports of that country proceed early in the spring as far west as Cape Clear, and the fish taken in their nets are salted in bulk on board the boats. They even obtain two or three full cargoes in the course of the summer ; which proves that more use is made of salted Mackerel in France than in this country." A small quantity is so preserved in Cornwall, which is consumed by the poorer classes.

The vignette, from a pen-and-ink sketch by Mr. Couch, represents the apparatus as used when fishing for Mackerel. The ascending line is that which hangs from the boat ; the line connecting the leaden plummet and the hook is called the snood or snoozing ; the bait is cut thick near the hook, and thinner backwards, that it may vibrate when drawn through the water. The number of fin-rays in the Mackerel are,—
D. 10 — 13 — V. : P. 13 : V. 6 : A. 11 — V. : C. 22.
The nose is pointed ; the under jaw the longest ; the teeth

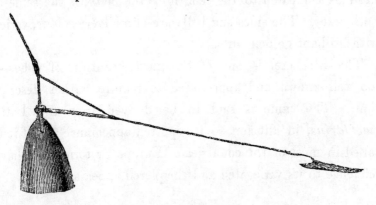

are alike in both jaws, resembling small pointed cones, curving slightly inward, and placed in a single row; the anterior edge of the eye one third of the distance from the point of the upper jaw to the edge of the operculum; the irides partly concealed by a membrane before and behind; the eye itself said to be more or less opaque during the colder months: preoperculum triangular; operculum large, rounded, and smooth: the pectoral and ventral fins both in advance of the first ray of the first dorsal fin, the pectoral fin the most so; the vent in a line under the first ray of the second dorsal fin; the five finlets above and below the fleshy portion of the tail, behind the second dorsal and the anal fins, placed vertically over each other: the tail crescent-shaped; the lateral line ascends gradually from the tail to its termination over the pectoral fin: the colour of the back above the lateral line is a fine green, varied with rich blue, and marked with broad, dark descending lines.   Mr. Donovan says, "the males have these dark transverse bands nearly straight; while in females these bands are elegantly undulated." The elongated gill-cover and more attenuated form of body of the males of fish generally, compared with the shorter gill-cover and deeper body of the females, are good sexual distinctions; and in consequence, the relative length of the head as compared to the length of the body, is the same in both sexes.   The sides and belly are of a silvery colour, varied with brilliant golden tints.

The Mackerel is one of the most beautiful of fishes— too well known and appreciated to require farther description.   The name is said to be derived from the Latin *macularius*, in allusion to its spotted appearance; and it is called in most of the countries of Europe by terms that have reference to its variegated and chequered appearance.

ACANTHOPTERYGII.   *SCOMBERIDÆ.*

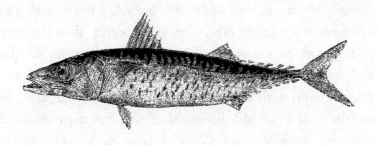

THE SPANISH MACKEREL.

Scomber colias,   GMELIN.
   ,,   ,,   CUV. et VALENC. Hist. Nat. des Poiss. t. viii. p. 39, pl. 209.
   ,,   ,,   *Coly Mackrel,*   TURTON, Brit. Fauna, p. 100, sp. 76.
   ,, *maculatus, Spanish Mackerel,* COUCH, Mag. Nat. Hist. vol. v. p. 22,
            fig. 8.

MR. COUCH and Dr. Turton appear to be the only British naturalists who have noticed this second species of
Mackerel on the British coast; and the description of this
fish by Mr. Couch, who states that a few of them are taken
every year on the Cornish coast, agrees so closely with the
account by Cuvier and M. Valenciennes in the Ichthyological work above quoted, who, with Mr. Couch, consider it
as the *colias* of Rondeletius, that I shall, by permission,
adopt his description, taken from a recent Cornish specimen.

" This fish attains the weight of four or five pounds; but
the specimen described measured no more than fourteen and
a half inches in length : the figure round and plump, six and
a half inches in compass near the pectoral fins; the thickness
of its figure being carried far towards the tail.   Mouth large ;
jaws of equal length ; teeth small ; tongue moveable and
pointed : head large and long ; eye large : from the snout

K 2

to the pectoral fin three and a half inches. Rays of the gill-membrane six, concealed. Lateral line at first slightly descending, then straight. Scales on the superior plate of the gill-covers, as well as on the body. First dorsal fin in a depression ; seven rays, the first shorter than the second or third, which are of equal lengths : spurious fins six above and below, the anterior not high : tail divided, and at its origin doubly carinated : vent prominent. Colour dark blue on the back ; striped like a Mackerel, but more obscurely, and with fewer stripes : a row of large dark spots from the pectoral fin to the tail ; sides and belly thickly covered with smaller dusky spots : the tail, gill-covers, sides, and behind the eye, bright yellow.

" From the Mackerel, which it resembles, this fish differs in the markings of the head, longer snout, larger eye and gape, longer head, and in having scales on the anterior gill-covers. The body is not nearly so much attenuated posteriorly ; the ventral fins are sharp and slender, those of the Mackerel wider and more blunt : in the former the pectorals lie close to the body, in the latter they stand off; in the latter, also, is a large angular plate, the point directed backward, close above each pectoral fin, which does not exist in the Spanish Mackerel.

" It seems to be the *Colias Rondeletii* of Ray (*Syn. Pisc.* p. 59). I have given it the name by which it is known to our fishermen."

" This fish is scarce, but some are taken every year. It does not often take a bait, although the fishermen inform me that this sometimes happens, and that its infrequency is owing to the difference of feeding rather than to want of rapacity. It is more frequently taken in drift-nets ; but even then it is only one at a time, and at considerable intervals. It is in no estimation as food."

The figure of this fish at the head of the page is from the first plate quoted.

Dr. Turton states, that the species he has described under the name of the Coly Mackrel is found frequently in the rivers about Swansea, and seldom exceeds six or seven inches in length : he also adds, that it is varied with rich green and blue ; spurious fins five above and below.   Although Dr. Turton has called his second species of Mackerel *S. colias*, it is possible that his fish may prove to belong to a third species, which also occurs in the Mediterranean.   Cuvier and M. Valenciennes have described a species of Mackerel of small size, which is decidedly more green in its colour, has five spurious fins above and below, and seldom exceeds eight or ten inches in length.   Mr. Couch describes *S. colias* as possessing six spurious fins above and below.

The name of this small-sized species of Mackerel is *S. pneumatophorus ;* so called by M. Laroche, on account of its possessing a swimming-bladder.   *S. colias* is also provided with a swimming-bladder : the common Mackerel, *S. scombrus,* Linn. is, as before mentioned, without any.

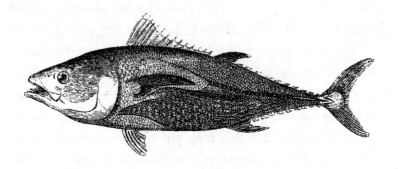

THE TUNNY.

Thynnus vulgaris, Cuv. et Valenc. Hist. Nat. des Poiss. t. viii. p. 58, pl. 210.
Scomber thynnus, Linnæus. Bloch, pt. ii. pl. 55.
    ,,   ,,   *Tunny*, Penn. Brit. Zool. vol. iii. p. 360, pl. 63.
    ,,   ,,   ,,   Don. Brit. Fish. pl. 5.
    ,,   ,,   ,,   Flem. Brit. An.

Generic Characters.—Form of the body like that of the Mackerel, but less compressed ; numerous scales surrounding the thorax : first dorsal fin extending nearly to the second ; second dorsal and the anal fin subdivided posteriorly, forming numerous finlets : sides of the tail decidedly carinated ; a single row of small pointed teeth in each jaw ; branchiostegous rays 7.

The Tunny was known to Aristotle ; and its goodness, in addition to its beauty, have caused this fish to be the praiseworthy theme of most of the writers on the fishes of the Mediterranean, ancient as well as modern. The fishery, also, is of great antiquity as well as value.

The Tunny is said to acquire a very large size. Although the specimens usually taken seldom exceed four feet in length, and frequently not more than three, Pennant saw one killed in 1769, when he was at Inverary, that weighed four hundred and sixty pounds, measuring seven feet ten inches long ; and they are recorded to have been taken of still greater bulk in the Mediterranean. There the habits of this fish have been studied with attention, the immense numbers affording great facilities. The roe is said to be of very rapid growth, and is deposited early in June. In July the young Tunnies do not weigh more than an ounce and a half; in August they weigh four ounces; and in October they weigh thirty ounces. In the months of May and June, when seeking a proper situation near the shore upon which to deposit their spawn, the adult fish rove along the coast in large shoals, and are known to be extremely timid, easily induced to take a new and apparently an open course to avoid any suspected danger. Advantage has been taken of these peculiarities to carry on a most extensive fishery against them at various places, which is as valuable as it is destructive. Cuvier and M. Valenciennes have described the two most common modes of effecting their capture. When the lookout sentinel, posted for that purpose on some elevated spot, makes the signal that he sees the shoal of Tunnies approaching, and the direction in which it will come, a great number of boats set off under the command of a chief, range themselves in a line forming part of a circle, and joining their nets, form an enclosure which alarms the fish, while the fishermen drawing closer and closer, and adding fresh nets, still continue driving the Tunnies toward the shore. When they have reached the shallow water, a large net is used, having a cone-shaped tunnel to receive the fish, which is drawn to the shore, bringing with it all the shoal. The fishermen carry out the young and small Tunnies in their arms; the larger ones are first killed with poles. This fishery, practised on the coasts of Languedoc, sometimes yields many hundred weight at each sweep of the nets.

Another mode of taking Tunnies is by the *madrague*, or, as the Italians call it, *tonnaro*. This is a more complicated engine, and somewhat expensive to set up. It consists of a series of long and deep nets fixed vertically by corks at their upper edges, and with lead and stones at the bottom. These are kept in a particular position by anchors, so as to form an enclosure parallel to the coast, sometimes extending an Italian mile in length : this is divided into several chambers by nets placed across, leaving narrow openings on the land side. The Tunnies, which in their progress, as before observed, proceed along the coast, pass between it and the tonnaro : when arrived at the end, they are stopped by one of the cross nets, which closes the passage against them, and obliges them to enter the tonnaro by the opening that is left for them. When once in, they are driven by various means from chamber to chamber to the last, which is called the chamber of death. Here a strong net placed horizontally, that can be raised at pleasure, brings the Tunnies to the surface, and the work of destruction commences. Sailors who have come off in boats for the purpose give unequal battle on all sides, striking the Tunnies with poles and all sorts of similar weapons. This imposing spectacle, which attracts a great number of curious people to witness it, is one of the great amusements of rich Sicilians, and, at the same time, one of the most considerable branches of the commerce of the island. When Louis the Thirteenth visited Marseilles, he was invited to a Tunny fishing at the principal *madrague* of Morgion; and found the diversion so much to his taste, that he often said it was the pleasantest day he had spent in his whole progress through the south.

The mode of curing the fish consists in taking out the whole of the inside, washing the flesh with brine, and cutting it in slices, which they cover with pounded salt. This is packed in layers in barrels, with alternate layers of salt. When sent to any distance, it is packed in smaller barrels with fresh salt.

The flesh of the Tunny is considered very delicious food ; but it is so solid, that it seems something between fish and meat : it is as firm as Sturgeon, but finer flavoured. " They dress this fish in France," says an author, " in a great variety of ways, and always excellent : it makes capital soup ; or it is served as a ragout, or plain fried or broiled : pies are made of it, which are so celebrated as to be sent all over France ; they will keep good for six weeks or two months. There is also a mode of preserving it to keep the whole year round, with salt and oil, called *Thon mariné :* this is eaten cold, as we eat pickled salmon." The flesh before it is cooked has the red appearance of beef, but when dressed it becomes more pale.

In the ocean, and on the western shores of the European Continent, the appearance of the Tunny is more rare,—almost accidental. Duhamel records having known it to be taken off Brest harbour. Mr. Couch has noticed their appearance on the Cornish coast, and will be referred to again. Mr. Donovan states that, in 1801, three Tunnies were taken near the entrance of the river Thames, and brought to Billingsgate market for sale. Mr. Paget says that small specimens are not unfrequently caught during the Mackerel fishery off Yarmouth. They have been taken among the islands west and north of Scotland, where they are called Mackrelsture or Mackerelstawr (Great Mackerel) ; a name derived from the Norwegians,—or, according to other authors, from the Danish word *stor*, which signifies ' great.'

Dr. Scouler has communicated to the Magazine of Natural History a notice (vol. vi. p. 559,) of a specimen of the Tunny taken in the Gair-loch, nearly opposite Greenock, in July 1831. It had entered the loch in pursuit of Herrings, got entangled among the nets, was sent by the fishermen to Glasgow, and is now deposited in the Andersonian Museum.

This specimen exceeded the average size, being nine feet in length.

Mr. Couch in his MS. states that " the Tunny appears on the Cornish coast in summer and autumn; but is not often taken, because it does not swallow a bait, or at least the fishermen use no bait that is acceptable to it; and its size and strength seldom suffer it to become entangled in their nets. It feeds on Pilchards, Herrings, and perhaps most other small fishes; but the Skipper, *Esox saurus,* seems to be a favourite prey; for it not only compels it to seek another element for safety, but will also spring to a considerable height after it,—usually across its course, at the same time attempting to strike down its prey with its tail. Osbeck says it feeds eagerly on the cuttle."

The fin-rays are as follows :—

D. 14—1+13—VIII : P. 31 : V. 1+5 : A. 2 + 12—VIII : C. 17 to 19.

The general form of the Tunny is similar to that of the well-known Mackerel, except that it is larger, more rounded, and that the jaws are shorter; the lower jaw is very little longer than the upper; the mouth is not deeply cleft; each jaw is furnished with a row of small teeth as sharp as pins, and slightly curved inwards; the tongue and the inside of the mouth very dark-coloured, almost black : the eye is surrounded by a membrane within the orbit, which covers part of its disk ; the cheeks are covered with long, narrow, pointed scales; the operculum smooth : the first dorsal fin is lodged in an elongated depression on the back, which conceals it when it is folded down ; a small spine before the commencement of the second dorsal fin, which fin is followed by nine finlets, which might be considered as ten, an apparent tenth being sometimes detached from the tail : the anal fin, preceded by two short spines, commences nearly on a line with

the origin of the last ray of the second dorsal fin, and is
followed by nine finlets : the tail crescent-shaped ; the membrane forming the lateral horizontal ridge on the fleshy portion of each side of the tail is produced, forming part of a
circle.

All the upper part of the body of the Tunny is very dark
blue ; the part of the corslet marked with scales is much
lighter ; the sides of the head white ; the whole of the belly
greyish white, spotted with silvery white ; these spots are
elongated towards the shoulders and flanks, but rounded over
the spaces between. The first dorsal fin, pectorals, and ventrals,
are black ; the tail paler ; second dorsal and anal fins almost
flesh colour, tinted with silver ; the finlets above and below
are yellowish, tipped with black.   The figure is taken from
the plate of Cuvier and M. Valenciennes, who consider most
of the figures of this fish more or less incorrect.

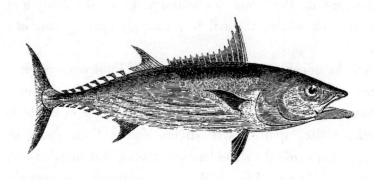

THE BONITO,

OR STRIPED-BELLIED TUNNY.

Thynnus pelamys, Cuv. et Valenc. Hist. Nat. des Poiss. t. viii. p. 113, pl. 214.
Scomber   ,,   Linnæus.
    ,,   ,,   *Bonito,* Couch's MS.

Specimens of this fish occasionally occurring on the
Cornish coast, the description of which by Mr. Couch will
here be inserted ; another, mentioned in the Magazine of
Natural History, vol. vi. p. 529, by Dr. Scouler ; besides a
notice by Dr. Fleming, on the authority of Stewart's Ele-
ments of Natural History,—the species is admitted among
British Fishes.

It should, however, be also stated that two distinct fishes
have been included under the term Bonito (*Scomber pelamys,*
Linn.) : the first, the *Thynnus pelamys* of Cuvier and M.
Valenciennes, the subject of the present article, has longitudinal bands on the sides of the belly, and very minute
teeth ; the second, the *Pelamys sarda* of Cuvier and M.
Valenciennes, has dark transverse bars reaching from the
ridge of the back to the lateral line, and large teeth ; and

this species has not, that I am aware, been taken on the shores of this country.

The Bonito is very similar to the Tunny in form, but is much smaller, seldom exceeding thirty inches in length. It inhabits the ocean, and is one of those species so well known to voyagers when within the tropics for the amusement they afford by their pursuit of the Flying Fish. Their attempts, however, to secure these unfortunate victims, sometimes lead to their own destruction. Sailors frequently amuse themselves by catching the Bonito with a hook fastened to a piece of lead shaped like the body of a small fish, to which a pair of wings made of feathers are attached, to give it the appearance of a Flying Fish. The food of the Bonito is fish, small cuttles, testaceous animals, and even marine vegetables. Though eaten with avidity by those who have been previously confined to salt provisions, the flesh has been considered dry, and by some even said to be disagreeable.

This fish is subject to several sorts of intestinal worms.

A specimen obtained by Mr. Couch on the Cornish coast " was twenty-nine inches long, and twenty inches round close behind the pectoral fins : head conical, ending in a point at the nose ; under jaw projecting ; teeth few and small ; tongue flat and thin ; nostrils obscure, not in a depression ; from the nose to the eye two and a half inches ; gill-covers of two plates. Body round to the vent, from thence tapering to the tail ; near the tail depressed ; lateral line at first descending and waved, becoming straight opposite the anal fin, from thence ascending and terminating in an elevated ridge, with another above and below the lateral line near the tail. Eye elevated, round ; iris silvery : from the nose to the pectoral fin eight and three-quarter inches, the fin pointed, four inches long, received into a depression. First dorsal fin seven inches long, four inches high, lodged in a groove ; the first two rays stout, the others low. The body is most solid opposite the second dorsal, which fin and the anal are falcate: tail divided and slender; ventral fins in a depression. Colour a fine steel blue, darker on the back; sides dusky, whitish below. Behind the pectoral fins is a bright triangular section of the surface, from which begin four dark lines, that extend along each side of the belly to the tail. Scales few, like the Mackerel.

This fish was taken in a drift-net in July, at which time the roe was abundant. It had no air-bladder; intestines simple; the muscle the colour of beef, greatly charged with blood. It rarely takes a bait, and is too wary to be often taken in a net.

Dr. Scouler states that a specimen of this tropical fish was taken in the Frith of the Clyde in July 1832. The specimen referred to by Dr. Fleming was taken in the Forth. The number of fin-rays are,—

D. 15—1 + 12. VIII : P. 27 : V. 1 + 5 : A. 2 + 12. VII : C. 35.

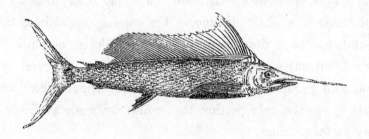

THE SWORDFISH.

Xiphias gladius, Linnæus. Bloch, pt. iii. pl. 76.
   ,,   ,,   Cuv. et Valenc. Hist. Nat. des Poiss. t. viii. p. 255, pl.
          225-6.
   ,,   ,,   *Swordfish,* Penn. Brit. Zool. vol. iii. p. 216, pl. 30.
   ,,   ,,   ,,   Flem. Brit. An. p. 220, sp. 169.

Generic Characters.—Body fusiform, covered with minute scales; a single
elongated dorsal fin; ventral fins wanting; tail strongly carinated; upper jaw
elongated, forming a sword; mouth without teeth; branchiostegous rays 7.

The Swordfish, inhabiting almost every part of the
Mediterranean Sea, was well known to the ancients, and was
called by various names, which had reference either to its
weapon, its supposed powers, or its imposing appearance.
It was first figured by Salvianus. When it issues from the
Mediterranean, it appears, like many other species from the
same sea, to take a course either north or south, but seldom
pursues the same parallel of latitude towards the west. It
has been found at Madeira and on the coast of Africa. In
the opposite direction, it has been taken on the coasts of
Spain and France. Daniel, in his " Rural Sports," states,
that " in the Severn, near Worcester, a man bathing was
struck, and actually received his death-wound from a Swordfish. The fish was caught immediately afterwards, so that

the fact was ascertained beyond a doubt." — In October 1834, a party of gentleman in their pleasure-boat fishing in the sea off the Essex coast, saw something bulky floating on the water at a short distance. On coming up with it, they found it to be a dead Swordfish, ten feet long, of which the sword measured three feet: decomposition, however, was going on so rapidly, that a skeleton of the bones, which were entire, was the only portion that could be made available to any useful purpose.

The Swordfish was first noticed in our seas by Sibbald; since which Dr. Leach, Mr. Pennant, Dr. Fleming, Dr. Knox, and Dr. Grant have each had opportunities of examining specimens obtained in different parts of Scotland. Still farther northward there is scarcely a writer on Ichthyology but mentions the Swordfish, several having been taken in various parts of the Baltic.

The Swordfish is supposed to entertain great hostility to the Whale, and accounts of conflicts that have been witnessed are recorded by mariners. Captain Crow, in a work lately published, relates the following as having occurred on a voyage to Memel :—" One morning during a calm, when near the Hebrides, all hands were called up at three A. M. to witness a battle between several of the fish called Thrashers, or Fox Sharks (*Carcharias vulpes*), and some Swordfish on one side, and an enormous whale on the other. It was in the middle of summer, and the weather being clear and the fish close to the vessel, we had a fine opportunity of witnessing the contest. As soon as the whale's back appeared above the water, the thrashers springing several yards into the air, descended with great violence upon the object of their rancour, and inflicted upon him the most severe slaps with their long tails, the sound of which resembled the reports of muskets fired at a distance. The Swordfish, in their turn, attacked the distressed whale, stabbing from below; and thus beset on all sides and wounded, when the poor creature appeared, the water around him was dyed with blood. In this manner they continued tormenting and wounding him for many hours, until we lost sight of him; and, I have no doubt, they in the end completed his destruction."

It is a commonly received notion, that it is in consequence of mistaking the hull of a ship at sea for a whale, that the Swordfish occasionally endeavours to thrust his sword-like beak into the vessel. Those who have been on board on such an occasion, found it difficult to believe that the vessel had not struck against some rock unseen below the surface, so great had been the violence of the shock, from the weight and power of the fish. Specimens of ships' planks and timbers, deeply penetrated by what appears to be the pointed upper jaw of a Swordfish, broken off by the concussion, are shown in various museums; the forms and structure of which indicate that, if they did belong to Swordfish, several species, some of them attaining a large size, must exist: some are evidently referrible to the allied genus *Istiophorus*, which is limited in its range to more tropical seas. Mr. Scoresby states an instance of a ship from the coast of Africa, the bow of which had been penetrated by a bone, which he considered was the snout of a Swordfish; and other instances are recorded.

Captain Beechey says, " When in the Pacific Ocean, near Easter Island, as the line was hauling in, a large Swordfish bit at the tin case which contained our thermometer, but fortunately failed in carrying it off."

The Swordfish are said to go in pairs, and would probably be captured more frequently, but that their great timidity and vigilance save them.

The mode of obtaining them, as practised in the Mediterranean, is reported to be still more amusing than that in use against the Tunny, which has been already noticed. A man, elevated on a mast or on a neighbouring rock, gives notice by signal of the approach of a fish. The fishermen row towards, and attack it with a small harpoon attached to a long line; and are so skilful, as often to strike the fish at a considerable distance. The struggle then commences; which is, in fact, whale-fishing in miniature. Sometimes they are obliged to follow a fish for hours, before they are able to get it into the boat. The fishing season is from May to August.

The length of the Swordfish is from ten to twelve feet; but they occasionally attain a larger size, and have been known to exceed four hundred pounds weight. Dr. Leach found small fish in the stomach of one: that examined by Dr. Fleming contained numerous remains of *Loligo sagittata*. The flesh of the adult is said to be hard but good; that of the young fish white, agreeable, and nourishing. At Genoa, young ones are sold and eaten; but the elongated jaw is cut off before the fish are brought to market. The fin-rays are,—

D. 3 + 40 : P. 16 . A. 2 + 15 : C. 17.

Body elongated, nearly round behind, but little compressed; upper part of the head nearly flat, slightly descending to the base of the sword, which is formed by an extension of the vomer, maxillary, and intermaxillary bones; the edges produced finely denticulated; the extremity pointed; upper surface finely striated; under surface smooth, with a slight groove along the middle. The sides of the head vertical; the eye round; nostrils placed near the upper surface, almost round, and close together, the posterior orifice the largest: the under jaw does not extend beyond the line of the curve

formed by the upper as it descends from the cranium, and
ends in a point; the opening of the mouth extends backwards beyond the line of the eye; no teeth; the branches of
the lower jaw only slightly rough; mouth divided by a
transverse membrane, with a smaller similar membrane to the
lower jaw; no true tongue; pharyngeal bones furnished with
very minute teeth: skin of the body rough. The pectoral
fins are elongated, and attached very low down on the body;
the first three rays are the longest, the last the shortest;
no vestige of ventral fins: the dorsal fin commences on a line
with the gill-opening; the first three rays spinous, the fourth
or fifth ray the longest; the rays then diminish rapidly to the
tenth or eleventh, where they become very slender, and are
connected by a very slight membrane as far as the thirty-
ninth or fortieth. Through a great part of this length the
dorsal fin is only about half as high as the pectoral fin is
long; the three or four last rays are rather longer, and the
fin attains more power. This is the state of the fin in a
young fish when it has been but little used; but the portion
of the fin intermediate between the two ends is so slight that
it is easily torn, or even entirely worn away by use during
life; and this will help to explain the representations of this
fish when adult, which exhibit only the two extreme ends of
this fin, and make it appear like two dorsal fins separated.
The two portions of the tail are elongated.

The whole of the body is covered with a rough skin; the
operculum smooth; lateral line scarcely visible; on each side
of the tail a membranous projection. The whole of the
underpart of this fish is of a fine pure silver colour, shaded
with bluish black on the upper part. Very young specimens
of twelve or eighteen inches long have the body covered with
small tubercles: these inequalities on the surface disappear as
the fish increases in size, first on the back, afterwards on the

L 2

belly, and, by the time it attains the length of three feet, are no longer apparent.

Most of the works on Ichthyology containing a figure of the adult fish, a young one, in which only a small part of the anal fin was worn away, has been selected for representation in the cut at the head of this article.

ACANTHOPTERYGII. *SCOMBERIDÆ.*

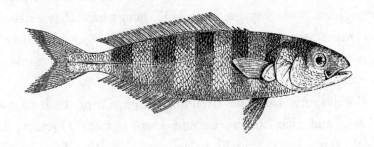

THE PILOT-FISH.

Naucrates ductor,	Cuv. et Valenc. Hist. Nat. des Poiss. t. viii. p. 312, pl. 232.
Gasterosteus ,,	Linnæus. Willughby, App. p 7, tab. viii. fig. 2.
Scomber ,,	Bloch, pt. x. pl. 338.
,, ,,	*Pilot-fish,* Couch, MS.

Generic Characters.—Body covered with small scales; dorsal fin single, elongated; free spinous rays before the dorsal and anal fins; sides of the tail carinated; teeth small, numerous; branchiostegous rays 7.

THE PILOT-FISH has been so often seen, and occasionally taken on our southern coast, as to be entitled to a place among British Fishes : it may be immediately recognised by its mackerel-like form of body and conspicuous transverse bands.

This fish was placed by Linnæus in his genus *Gasterosteus* on account of the free spines anterior to the dorsal fin ; but the form of the body, the minute scales, and the cartilaginous horizontal keel on the sides of the fleshy portion of the tail, indicate the family to which it belongs. Raffinesque considered this fish entitled to generic distinction, and assigned to it the name of *Naucrates :* Cuvier and M. Valenciennes coincide in this separation, and have adopted the name.

The Pilot-fish is supposed to have been the Pompilius of
the ancients ; a fish which is said to have pointed out the
desired course to doubtful navigators, accompanied them
throughout their voyage, and left them when they reached
the wished-for land.   The fish was therefore considered
sacred, and was invested with a Greek name, which signifies
' a companion.'

Besides this habit of attending ships during their course
at sea, and that for weeks and even months together, of
which some instances will be quoted, the Pilot-fish also ac-
companies large Sharks : but their motives for this association
have been variously interpreted.   By some it has been considered that the Pilot-fish acted as a guide to direct the
Shark to his food ; while others state, that when a Shark and
his Pilot were following a vessel, if meat was thrown overboard cut into small pieces, and therefore unworthy the
Shark's attention, the Pilot-fish showed his true motive of
action by deserting both Shark and ship to feed at his leisure
on the morsels.

M. Geoffroy relates an instance of two Pilots that took
great pains to direct a Shark towards a bait.   On the other
hand, Colonel Hamilton Smith has furnished an account of
an opposite character, which is thus related in Griffith's
Animal Kingdom, Fishes, vol. x. page 636.   " Captain
Richards, R. N., during his last station in the Mediterranean,
saw on a fine day a blue Shark which followed the ship,
attracted perhaps by a corpse which had been committed to
the waves.   After some time a shark-hook, baited with pork,
was flung out.   The Shark, attended by four Pilot-fish *Scomber ductor*, repeatedly approached the bait ; and every time
that he did so, one of the Pilots preceding him was distinctly
seen from the taffrail of the ship to run his snout against the
side of the Shark's head, to turn it away.   After some far-

ther play, the fish swam off in the wake of the vessel, his dorsal fin being long distinctly visible above the water. When he had gone however a considerable distance, he suddenly turned round, darted after the vessel, and, before the Pilot-fish could overtake him and interpose, snapped at the bait and was taken. In hoisting him up, one of the Pilots was observed to cling to his side until he was half above water, when it fell off. All the Pilot-fishes then swam about awhile, as if in search of their friend, with every apparent mark of anxiety and distress, and afterwards darted suddenly down into the depths of the sea. Colonel H. Smith has himself witnessed, with intense curiosity, an event in all respects precisely similar."

In the year 1831, two specimens of the Pilot-fish were caught on the opposite side of the British Channel, and more than one instance has occurred of their following ships into Guernsey. A few years since, a pair accompanied a ship from the Mediterranean into Falmouth, and were both taken with a net. In January 1831, the Peru, Graham master, put into Plymouth, on her voyage from Alexandria for London, after a passage of eighty-two days. About two days after she left Alexandria, two Pilot-fish, *Gasterosteus ductor,* made their appearance close alongside the vessel, were constantly seen near her during the homeward voyage, and followed her into Plymouth. After she came to an anchor in Catwater, their attachment appeared to have in-creased ; they kept constant guard to the vessel, and made themselves so familiar, that one of them was actually cap-tured by a gentleman in a boat alongside, but, by a strong effort, it escaped from his grasp, and regained the water. After this the two fish separated ; but they were both taken the same evening, and, when dressed the next day, were found to be excellent eating. In October 1833, nearly one hun-

dred Pilot-fish accompanied a vessel from Sicily into Catwater ; but they were not taken.

The usual length of the Pilot is about twelve inches : the
stomach has been found full of small fish : the flesh is delicate, and said to resemble that of the Mackerel. The fin-
rays are,—

D. IV. 26 : P. 18 : V. 1 + 5 : A. II. 16 : C. 17.

The nose is rounded ; the under jaw rather the longest ; the
diameter of the eye one-fifth of the whole head, and placed
at one-third of the distance from the nose to the end of the
operculum ; irides golden yellow ; nostrils placed near the
line of the profile, and rather nearer the point of the nose
than the eye ; mouth not very deeply divided ; teeth very
small, numerous, forming a band on each jaw ; a narrow band
on each palatine bone ; one single, short, but strong tooth,
on the front of the vomer, and one on the tongue ; the
tongue large, thin, and free ; ventral fins attached to the
abdomen by a membrane through one-third of their length.
The dorsal and anal fins end on the same line.

The body is covered with small oval scales, except one
triangular spot above the base of the pectoral fin ; the cartilaginous keel-like projection on each side the fleshy portion of
the tail reaches from the origin of the caudal rays forward
beyond the line of the base of the last rays of the dorsal and
anal fins.

The general colour of the fish is a silvery greyish blue,
darkest on the back, much paler on the belly ; the five dark
blue transverse bands pass round the whole of the body :
there are also indications of two other bands, one on the
head, the other on the tail : pectoral fins clouded with white
and blue, ventrals nearly black.

Individuals from various localities exhibit but very trifling differences.

In the Linnean Transactions, vol. xiv. page 82, Mr. Couch, in his paper on the Fishes of Cornwall, says of the Albacore, " I believe this fish is not uncommon in summer ; but keeping at a distance from the shore, and seldom taking a bait, it is but rarely caught."

Under the name of Albacore, like that of Bonito, two species have been included. The *Scomber glaucus* of Linn. will be found described and figured by Cuvier and M. Valenciennes, t. viii. p. 358, pl. 234.

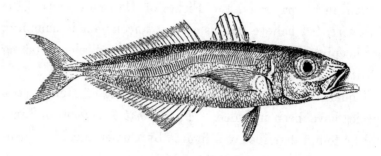

THE SCAD, OR HORSE-MACKEREL.

Caranx trachurus,   L<small>ACEPEDE</small>.
     ,,   ,,   C<small>UV</small>. et V<small>ALENC</small>. Hist. Nat. des Poiss. t. ix. p. 11, pl. 246.
Scomber   ,,   L<small>INNÆUS</small>. B<small>LOCH</small>, pt. ii. pl. 56.
     ,,   ,,   *Scad,* P<small>ENN</small>. Brit. Zool. vol. iii. p. 363, pl. 62.
     ,,   ,,   ,,   D<small>ON</small>. Brit. Fish. pl. 3.
Trachurus vulgaris, ,,   F<small>LEM</small>. Brit. An. p. 218, sp. 163.

Generic Characters.—Body covered with small scales, with the exception of the lateral line ; lateral line armed with a series of broad scales, those on the posterior half of the body having an elevated horizontal keel in the centre, form-ing a continuous ridge, each scale ending in a point directed backwards ; two distinct dorsal fins ; free spines before the anal fin ; teeth exceedingly minute ; branchiostegous rays 7.

T<small>HE</small> S<small>CAD</small>, or H<small>ORSE-</small>M<small>ACKEREL</small>, as it is commonly called, in reference to its supposed coarseness and consequent inferiority, rather than to its size, is occasionally abundant on particular parts of our southern shore, and may be traced nearly all round the British coast. Communications from various sources will supply a better history of this species than any materials I could myself furnish.

This fish occurs on the coast of Antrim in Ireland, at Belfast bay in the north, along the shore of the county of Cork in the south, and probably at many intermediate

points. Part of a letter from my friend Mr. Bicheno, residing on the coast of Glamorganshire, is as follows :—" On
Tuesday, the 29th of July 1834, we were visited by immense shoals of Scad, or, as they are also called, Horse-
Mackerel. They were first observed in the evening; and
the whole sea, as far as we could command it with the eye,
seemed in a state of fermentation with their numbers. Those
who stood on some projecting rock, had only to dip their
hands into the water, and with a sudden jerk they might
throw up three or four. The bathers felt them come against
their bodies; and the sea, looked on from above, appeared
one dark mass of fish. Every net was immediately put in
requisition; and those which did not give way from the
weight, were drawn on shore laden with spoil. One of the
party who had a herring-seine with a two-inch mesh was the
most succesful : every mesh held its fish, and formed a wall
that swept on the beach all before it. The quantity is very
inadequately expressed by numbers,—they were caught by
cart-loads. As these shoals were passing us for a week, with
their heads directed up channel, we had the opportunity of
noticing that the feeding-time was morning and evening.
They were pursuing the fry of the Herring, and I found
their stomachs constantly full of them."

According to Mr. Couch, the Scad " regularly visits the
coast of Cornwall and Devon, commonly in scattered quantities, but occasionally in considerable schulls. The first
appearance of this fish in spring is not until towards the
end of April; they are not abundant before the warmer
months of the year, when some may be found on board of
every fishing-boat. They are rarely brought to market, and
in many places even the fishermen are not in the habit of
eating them : in the west of Cornwall, however, they are
salted in the same way as Mackerel, and in this state meet

with a ready sale in winter. The usual habit of this fish is
to keep near the ground ; but when they assemble in pursuit
of sandlaunce or other favourite food, as they sometimes do
in innumerable multitudes, they become so eager as to thrust
each other in heaps on the land.

" On Tuesday evening, August, upwards of ten thousand
Scads were taken by a foot-sean near Marazion. These fish
frequently come so near the shore as to enable persons to
take them by hand. On Wednesday evening another shoal
appeared, when a number of men, women, and children went
into the water to catch them, while others stood on the sand
to see them throw the fish on shore ; and by this means a
vast quantity were obtained. The young keep near the
shore after the larger fish have retired to deep water."

Montagu found this species common on the Devonshire
coast, and well known to the fishermen by the names before
given. In one week, at the latter end of August, he obtained several, varying in length from three to fifteen inches ;
but the most common size was about nine inches. In West-
bay and at Weymouth this fish is common. I saw about
a score in the London market at the end of May 1834, and
purchased two. They possessed a portion of the flavour of
Mackerel, but were not so fine. These were about twelve
inches long, and would have spawned about the same time
as the Mackerel. They have been taken off Yarmouth, in
Berwick bay, in the Frith of Forth ; and Dr. Fleming found
part of one in the estuary of the Tay. Professors Reinhardt
and Nilsson have ascertained their existence, also, as far north
as the coast of Denmark and the west coast of Norway.

Montagu's description of a fresh specimen fifteen inches
long is as follows : — " The depth behind the gills three
inches ; the mouth large ; the upper lip capable of considerable projection ; the teeth minute, not discernible without a

lens ; the eye very large, equal nearly to one-half the depth of the head, part silvery, part dusky ; operculum rounded ; the last ray of the first dorsal fin connected by a membrane to the first ray of the second dorsal fin ; the two spines anterior to the anal fin slightly united by a membrane to each other, and to the base of the first ray of the anal fin. The ventral fins are placed in depressions ; the two spines and the anterior part of the anal fin are lodged in a groove. The curve of the lateral line is over the vent ; the body from thence to the tail becomes quadrangular, on account of the bony plates of the lateral line, which are terminated by a spine pointing backwards, and forming a strong carina on each side quite to the tail.

The number of fin-rays are as follows :—

D. 8. 1 + 32 : P. 21 : V. 1 + 5 : A. II. 1 + 26 : C. 17.

" The colour is a dusky olive above, changing to a resplendent green, with a bluish waved gloss ; sides of the head, and beneath the lateral line, silvery, with waved reflections ; dorsal fins dusky, the lower fins quite pale ; on the margin of the gill-cover, above the pectoral fin, a large black spot ; the throat and under part of the jaw is also black. This specimen was caught on a whiting-hook baited with sandlaunce."

The Scad, or Horse-Mackerel, occurs in the Mediterranean and at Madeira : some variation, however, is found in the number of lateral plates.

In a specimen of the Scad of our seas, the number of these lateral plates was seventy-four.

THE BLACKFISH.

Centrolophus pompilus,	Cuv. et Valenc. Hist. Nat. des Poiss. t. ix. p. 334.
,, *morio,*	,, ,, ,, t. ix. p. 342.
Holecentrus niger,	Lacepede, t. iv. p. 441.
Centrolophus niger,	,, t. ix. p. 347, 8vo. edit. Paris, 1831.
,, ,,	*Black Perch,* Penn. Brit. Zool. vol. iii. p. 351.
,, ,,	*Blackfish,* Couch, MS.

Generic Characters.—Body covered with minute scales ; one dorsal fin elongated, the rays short ; teeth very small, numerous ; branchiostegous rays 5.

Lacepede, when describing this rare fish, considered it unknown to naturalists, and that its singular form required generic distinction. Mr. Couch has had the good fortune to see two specimens of it, and I avail myself of his kind permission to give his account in full.

" The specimen described was fifteen inches long; blunt and rounded over the snout, flattened on the crown ; mouth small ; tongue rather large ; teeth in the jaws fine ; nostrils double, that nearest the eye large and open ; eye prominent and bright ; five gill-rays : though soft, the membrane of the preoperculum had a free edge, somewhat incised. Body compressed, about three inches deep ; a thin elevated ridge, which makes it appear deeper on the back, on which the dorsal fin is seated. This fin begins at four and a half inches from the snout, and reaches to the distance of twelve inches from it; the rays fleshy at the base, many of them obsolete; vent six and a half inches from the lower jaw; pectoral fins pointed; ventral fins bound down by a membrane; tail forked; lateral line somewhat crooked at its commencement. Body covered with minute scales, which when dry appear curiously striated. Colour of the whole black, the fins intensely so, very little lighter on the belly; somewhat bronzed at the origin of the lateral line. While employed in drawing a figure, the side on which it lay changed to a fine blue.

" Another specimen measured two feet eight inches in length, and weighed nearly fourteen pounds. The skin was observed to be so tough, as to be stripped from the fish like that of an Eel: no air-bladder was found. The taste was delicious.

" This fish, first described as British by Borlase from the papers of Mr. Jago, of East Looe, has been a stumbling-block to naturalists for the greater part of a century. Stewart and Turton fixed it in the genus *Perca*, under the name of *P. nigra;* and Stewart supposed it a variety of the Ruffe, in which opinion he was joined by Dr. Fleming. All this, however, is to be traced to an original mistake of the Cornish historian, who, in copying Jago's description, represents it as three-fourths of an inch broad, which would make it as slender as a Tapefish, where he should have read three or four inches, which was the exact dimensions of my specimen,—a little more than three behind the head, a little less than four at the commencement of the dorsal fin, and the precise measurement of Jago's fish. The difference of colour in the four specimens now recorded as taken in Cornwall, (Jago's two were caught in one net,) and those described by other authors, is easily explained by what is known to occur in reference to other species. The Tunny, like the Pompilus, is beautifully variegated in the Mediterranean Sea; but with us both of them assume an intense black."

" The great strength and velocity of this fish have been spoken of in terms of admiration by several authors; and the larger individual above mentioned, that fell into the hands of my friend Mr. Jackson, of East Looe, afforded a corroboration of the truth of the observation. It was caught in a net set for Salmon, at the mouth of the river, in the last week in November 1830; and such was the force with which it struck the bottom of the net, that it carried it before it over the head-rope. Jago found oreweed in the stomachs of his fishes; Ruysch says they feed on seaweed, though chiefly on flesh; and in my own specimen were found a muscle without a shell, and a piece of a Sea Bream *Pagellus centrodontus*, both, as I suppose, snatched as bait from the fishermen's hooks, but was captured baited with the lask or slice cut from the side of a Mackerel."

Cuvier and M. Valenciennes appear to believe that the two fishes they have designated by the terms *C. pompilus* and *C. morio* are only different states of the same species; and the descriptions and remarks of different authors in reference to the colour, as well as other particulars of specimens taken in the Mediterranean Sea, and in the ocean on the western coast of Europe, go far to confirm their view: the two names have therefore been brought together at the head of this subject.

The representation of this fish is derived from Mr. Couch's drawing of the smaller specimen of the two examples recorded by him.

The number of fin-rays in several specimens as stated by

different authors agree so closely, that they may be considered as amounting to—

$$D. 38 : P. 20 : V. 1 + 5 : A. 22 : C. 17.$$

M. Laurillard, not long since, obtained a specimen, at Nice, twenty-seven inches long. The Blackfish has now been taken of various sizes, from thirteen to thirty-two inches.

A figure of this fish occurs in the *Traité Général des Pêches* of Duhamel, *deuxième partie*, sec. iv. pl. vi. fig. 2, under the name of *Serran de Provence*, and another figure is also given in a recent Paris edition of the Works of the Comte de Lacépède, in eleven volumes, octavo, edited by M. Desmarest, plate xcv. fig. 3.

THE DORY, OR DOREE.

Zeus faber,	Linnæus. Bloch, pt. ii. pl. 41.	
,, ,,	Cuvier, Règne Animal, edit. 1829, t. ii. p. 211.	
,, ,,	*Doree,* Penn. Brit. Zool. vol. iii. p. 296, pl. 45.	
,, ,,	,, Don. Brit. Fish. pl. 8.	
,, ,,	,, Flem. Brit. An. p. 218, sp. 164.	

Generic Characters.—Body oval, very much compressed, surface smooth, without scales ; spinous portions of the dorsal and anal fins separated from the flexible portions by a depression, dorsal spines with long filaments ; spinous scales along the line of the dorsal and ventral edges ; mouth capable of considerable protrusion ; teeth numerous ; branchiostegous rays 7.

The Dory was first described by Pliny ; unless, indeed, it be the *Chalceus* of Oppian and Athenæus. The ancients must have entertained a high regard for it, since they gave it the name of Jupiter, *Zeus.*

The Dory, or Doree, contends with the Haddock for the honour of bearing the marks of St. Peter's fingers, each being supposed to have been the fish out of whose mouth the Apostle took the tribute money; leaving on its sides, in proof of the identity, the marks of his finger and thumb. Another origin for the spots on the sides of the Dory has also been assigned. St. Christopher, in wading through an arm of the sea, bearing the Saviour, whence his name of Christophorus, is reported to have caught a Dory, and to have left those impressions on its sides, to be transmitted to all posterity as an eternal memorial of the fact. The name of Doree was therefore said to be derived from the French, *adoree*, ' worshipped.'

Our common appellation of John Dory is also said to be of foreign derivation, and even with a second reference to St. Peter. The fishermen of the Adriatic call this fish *il janitore*, 'the gatekeeper,' in allusion to the supposed keys of the gates of heaven, of which the Apostle is pictured to be the bearer; and in several countries of Europe the Dory is called St. Peter's fish. The real origin of the English name for this fish may be questioned; but it is probably derived from the French, *dorée*, or *jaune dorée*, in reference to its peculiar golden yellow colour.

At what precise time the epithet of John became prefixed to the simple name of this fish, it might be difficult to ascertain: its name of Doree is at least as old as Merrett, who, in his *Pinax Rerum Naturalium Britannicarum*, 1666, speaks of it as a Doree, or a Dorn.

The Dory is considered a rare fish in the northern counties. It has been taken on the coast of Cumberland. In Ireland it occurs on the coast of Londonderry and Antrim; and, on the south, along the coast of Waterford. It is taken on the Cornwall and Devonshire coasts, sometimes even in profusion; and, onwards to the east, on the Hampshire and Sussex shores; but on the north-east coast it

M 2

is again considered rare. Mr. Paget says that several were caught during the summer of 1834 by the Yarmouth fishermen when taking Turbot on the Knowl. The food of the Dory is the fry of other fishes, molluscous animals, and shrimps. The largest specimens that come to the London fish-market weigh from ten to twelve pounds ; but the average weight is scarcely half as much. Pennant says the largest are from the Bay of Biscay.

Mr. Couch considers the Dory as " rather a wandering than a migratory fish ; and its motions are chiefly regulated by those of the smaller kinds on which it preys. When the Pilchards approach the shore, the Dory is often taken in considerable numbers. In the autumn of 1829, more than sixty were hauled on shore at once in a net, some of them of large size, and yet the whole were sold together for nine shillings. It continues common until the end of winter ; after which it is more rare, but never scarce. The form of the Dory would seem to render it incapable of much activity ; and it is sometimes seen floating along with the current, rather than swimming ; yet some circumstances favour the idea that it is able to make its way with considerable activity. It keeps pace with schulls of Pilchards, so that some are usually enclosed in the sean with them ; it also devours the common Cuttle, a creature of vigilance and celerity ; and I have seen a Cuttle of a few inches long taken from the stomach of a Dory that measured only four inches. It takes the hook, but gives the preference to a living bait ; and a Chad,* hooked through the back, with the prickly dorsal fin cut off, is sure to entice it."

" It is now," says Colonel Montagu,† " about sixty years since the celebrated Mr. Quin, of epicurean notoriety,

---

* The young of the Sea Bream, *Pagellus centrodontus.*
† Colonel Montagu died in July 1815.

first discovered the real merit of the Doree ; and we believe from him originated the familiar, and we may say national, epithet of John Dory, as a special mark of his esteem for this fish ; a name by which it is usually known in some parts, especially at Bath, where Quin's celebrity as the prince of epicures was well known, and where his palate finished its voluptuous career."

" Notwithstanding the numerous anecdotes recorded of this gentleman, as famous for his love of good living as for his excellence as a comedian, and who equally shone as a *bon-vivant* or in the character of Falstaff, we may be allowed to record one more in honour of both the person who brought the Doree into such high estimation and of the fish itself.

" An ancestor of ours, a Mr. Hedges, was an intimate friend of Quin's, and was induced by him to take a journey from Bath to Plymouth, on purpose to eat John Dory in the highest perfection,—not only from procuring it fresh, but with the additional advantage of having it boiled in sea-water, a matter of very great importance to the palate of Quin.

" As this journey was purposely taken to feast on fish, their stay at Plymouth was not intended to exceed a week, by which time they expected to have their skins full of Doree ; but that no opportunity might be lost, Quin left strict charge with the host at Ivybridge to procure some of the finest Doree he could get, for his dinner on his return, fixing the day. Whether our celebrated epicure was disappointed in his expectations at Plymouth, is not recollected ; but that he might have the provided fish at Ivybridge in the highest perfection, and remarking that the place was too remote from the coast to obtain sea-water for dressing the Dorees anticipated, he ordered a cask of sea-water to be tied behind his carriage. Unfortunately, the weather had been stormy, and no fish of note could be procured. Every apology was made by the host, who assured him that an excellent dinner was provided, which, he had no doubt, would be to his taste; but no fish. The disappointment, however, was too great to be borne with patience; after having made a water-cart of his carriage, and the appetite having been set for John Dory boiled in sea-water, no excuse, no apology, would satisfy Quin; and he declared he would not eat in his house, but, like a ship in distress, threw his water-cask overboard, and pursued his journey not a little sulky, till some fortunate stroke of wit, or some palatable viand roused him to good humour.

" This western tour of Quin's did not appear to have given him much satisfaction, as may readily be imagined by his reply to a friend on his return to Bath. Being asked if he did not think Devonshire a sweet county,—" Sir," said Quin, " I found nothing sweet in Devonshire—— but the vinegar."—*Montagu's MS*.

The body of the Dory is oval, very much compressed; the head large; the mouth capable of great protrusion, so much so, that from the point of the lower jaw when extend-ed, to the posterior angle of the operculum, is as long as from that angle to the base of the caudal rays. The length of the head when the mouth is not projected is nearly as long as the body is in depth. The mouth large; the teeth small and numerous, placed in a single row in each jaw, and curving inwards; the eyes large, situated laterally, and high up on the head; irides yellow; a spine behind and over each orbit about halfway between the eye and the first ray of the spinous portion of the dorsal fin; the spines of the first dorsal fin very long, the longest half as long as the body is deep; the membrane between the spines ending in a filament three times as long as the rays. The base of the second dorsal fin about as long as that of the first; the rays flexible, and only half as high as those of the first: the pectoral fin small and short, ending on a line with the anterior edge of the dark spot on the side; the ventrals very long and slender, arising in advance of the pectorals, the rays reaching as far back as the first flexible ray of the anal; the first spinous ray of the anal fin is on a line with the posterior edge of the dark spot, and with the sixth spinous ray of the dorsal; the flexible portion commences and ends nearly on the same planes as the flexible dorsal: the tail is narrow, long, and slender; the lateral line advancing at first straight, afterwards rises in an elevated arch over the dark spot, which is placed at about the diameter of its own breadth behind the posterior angle of the operculum. A row of spiny scales pointing backwards are ranged along the base of the dorsal and anal fins on both sides.

The number of fin-rays are—

D. 9. 22 : P. 13 : V. 9 : A. 5. 21 : C. 13.

The prevailing colour of the body is an olive-brown, tinged with yellow, and reflecting in different lights, blue, gold, and white; when the living fish just taken from the net is held in the hand, varying tints of these different colours pass in rapid succession over the surface of the body. The membranes of the flexible portions of the fins are light brown: those of the spinous portions are much darker.

A large portion of the Dorees supplied to the London fish-market is brought by land-carriage from Plymouth, and some other parts of the Devonshire coast. Being a ground fish,* they are little or none the worse for keep-

* See page 22.

ing till the second or third day. Montagu, disliking the
toughness of a fresh-caught Dory, says, they are most
palatable after keeping two days. Fish for the supply
of the London market was not brought by land-carriage
until the year 1761. Steam-boats seem likely to effect
another change. In the summer of 1834, a cargo of Salmon from Scotland was deposited in the London market
within forty hours.

ACANTHOPTERYGII. *SCOMBERIDÆ.*

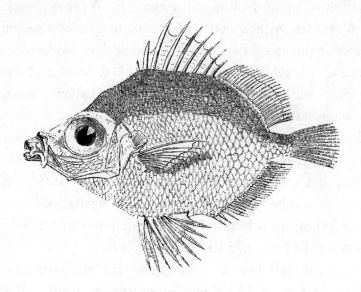

THE BOAR-FISH.

Capros Aper.	Lacepede.
,, ,,	Cuvier, Règne An. 1829, t. ii. p. 211.
Aper Rondeletii,	Willughby, p. 296, p. T. iv. fig. 4.
Zeus Aper,	Linnæus.
Perca pusilla,	Brunnich, p. 62, sp. 79.
Capros Aper,	Risso, t. iii. p. 380. sp. 296.
,, ,,	Proceedings Zool. Soc. 1833, p. 114.

Generic Characters.—Body oval, compressed; two dorsal fins; no spines at the base of the dorsal or anal fins; scales on the body small, adherent, ciliated; mouth capable of protrusion; teeth small, numerous; branchiostegous rays 6.

A SPECIMEN of the *Capros Aper* of authors having been taken in Mount's Bay, in October 1825, and a notice of the occurrence forwarded to the Zoological Society, with a drawing and description, by Dr. Henry Boas, a figure and a short account of the species necessarily belong to a History of British Fishes. Its right to rank among them is confirmed by the fact, that at the moment of preparing this account, I am favoured, by the united kindness of W. C. Trevelyan, Esq. of Nettlecombe, and Mr. William Baker, of Bridgewater, with a notice of the occurrence of a second example of this rare fish, which was obtained in Bridgewater fish-market on the 18th of April 1833. A drawing of the fish, made while this specimen retained its natural colour, also accompanied the communication.

The *Capros*, according to the ancient authorities, was known to Aristotle. It is figured and described by Rondeletius, and again by Willughby, as quoted. It is said to have been called by several names that signify wild boar and marine boar, on account of its projecting nose and mouth, the form of the head, and its bristling spines.

It is a fish well known as inhabiting the Mediterranean, where, according to M. Risso, it spawns in April. The flesh is hard, and Lacépède says it emits also an unpleasant odour.

The Zoological Society having received three specimens of it from Madeira, presented by the Rev. R. T. Lowe, and the use of one of these specimens having been immediately granted to me on my making the request, I am thus enabled to supply both figure and description from the fish.

I am not aware that any figure from nature of the Boarfish has hitherto been published, except the one originally given by Rondeletius. While referring to this representation, I may be excused reminding the reader who possesses a copy of the work of Rondeletius on the fishes of the Mediterranean, of the opportunity now afforded of comparing the representations of fishes cut in wood at the present time, with others also cut in wood nearly three hundred years ago. Many of those contained in the work referred to, although coarse in their execution, are by no means deficient in character or spirit; but the name of the artist who engraved them at that distant period is unknown.

The form of the body is a shorter oval than that of the Dory; the mouth protrudes; a band of minute teeth considerably within each jaw; the eye very large, placed at the distance of its own diameter from the end of the nose when the mouth is shut; the nostrils large, just anterior to the edge of the orbit: the origin of the first dorsal, pectoral, and ventral fins is nearly on the same plane; the base of the first dorsal about as long as its third spine, which is the longest; the base of the second dorsal fin equal to that of the first, the rays very slender and flexible, the membrane only extending up one-third of the length of the rays; the pectoral fin as long as the third ray of the first dorsal fin, slender and delicate in structure; ventral fin with one strong spine, the other rays flexible and branched, the membrane not extending the whole length of the rays; anal fin with all the characters observable in the second dorsal fin, and ending at the same distance from the tail; the caudal rays slender, and twice as long as the fleshy portion of the tail. The number of fin-rays are—

D. 9. 24 : P. 14 : V. 1 + 5 : A. 3 + 24 : C. 12.

No lateral line is observable; the body is quite smooth when the finger is passed from before backwards, but rough to the touch in the contrary direction, from numerous small scales which are minutely ciliated. The specimen belonging to the Zoological Society is five inches long from the point of the nose to the end of the tail; and the colour, probably altered from having been kept two or three years in spirit, of a uniform pale yellow brown.

The specimen of this fish taken in Mount's Bay measured

six and a half inches.   Mr. Baker's example was seven inches.
In both these last the irides were orange colour, the pupil
blueish black ; the upper part of the back and sides pale
carmine, still lighter below, and passing to silvery white on
the belly; the body divided by seven transverse orange-
coloured bands reaching three-fourths of the distance from
the back downwards.   The Mount's Bay specimen, according to Dr. Boas, had no bands.   All the fin-rays the same
colour as the back ; the membranes much lighter.

ACANTHOPTERYGII. SCOMBERIDÆ.

THE OPAH, OR KING-FISH.

Lampris guttatus,	Retz.	Cuvier, Règne Animal, t. ii. p. 211.
Zeus Luna,	Gmfl.	Linn.
,, ,,		*Opah,* Penn. Brit. Zool. vol. iii. p. 299, pl. 46.
,, ,,	,,	Don. Brit. Fish. pl. 97.
,, *Imperialis,*	,,	Shaw, Nat. Misc. pl. 140.
Lampris Luna,	,,	Flem. Brit. An. p. 219, sp. 168.

Generic Characters.—Body oval, greatly compressed, scales small; a single elevated and elongated dorsal fin; sides of the tail carinated; teeth wanting; branchiostegous rays 7.

THE OPAH, or KING-FISH, originally included in the genus *Zeus*, has been removed by some authors, on account of its possessing but a single dorsal fin; and the generic term *Lampris* has been applied to distinguish it.

This fish is as beautiful as it is rare. At the date of the first edition of Pennant's British Zoology, only about five examples were recorded as having been taken in different parts of the British Islands; four of them in the north, and one at Brixham. Since that time three others have been obtained, one of which is now preserved in the British Museum, and from that example the representation above was drawn and engraved. A specimen taken in the Clyde some years since is now preserved in the Andersonian Museum at Glasgow. It has also appeared still farther north, since Nilsson includes it in his Prodromus of the Fishes of Scandinavia.

Professor Reinhardt has recorded that within the last thirty years three examples have been taken on the coast of Denmark ; and, what is remarkable, they were all caught very near the same spot.

This fish was first described by Dr. Mortimer, in the Philosophical Transactions, from a specimen taken at Leith in the year 1750 : the preserved fish was exhibited at a meeting of the Royal Society. To his account of it Dr. Mortimer has added " that the Prince of Anamaboo, a country on the west coast of Africa, being then in England, recognised the fish immediately as a species common on that coast, which the natives called Opah, and said it was good to eat."

Little or nothing is ascertained of the habits of this fish : one exhibited at Dieppe was unknown to the oldest fishermen there. The specimen before referred to as taken at Brixham, measured four feet six inches in length, and weighed one hundred and forty pounds.

By the evidence of Chinese drawings, it would appear that the Opah is also a native of the eastern seas ; and it is certainly not a little singular, as observed by Mr. Couch, that by a people so distant and secluded as the Japanese, a fish, considered originally as belonging to the same genus as the Doree, should also be regarded as devoted to the Deity, and the only one that is so. The Opah is by them termed Tai ; and is esteemed as the peculiar emblem of happiness, because it is sacred to Jebis or Neptune.*

The number of fin-rays are—

D. 2 + 52 : P. 28 : V. 1 + 9 : A. 1 + 25 : C. 30.

The length of the body including the tail is to the depth of the body without the fins as two to one ; the form of the body oval; the profile of the head, both above and below, falling in with the outline of the body; the mouth small, without teeth ; tongue thick, with rough papillæ pointing backwards, and well calculated to assist in conveying food towards the pharynx. The base of the dorsal fin is rather longer than the depth of the body, the first eight or nine rays elongated, the longest four times as long as the rays of the posterior portion ; pectoral and ventral fins very long, full one-third the whole length of the body and tail; the anal fin, which is preceded by a triangular scale pointing backwards, equal in length to half the length of the base of the dorsal; the tail in shape lunate. The ventral, pectoral, and anterior part of the dorsal fins falciform ; the lateral line forms an elevated arch over the pectoral fin, its highest part being immediately under the longest ray of the dorsal fin.

The upper part of the back and sides are of a rich green, reflecting both purple and gold in different lights, passing into yellowish green below; above and beneath the lateral line are various round yellowish white spots, from which the fish received the name of *Luna ;* the irides are scarlet ; all the fins bright vermilion.

The showy colours with which the Opah is ornamented induced an observer to remark, that it looked like one of Neptune's lords dressed for a court-day.

* Kœmpfer. History of Japan, folio, vol. i.

ACANTHOPTERYGII.   *RIBAND-SHAPED.**

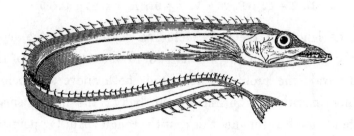

THE SCABBARD-FISH.

Lepidopus argyreus,   Cuv. et Valenc. Hist. Nat. des Poiss. t. viii. p. 223,
            pl. 223.
Vandellius Lusitanicus, Shaw, vol. iv. p. 99.
Xipotheca tetradens,   Montagu, Mem. Wern. Soc. vol. i. p. 81, and 623,
            pl. 2 and 3 ; and vol. ii. p. 432.
Lepidopus Lusitanicus, Leach, Zool. Misc. pl. 62.
            Scabbard-fish, Penn. Brit. Zool. vol. iii. p. 210.
            Scalefoot,   Flem. Brit. An. p. 205, sp. 116.

     Generic Characters.—Head pointed ; body without scales, elongated, com-
pressed, thin, riband-shaped : dorsal, anal, and caudal fins distinct ; the dorsal
fin extending the whole length of the body : two small scales on the abdomen
in the place of ventral fins : teeth in a single row in each jaw, compressed,
cutting, and pointed ; others very small on the palatine and pharyngeal bones,
and on the branchial arches : branchiostegous rays 8.

     Of the family of riband-shaped fishes, not more than six
species, belonging to five genera, have been obtained on the
British shores ; and these so rarely, that little is known of
their habits.

     Colonel Montagu first described the Scabbard-fish as a
British species under the name of *Xipotheca tetradens,* from
its sword-like form and four elongated teeth in front, be-

     * The family of riband-shaped fishes.

lieving it to be then entirely unknown to naturalists; but this fish appears to be an inhabitant of the Mediterranean as well as the European seas, and has been taken occasionally in several different parts of southern and western Europe.

One specimen taken at the Cape of Good Hope is described and figured by Euphrasen, in the new Memoirs of Stockholm for 1788, t. ix. p. 48, pl. 9, fig. 2; and other descriptions and figures were equally known.

Four examples of this fish have occurred on the southern shores of England: two fortunately came into the possession of Colonel Montagu, and are still preserved in the British Museum. In the summer of 1787, a specimen came ashore near Dawlish; and notes with a drawing of this fish were sent by Mr. Matthew Martin to his friend and correspondent John Walcott, Esq. for his then projected work on British Fishes. A fourth example was received a few years back by the Linnean Society.

Colonel Montagu's first and largest specimen measured five feet six inches in length; the depth at the gills four and a half inches; the weight, without the intestines, six pounds one ounce. This fish was taken in Salcombe harbour, on the coast of South Devon, in June 1808. It was swimming with astonishing velocity, with its head above water,—to use the fisherman's expression, "going as swift as a bird,"—and was killed by a blow of an oar.

"The specimen was considered so rare, that a public show was made of it at Kingsbridge, where, in one day a guinea was taken for its exhibition, at one penny each person. It was embowelled when I first saw it. In preparing it, I observed within the skin, on the abdominal parts, a great many small ascarides, pointed at each end, and of a whitish colour: they were all coiled up in a spiral manner.

VOL. I.   N

On the head, beneath the skin, and along the root of the dorsal fin, were several of a species of *Echinorhynchus,* of a yellow colour, nearly two inches in length, and more than one-eighth of an inch in diameter : the proboscis short, with a round termination furnished with spines : the anterior end of the body sub-clavate, with a groove on each side ; posterior part wrinkled, and obtusely pointed. These *vermes* had formed sinuses under the skin, and were firmly attached by one end." This fish has been observed by other authors to be infested with worms.

Not to multiply the description of Montagu, an abridgment of that of Cuvier is here given.

The head is pointed and slender ; the edge of the back thin ; the dorsal fin low, the rays of nearly equal length throughout, and the fin occupies the whole length of the back ; the edge of the belly is rounder, and has but a small anal fin at the posterior end ; the tail is small and forked. Its remarkable characters are, the pointed and cutting teeth, the two rounded scales in the place of the ventral fins, and in a third triangular scale situated behind the vent. These are the only scales, for the skin is smooth.

The head is about one-seventh of the whole length of the fish, and in height about equal to half the length of the head ; the thickness of the body one-fourth of its height. The eye is placed about half way between the end of the lower jaw, which is the longest, and the hinder edge of the operculum ; the nostrils ovate, and just before the eyes. Each intermaxillary bone has a row of twenty to twenty-two compressed, cutting, sharp-pointed teeth : in front, just within, are two or three teeth four times as large and as long as the others, slightly bent inwards ; six of these are the correct number, but two or three are generally observed to be broken. The under jaw has also one entire row of teeth, with two longer ones. The vomer is not furnished with any teeth, but the long external edge of each palatine bone has one row of very minute teeth ; the pharyngeal bones and the branchial arches are also furnished with teeth, but they are exceedingly minute.

The pectoral fin is about one-fifteenth part of the whole length of the body, and the lower rays are the longest ; the two upper rays are short and simple, the other ten rays are branched and articulated. The two half-circular scales in the place of ventral fins are situated rather nearer the end than the origin of the pectorals, and are connected to each other at the base. The dorsal fin commences at the nape ; the height one-fourth that of the body, the rays simple and flexible. The vent is at an equal distance from each extremity of the fish, with a moveable triangular scale behind it. The anal fin commences far behind the scale ; the tail is forked ; all the membranes of the fins are slender and easily injured. The lateral line is a narrow depression, which descending gradually from the upper edge of the operculum, afterwards passes along the middle of the body to the centre of the tail.

The irides are silvery, the fins greyish yellow ; the colour of the skin of the body, which is quite smooth and destitute of scales, is like burnished silver, with a bluish tint. The fin-rays are—

D. 105 : P. 12 : A. 17 : C. 17 : vertebræ 111.

The difference in the number of fin-rays, according to authors, leads to the supposition that more than one species will yet be defined.

The flesh is eaten, and, according to Risso, it is firm and delicate.

N 2

The females are full of ova in spring; they approach the shore in May.

A very young specimen of this fish was found alive on the shore in Slapton bay, on the south coast of Devon, about four miles east of the Start Point, in Feburary 1810. " I regretted," says Colonel Montagu, " not having seen it alive; but it was quite fresh and perfect when brought to me the day after it was taken, and is now in high preservation in spirits. It measures about ten inches in length, and half an inch in breadth, at the broadest part, just behind the head, and where its thickness does not much exceed one-eighth of an inch. It differs in nothing but size from that before described : the characteristic larger teeth are conspicuous, and the two ventral scales are also obvious by the assistance of a glass : the dorsal and anal fins are so fine in this young specimen, and lie so close, that they are not easily discovered, unless they are lifted up by some pointed instrument : the caudal fin is very small, but perfect : the under jaw projects full as much in proportion as in the larger fish : the whole skin is covered with a silvery cuticle, which is easily separated by gentle friction, and adheres to the fingers ; it is not of that high polish observed in some of the scaly fishes, and is a little wrinkled; there are also several slight lon-gitudinal depressions on the sides, that give a striped ap-pearance in some points of view.

" How are we to account for this very young speci-men being found in our seas, unless the spawn had been deposited on our coast ? And if, as we may now conclude, this fish actually inhabits our seas, it is curious that it should never before have been discovered."

This small specimen is still in good preservation at

the British Museum, the depository of Montagu's collection of fishes and shells, as well as of his birds. By the kindness of the Zoological officers of that establishment, the vignette at the foot of the page represents exactly a portion of this small specimen of the natural size.

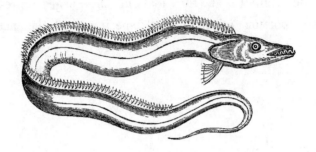

THE SILVERY HAIRTAIL.

Trichiurus lepturus, Linnæus. Bloch, pt. v. pl. 158.
,,   ,,   Cuv. et Valenc. Hist. Nat. des Poiss. t. viii. p. 237.
,,   ,,   Hoy, Linn. Trans. vol. xi. p. 210.
,,   ,,   *Blade-fish,* Flem. Brit. An. p. 204, sp. 115.

Generic Characters.—Head and body very much resembling those of the fishes of the genus *Lepidopus* last described ; no ventral fins, nor scales instead ; no anal fin ; tail without rays, ending in a single elongated hair-like filament (from which the generic name is derived) ; branchiostegous rays 7.

Two specimens of fishes regarded as belonging to the genus *Trichiurus* have been found dead on the shore of the Moray Frith, both of which were examined by Mr. James Hoy, a Fellow of the Linnean Society, and an account of them published in the Transactions of the Society as quoted.

" On the second of November 1810, after a high wind from the north, a specimen of the *Trichiurus lepturus,* Linn. was cast upon the shore of the Moray Frith, near the fishing village of Port Gordon, about three miles east from the mouth of the river Spey; and it was brought to me the next day, as a kind of fish which had never been seen before by any of the fishermen in this part of the country.   They said, that in seeking for lobsters cast ashore by the storm, they found it lying dead upon the sandy beach."

" Its head was much broken, probably by being dashed upon the rocks above low-water mark : the bones of the upper part of the head still remained, and the sockets of the eyes were distinguishable, very near to each other : the extremity of the upper jaw, or upper part of the mouth, was entire ; upon either side of which was an operculum. The length of the head could not be measured exactly, but was about eight or nine inches : the body, from the gills to the point of the tail, was three feet two inches long ; its greatest breadth six inches and a quarter, and its greatest thickness only an inch : the vent was two inches from the gills ; these were much broken, and partly gone, so that the number of the rays could not be ascertained. Both sides of the fish were wholly white, without a spot upon them ; the dorsal fin was the only part of a different colour, being a blackish green : this fin ran all along the back from the gills to the tail, consisting of a great number of rays, soft, and little more than an inch long. Each of the pectoral fins had six double rays. There were no ventral nor anal fins ; but the belly was a sharp, smooth, and entire edge. The tail ended in a point, consisting of three or four soft spines or bristles of different lengths, not exceeding two inches. The body was nearly of the same breadth for one half of its length, and then its breadth diminished gradually till within three inches of the tail, when the diminution became more quick. The lateral line was straight, and strongly marked along the middle of the two sides."

" This was the first individual of the genus *Trichiurus*, as far as I know, that had ever been found on the British coast. But although the fishermen have not found out the means to catch them, it now appears that these fish inhabit

our seas ; for on the 12th of November 1812, another of
them was found on the beach, hard by the same fishing village as the former, but of a much larger size : it was brought
next day in a cart to the Duke of Gordon, at whose desire
I made the following observations :"—

" Its head had been broken off, and was quite gone ; a
small bit of the gills only remained about the upper part of
the throat ; from whence to the extremity of the tail its
length was twelve feet nine inches : its breadth, eleven inches
and a quarter, was nearly equal for the first six feet in length
from the gills, diminishing gradually from thence to the tail,
which ended in a blunt point, without any of those kind of
bristles which projected from the tail of the one found formerly : its greatest thickness was two inches and a half :
the distance from the gills to the anus forty-six inches.
The dorsal fin extended from the head to the tail, but was
much torn and broken : the bones and muscles to which the
pectoral fins had been attached, were perceivable very near
the gills.   There were no ventral nor anal fins ; but the
thin edge of the belly was closely muricated with small hard
points, which, although scarcely visible through the skin,
were very plainly felt all along it.   Both sides of the fish
were white, with four longitudinal bars of a darker colour ;
the one immediately below the dorsal fin was about two
inches broad, each of the other three about three-fourths
of an inch.   The side line straight along the middle."

Dr. Fleming has remarked, that " from the preceding
descriptions, it appears probable that the two fishes examined
by Mr. Hoy belong to different species.   The difference
in the position of the vent, the structure of the tail, and the
condition of the ridge of the belly, seem too great to justify
the inference of their being only varieties.   The latter fish

appears identical with the *lepturus* of Artedi, and consequently of Linnæus."

Cuvier and M. Valenciennes, in their description of *T. lepturus*, state the situation of its lateral line to be but one-third of the space above the line of the edge of the abdomen: Mr. Hoy states that the side line went straight along the middle : in other respects, Mr. Hoy's second fish agrees nearly with *T. lepturus*, as described in the *Histoire Naturelle des Poissons*, already referred to. It would seem, however, that it must have been comparatively a deeper fish : the barring of the sides does not occur in *T. lepturus ;* and the latter has never yet been recorded as arriving at the gigantic size of Mr. Hoy's specimen, which could not have been less than fourteen feet and a half in length : the largest in the Paris Museum is stated to measure only three feet. It is evident that more information on the subject is required : the result of it may be the establishment of Mr. Hoy's second fish as a new species of *Trichiurus*, and of his first fish, which is evidently distinct from the second, as the type of a new genus,—if, as Dr. Fleming has suggested,* it was not a mutilated example of the Dealfish of the Orcadians, *Gymnetrus arcticus*, the fish described next but one in this work.

Specimens of *Trichiurus* have been taken at New York, Cuba, Jamaica, Porto Rico, St. Bartholomew's, Cayenne, Rio Janeiro, and Monte Video. Cuvier thinks it may cross the Atlantic ; and adds, that specimens received from Senegal in no way differed from those received from America.

Two species at least, if not more, inhabit the Indian Sea ; and all the species are truly marine. The differences, however, which characterise the various species, are as yet not

* Loudon's Magazine of Natural History, vol. iv. p. 219.

sufficiently known.   The work of Cuvier and M. Valenciennes contains the characters of three species,—*lepturus*,
haumela, and *savala*.   Mr. J. E. Gray has published the
characters of three species in the collection at the British
Museum, under the names of *armatus*, *intermedius*, and
muticus, in the first part of his Zoological Miscellany, pages
9. and 10 ; and representations of three species will be found
in that part of the Animal Kingdom, by Edward Griffiths,
Esq. and others, which is devoted to Fishes, plate 9.

The number of fin-rays in *T*. *lepturus* are—

D. 135 or 6 : P. 11.

Mr. Hoy remarks, that as the second fish appeared to be
very fresh, a cut of it was boiled, which he tasted, and found
to be very good, approaching nearly in taste to the Wolf-
fish, *Anarhichas lupus*, which he had an opportunity of
tasting only a few days before.

The figure at the head of the present article, which will
assist an observer in determining correctly the true *Trichiurus lepturus* in the event of its occurring on the coast, is
derived from Bloch ; and subjoined is an abridged description of this fish from the work of Cuvier and M. Valenciennes.

The height of the body at the deepest part is to the
whole length, reckoning from the point of the nose to the
end of the hair-like tail, as one to sixteen or seventeen :
at about one-half of the whole length of the fish the body
begins to diminish in size, and continues declining, the
latter fifth portion being little more than the slender tail :
the length of the head, from the point of the lower jaw,
which is the longest, to the end of the operculum, is equal
to one-eighth of the whole length of the body ; the descending line of the profile from the nape to the nose is straight ;

the face and crown flat, sides of the head vertical ; the eye placed high up near the line of the profile, the posterior edge of the orbit dividing the length of the head, the diameter one-sixth of the whole head ; nostril oval, and near the anterior edge of the orbit : the mouth furnished with a single row of about fifteen teeth on each side of each jaw, compressed, cutting, and pointed ; of which those towards the front are the smallest, except that there are two on each side of the upper jaw long and curved with a slight barb, and two or three rather longer than the others on the lower jaw : the vomer is without teeth, but the palatine bones have each a row of very minute teeth, more easily felt than seen ; tongue long, pointed, free, and perfectly smooth : the edge of the preoperculum forms a half-circle.

The pectoral fin is small, not so long as the body of the fish is deep, the second and third rays the longest, eleven rays in all ; no vestige of ventrals : the dorsal fin commences on a line with the superior angle of the operculum, the rays uniform in height throughout the greater part of its length, diminishing towards the end : the anal orifice at one-third of the length of the fish from the head ; behind it are numerous small spiny points, to the number of one hundred and fifteen, or one hundred and eighteen. No scales on the body visible ; the skin covered with a delicate silvery membrane : the lateral line, commencing at the upper edge of the operculum, descends to the line of the lower third of the body, and follows that parallel to its termination.

The colour of the fish a bright and shining silver : the fins greyish yellow ; the edge of the dorsal speckled with black, forming a spot between the first rays : the irides golden.

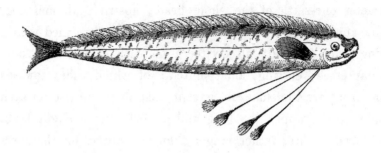

HAWKEN'S GYMNETRUS.

CEIL CONIN. *Cornwall.*

Gymnetrus Hawkenii,	Bloch, pt. xii. pl. 423.		
,,	,,	Cuvier, Règne An. t. ii. p. 220.	
,,	,,	*Ceil Conin,*	Couch, Trans. Linn. vol. xiv. p. 77, and MS.

Generic Characters.—Body elongated, compressed ; a single dorsal fin, extending the whole length of the back ; no anal fin ; jaws capable of considerable extension ; teeth pointed.

THE species of the genus *Gymnetrus* of Bloch have very rarely been obtained entire ; more or less mutilation has hitherto been found to have happened to the few specimens that have occurred ; and authors have consequently taken very different views of many of their characters.

Of this genus, instituted by Bloch for the reception of a fish sent to him from India with a drawing by Hawken, the species have been ranked under four other different names : viz. *Trachypterus* of Gouan and Bonelli ; *Bogmarus* of Schneider ; *Epidesmus* of Ranzani ; and *Argyctius* of Rafinesque.

Cuvier adopts the name proposed by Bloch.

Of the genus *Gymnetrus*, three species probably belong

to the Mediterranean, two to the seas of the North of
Europe, and two to India. One Northern species, besides
one of those apparently belonging to India, has been taken
on the shores of this country. That of the North has occurred more than once in Scotland ; that of India, once on
the coast of Cornwall. Figures of both are inserted in this
work, rather with a view to invite investigation of such rare
species should they again occur, than from any novelty now
to be communicated.

A fish apparently of the species called by Bloch *Gymnetrus Hawkenii* was drawn on shore dead in a net at Newlin,
on the western side of Mount's bay, on the south coast of
Cornwall, in February 1791 ; from a large original drawing
of which, with notes, in the possession of William Rashleigh,
Esq. of Menabilly in that county, Mr. Couch's account of
it is derived. It is as follows :—

" The length, without the extremity of the tail, which
was wanting, was eight and a half feet ; the depth ten and a
half inches ; thickness two and three-quarter inches ; weight
forty pounds. In the drawing, the head ends in a short
and elevated snout ; eye large ; pectoral fin round ; no anal
fin ; the dorsal fin reaches from above the eye to the tail.
In the drawing, as well as in Bloch's engraving, the caudal
fin is supplied. The ventrals are formed of four long red
processes, proceeding from the thorax, and ending in a
fan-shaped appendage, of which the base is purple, the expansion crimson. The back and belly are dusky green, the
sides whitish ; the whole varied with clouds and spots of a
darker green ; the fins crimson."

The account given by Bloch is as follows :—

" This fish was sent to me by Mr. Hawken : from him
also I received the drawing. He wrote me at the same time,
that the fish was caught near Goa, in the Indian Sea, on

the 23rd of July 1788. The specimen was two and a half feet long, six inches deep, and weighed ten pounds."

The number of fin-rays, according to Bloch, are—

D. 17 ? 117 : P. 8 : V. 2 : C. 13.

The woodcut represents this fish as shown in Mr. Couch's drawing, but reduced to one-fourth, and differs a little in the form of the head from the figure in Bloch's work.

ACANTHOPTERYGII.   *RIBAND-SHAPED.*

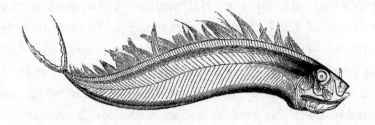

THE VAAGMAER, OR DEALFISH.

Gymnetrus arcticus,	Cuvier, Règne An. t. ii. p. 219.
Gymnogaster ,,	Brunnich.
Bogmarus Islandicus,	Schneider, p. 518, pl. 101.
Gymnetrus arcticus,	*Vaagmaer, or Dealfish,* Flem. Mag. Nat. Hist. vol iv.
	p. 215, fig. 34.
Vogmarus Islandicus,	Reinhardt's *Vaogmære.*

Dr. Fleming has published, in the Magazine of Natural History above quoted, an account and description of this interesting addition to the catalogue of British Fishes, and appears to be the only British naturalist who has made known its occurrence in Scotland. The specimens obtained, however, were either so mutilated, or so imperfectly preserved, that the author in his paper was induced to doubt the propriety of retaining this species in the genus *Gymnetrus,* and proposed to restore it, as a fish having no ventral fins, to its original station in the genus *Gymnogaster* of Brunnich.

A recent notice of the Vaagmaer, or Vaogmære, as it is there called, appeared in the *Institut,* or *Journal Général des Sociétés et Travaux Scientifiques,** a French periodical

* Paris, tome ii. 1834, p. 158 and 193.

publication devoted to giving reports of the proceedings of Societies, of which the following is a free translation :—

" Professor Reinhardt communicated to the Royal Society of Natural History and Mathematics of Denmark a continuation of his Ichthyological memoirs. It contained descriptions of two genera which up to the present time have not been perfectly understood ;—the *Macrourus (Berglax)*, and the *Vogmarus* (Vaogmære), the species of which are found in the Polar Seas, as well as in the Mediterranean.

" The Ichthyologists of the North, it is stated, have inaccurately described the *Vogmarus Islandicus :* their specimens were mutilated, or badly preserved. A specimen, almost entire, was thrown ashore during last year on the coast of Skagen, which is now in the zoological collection of the university : another was caught at the Feroe Islands, and is preserved in the Royal Museum. These specimens have been carefully examined, and prove that the Vaogmære does not belong, as Linnæus believed, to the apodal fishes, but to the thoracic; although neither of these two specimens are sufficiently perfect to admit measurement of the fin-rays."

This Northern species differs from those of the Mediterranean.

In Dr. Fleming's paper above referred to, one specimen caught alive in Sanday, in Orkney, is thus described :— " Length three feet ; body excessively compressed, particularly towards the back, where it does not exceed a table-knife in thickness ; breadth nearly five inches, tapering to the tail. Colour silvery, with minute scales ; the dorsal fin of an orange colour, occupying the whole ridge from the head to the tail, with the rays of unequal sizes. Caudal fin forked, the rays of each fork about four inches long. Pectoral fins very minute : no ventral nor anal fins whatever. Vent immediately under the pectoral fins, and close to the gill-openings. Head about four and a half inches long, compressed like the body, with a groove on the top. Gill-lids formed of transparent porous plates. Eyes one inch and a quarter in diameter. Both jaws armed with small teeth. Lateral line rough, and, towards the tail, armed with minute spines pointing forwards ; and these are the only spines on the body."

Another specimen found on the beach of Sanday is described as follows :—" Length four and a half feet ; breadth eight inches ; thickness one inch, thin at the edges of the back and belly. Length of the head five inches, terminating gradually in a short snout. Tail consists of eight or nine fin-bones or rays, the third ray seven inches long, the rest four inches. The dorsal fin reaching from the neck to the tail, rays four inches long. On each side of the fish, from head to tail, a row of prickles pointing forward ; distance between each half an inch. Under edge fortified by a thick ridge of blunt prickles. Pectoral fins one inch long, lying upwards. Skin rough. Colour a leaden or silvery lustre ; dorsal fin and tail blood colour. The skin or covering of the head like that of a Herring : several small teeth ; gills red, consisting of four layers. Heart half an inch ; liver two and a half inches ; stomach four and a half inches, full of a gelatinous substance. Flesh perfectly white. Spine in the middle of the fish. Body thin towards the back and belly, and wears very small towards the tail. Eyes and brain wanting."

Various specimens, probably to the number of twelve or more, appear to have been obtained on the island of Sanday between the years 1817 and 1829. Some of the natives were sufficiently acquainted with it to induce a belief that they had even eaten it. Most of the specimens, varying in

VOL. I. o

size from one to six feet, were driven on shore by bad weather.

Olafsen, in his Voyage to Iceland, states that this fish is rare even in Iceland : it seems to approach the shore at flood-tide, in those places where the bottom is sandy and the shore not steep, and where it remains till left dry. The inhabitants, he adds, consider the fish as poisonous, because the ravens will not eat it.

To assist observers in identifying this species, the representation at the head of this article is, reduced in size, from the figure in the Magazine of Natural History, which contains Dr. Fleming's paper at length, parts only of which are here extracted. To this is added, as a vignette, an outline of the fish, and the form of the lateral spine, from Schneider.

A good figure of this species is still wanting.

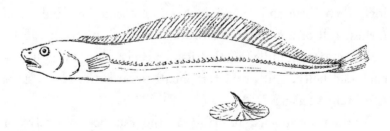

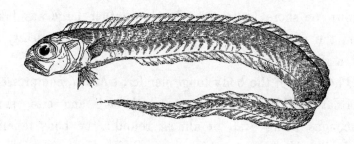

THE RED BANDFISH, OR RED SNAKEFISH.

Cepola rubescens, Linnæus.
  ,,   ,,   Cuvier, Règne An. t. ii. p. 221.
  ,,   ,,   Montagu, Linn. Trans. vol. vii. p. 291, tab. 17.
  ,,   ,,   *Red Bandfish,* Penn. Brit. Zool. vol. iii. p. 285.
  ,,   ,,   ,,   Don. Brit. Fish. pl. 105.
  ,,   ,,   *Bandfish,*   Flem. Brit. An. p. 204, sp. 114.
  ,,   ,,   *Red Snakefish,* Couch, Linn. Trans. vol. xiv. p. 76, and MS.

Generic Characters.—Head short, rounded; body elongated, compressed, lanceolate; dorsal and anal fins extending very nearly the whole length of the body; teeth prominent, curved, and sharp; branchiostegous rays 4.

Colonel Montagu first described the Red Bandfish as a British species in 1803. Two specimens were taken in Salcombe bay, on the south coast of Devonshire,—the first in February, the second in March; and an account, with a description and figure, appeared in the Transactions of the Linnean Society, vol. vii. In 1822, Mr. Couch included this species in his paper on the Fishes of Cornwall, in the fourteenth volume of the Transactions of the same Society; and referring to his MS., I find the following additional information :—" Until within a few

o 2

years the Red Snakefish had not been recognised as a British species; yet it is not uncommon on the western coast. No less than nine specimens have fallen into my hands, of which three were at different times killed and thrown on shore by tempests. One rather large was taken from the stomach of a hake; and one more, at least, was taken with a line."

The form of the body long, slender, smooth, compressed; this latter character increasing with age and size, small specimens being oval, or almost round: the body tapering gradually, both as to thickness and depth, from the head to the tail; head not larger than the body; both jaws sloping equally towards each other; the lower jaw the longest when the mouth is opened; the line of the upper jaw ascending obliquely; the mouth large, the angle depressed; the tongue short and smooth: both jaws furnished with a single row of conical, curved, pointed teeth, not set close together; the teeth ranged along the outer edges of the jaws, and projecting considerably, particularly those of the lower jaw: the eyes large; the nose short; gill-covers of two pieces: pectoral fins small and rounded; ventrals placed rather before the line of the origin of the pectorals, the first ray spinous, the inner ray of each united at the base; the dorsal fin commences at the nape and extends to the tail, the anterior rays shorter than the others; the vent is about an inch behind the ventral fins; the anal fin commences immediately behind the vent, extending like the dorsal fin to the tail, and having also the anterior rays rather shorter than the others; caudal fin lanceolate, middle ray the longest: the distinction between the rays of the dorsal, anal, and caudal fins, is lost by union, and the tail ends in a point. The lateral line, not very obvious on some parts of the body, is a little curved near the head, and afterwards runs quite straight to the tail : skin smooth, but when examined with a lens, appears finely and regularly punctured. A specimen seven and a half inches long, for which I am indebted to the kindness of Mr. Couch, exhibits here and there an occasional thin, oval, semi-transparent scale. The irides are silvery with a tinge of crimson, pupils bluish black ; gill-plates silvery. The body appears subject to some variation in colour. One of Colonel Montagu's specimens was pale carmine, the second darker. Mr. Couch had specimens of a pale red. A dried example from the Mediterranean, now before me, is orange red : the Cornish specimen, preserved in spirits, has lost colour, and is now greyish orange. Brunnich, describing the colour of his *Cepola rubescens*, calls it *pallide carneum*, pale flesh colour ; and M. Risso says it is the colour of the red oxide of mercury. In the first edition of his work, M. Risso includes two species of this genus, *C. tænia* and *C. rubescens* ; in the second edition, *rubescens* only is retained. Brunnich, in a note at the end of his description of *rubescens*, asks, Is this fish distinct from the *tænia* of Linnæus, and how ? The latter is said to be distinguished by a row of hard points along the side, above the lateral line, and by an inner second row of teeth on the lower jaw. My Mediterranean specimen, thirteen inches long, has the rough line just below the base of the dorsal fin, and a second row of six small teeth within the lower jaw.

In reference to the first of these distinctions, it is essential to remark, that Mr. Couch, in his description in the Linnean Transactions of a Cornish specimen fifteen inches long, says, " Besides the lateral line, there was a row of small bony prominences near the dorsal fin ;" and that in the smaller Cornish specimen sent to me by Mr. Couch, there is a single tooth in the lower jaw on the line of the second row of teeth in the larger Mediterranean fish. May not the rougher dorsal line and the six additional teeth be the consequence of age ?

The numbers of fin-rays agree very nearly : in the small specimen preserved in spirits, they are—

D. 69 : P. 16 : V. 1 + 5 : A. 61 : C. 13.

Of the habits of this fish but little is known. M. Risso says, that when moving in the water, its appearance has suggested the epithets of Fire-flame and Red-riband, by both of which names it is known at Nice. He adds, also, that it lives principally among seaweed near the shore, and though it feeds on crustaceous and molluscous animals, yet its flesh is not esteemed for its flavour.

"The air-bladder of this fish," says Mr. Couch, "is remarkable for its large size, and the chief part, not in the abdomen, but behind it, occupying the space from the spine behind the vent and along the anal fin."

It may be considered worth noticing here, that a large proportion of the examples of the family of riband-shaped fishes that have been obtained in this, as well as in other countries, have been left on the shore after stormy weather.

Does their elongated form prevent their swimming with ease at mid-water, and inducing a habit of keeping near the ground, or occasionally seeking cavities among rocks for shelter, thus render them liable to be left dry by the retiring tide, or destroyed by the force of waves dashing them against such opposing substances ? The combination of great length with extreme tenuity of body, by diminishing the quantity of muscle, and at the same time preventing its being brought into concentrated action upon a single centre of motion, must necessarily leave them at all times much at the mercy of the currents, amid which they may wriggle or float, but against which they are evidently incapable of swimming with any vigorous effort : by their struggles in the ocean, they cannot fail to become speedily exhausted, and they are rejected by the waves like inanimate matter, upon any coast toward which the winds may have driven them.

THE GREY MULLET.

Mugil capito,   Cuvier, Règne An. t. ii. p. 232.
  ,,   *cephalus,* Willughby, p. 174, tab. R. 3.
  ,,   ,,   *Grey Mullet,*   Penn. Brit. Zool. vol. iii. p. 346, pl. 77.
  ,,   ,,   ,,   ,,   Don. Brit. Fish. pl. 15.
  ,,   ,,   *Common Mullet,* Flem. Brit. An. p. 217, sp. 159.

Generic Characters.—Body nearly cylindrical, covered with large scales; two dorsal fins, widely separated, the rays of the first fin spinous, those of the second flexible; ventral fins behind the pectorals; middle of the under jaw with an elevated angular point, and a corresponding groove in the upper; teeth small; branchiostegous rays 6.

Baron Cuvier, in the last edition of his *Règne Animal,* states, in a note at the foot of page 231, that Linnæus and several of his successors have confounded all the European Grey Mullets under one common name,—that of *Mugil cephalus.* He has, however, distinguished among them several species: and according to him, the description of the *cephalus* of Willughby and the figure of the *cephalus* of

Pennant both appear to belong to the *M. capito* of the *Règne Animal.*

This opinion, that the *cephalus* of Linnæus is not the true *cephalus*, receives support from other authors who have attended to fishes. Professors Reinhardt and Nilsson each refer the Grey Mullet of the Baltic and the coast of Norway to the *capito* of Cuvier; and the Prince of Musignano, who has described and figured in his *Fauna Italica* five species of Grey Mullets as belonging to the Mediterranean, including both *cephalus* and *capito*, makes no reference to Linnæus in his account of *cephalus*, and considers his *capito* as identical with the *cephalus* of Pennant.

Mugil cephalus is distinguished by having its eyes partly covered with a semi-transparent membrane adhering to the anterior and posterior edges of the orbit, and also by a large elongated triangular scale pointing backwards, placed just over the origin of the pectoral fin on each side. A dried specimen of this fish from the Mediterranean, now before me, exhibits both these peculiarities, which *M. capito* does not possess. The vignette accompanying this article represents the appearance of the pectoral fin, and the superposed triangular scale of *M. cephalus*, both for the purpose of supplying the means of comparison with our common Grey Mullet, in which the pectoral fin-scale is short and blunt, and to enable observers to identify the true *cephalus*, should it occur on our coast; which is not improbable, when it is recollected how many Mediterranean species have been recorded as occurring along the line of our southern shore.

Our most common Grey Mullet may therefore be considered as the *M. capito* of Cuvier, an inhabitant not only of the Mediterranean, but also of all the western shores of the more temperate part of Europe.

In Ireland this fish occurs on the coast of the northern counties of Londonderry and Antrim; in the south, on those of Cork and Waterford; and probably at many intermediate points. It is found plentifully in Cornwall and Devonshire, and along the whole line of our south coast. It occurs constantly on the Kentish and Essex coast; is taken at Yarmouth: Mr. Neill has met with it at the mouth of the Esk; and it has been traced to the Baltic and the west coast of Norway, as previously quoted.

Mr. Couch, in his MS., has described the habits of this fish so much better than any account I could offer of my own, that I shall be excused quoting his remarks at some length.

" This fish never goes to a great distance from land, but delights in shallow water when the weather is warm and fine; at which time it is seen prowling near the margin in search of food, and imprinting a dimple on the placid surface as it snatches beneath any oily substance that may chance to be swimming. It ventures to some distance up rivers, but always returns with the tide. Carew, the Cornish historian, had a pond of salt water, in which these fish were kept; and he says, that having been accustomed to feed them at a certain place every evening, they became so tame, that a knocking like that of chopping would certainly cause them to assemble. The intelligence this argues may also be inferred from the skill and vigilance this fish displays in avoiding danger, more especially in effecting its escape in circumstances of great peril. When enclosed within a a ground-sean or sweep-net, as soon as the danger is seen, and before the limits of its range are straitened, and when even the end of the net might be passed, it is its common habit to prefer the shorter course, and throw itself over the head-line, and so escape ; and when one of the company passes, all immediately follow."

" This disposition is so innate in the Grey Mullet, that young ones of minute size may be seen tumbling themselves head over tail in their active exertions to pass the head-line. I have even known a Mullet less than an inch in length to throw itself repeatedly over the side of a cup in which the water was an inch below the brim."

" Mullets frequently enter by the floodgate into a salt-water mill-pool at Looe, which contains about twenty acres ; and the larger ones, having looked about for a turn or two, often return by the way they had come. When, however, the turn of the tide has closed the gates and prevented this, though the space within is sufficiently large for pleasure and safety, the idea of constraint and danger sets them on effecting their deliverance. The wall is examined in every part ; and when the water is near the summit, efforts are made to throw themselves over, by which they are not uncommonly left on the bank to their own destruction."

" When, after being surrounded by a net, two or three have made their escape, and the margin of the net has been secured and elevated above the surface to render certain the capture of the only remaining one, I have seen the anxious prisoner pass from end to end, examine every mesh and all the folds that lay on the ground, and at last, concluding that to pass through a mesh, or rend it, afforded the only though desperate chance of escape, it has retired to the greatest possible distance, which had not been done before, and rushed at once to that part which was most

tightly stretched. It was held, however, by the middle;
and conscious that all further effort must be unavailing, it
yielded without a further struggle to its fate."

" The Grey Mullet selects food that is soft and fat, or such
as has begun to suffer decomposition; in search of which it
is often seen thrusting its mouth into the soft mud; and, for
selecting it, the lips appear to be furnished with exquisite
sensibility of taste. It is, indeed, the only fish of which
I am able to express my belief that it usually selects for
food nothing that has life; although it sometimes swallows
the common sand-worm. Its good success in escaping the
hook commonly proceeds from its care not to swallow a
particle of any large or hard substance; to avoid which,
it repeatedly receives the bait into its mouth, and rejects
it; so that when hooked it is in the lips, from which the
weight and struggles of the fish often deliver it. It is
most readily taken with bait formed of the fat entrails of a
fish, or cabbage boiled in broth."

" The Grey Mullets shed their spawn about Midsummer;
and the young in August, then an inch long, are seen
entering the fresh water, keeping at some distance above
the tide, but retiring as it recedes. The change and re-
change from salt water to fresh seems necessary to their
health, as I judge from having kept them in glass vessels."

The Grey Mullet is frequently an object of sport to
the angler. They rise freely at the flies used for Trout,
and even at the larger and more gaudy flies used for
Salmon. They are reported to be strong in the water,
and require care in the management of them, as they plunge
violently. The best time for angling for them is when
the tide is coming in; as, when it ebbs, they return to salt
water.

The county of Sussex is proverbially celebrated for six

good things ; viz. a Chichester lobster, a Selsey cockle, an Arundel mullet, a Pullborough eel, an Amberley trout, and a Rye herring. In reference to the Mullet, I may notice, that during the summer of 1834, probably owing to the warmth of it, the Grey Mullet migrated much farther up the river than usual, and were caught above even where the spring-tides flow, as high up as Amberg Castle, which is by the river nearly ten miles above the town of Arundel, and nearly twenty miles from the sea.

The partiality exhibited by the Grey Mullet for fresh water has led to actual experiment of the effect of confining them to it entirely. Mr. Arnould put a number of the fry of the Grey Mullet about the size of a finger into his pond at Guernsey, which is of about three acres area, and has been before referred to under the article Basse. After a few years, Mullet of four pounds' weight were caught, which proved to be fatter, deeper, and heavier for their length, than others obtained from the sea. Of all the various salt-water fishes introduced, the Grey Mullet appeared to be the most improved. A slight change in the external colour is said to be visible.

The length of the head in this fish, compared with the length of the body and tail, is as one to four : the depth of the body is equal to the space from the anterior edge of the orbit to the end of the operculum, and the body does not decrease in size till the commencement of the anal fin : the fleshy portion of the tail is equal to half the depth of the body.

The form of the mouth is different from that of most other fishes. The lower jaw is divided in the middle by an ascending angular point, which, when the mouth is closed, passes within the upper jaw : the upper jaw, also, if viewed from below, is angular ; each jaw is furnished with a single row of minute teeth ; the nostrils are double on each side, placed near together, both pierced in the same depression, the anterior aperture round, the posterior orifice oblong and vertical; the operculum large and broad.   The number of fin-rays are—

D. 4. 1 + 8 : P. 17 : V. 1 + 5 : A. 3 + 9 : C. 13.

The first dorsal fin commences behind the nape at a distance equal to the length of the head, and nearly in a line dividing the distance between the origin of the ventral and anal fins ; the second dorsal fin begins on a line a little behind the origin of the anal fin, and being shorter than that fin, ends on the same line with it.   The general lengths of the longest of all the various fin-rays are nearly equal to each other, and about equal to three-fourths the length of the head ; except the caudal fin, the rays of which are longer, and the tail considerably forked.

The colour of the top of the head and back is dusky grey tinged with blue ; the sides and belly silvery white, marked with longitudinal parallel dusky lines ; membranes of the fins dull white ; cheeks and operculum silvery white ; irides reddish brown, pupil black, surrounded by a silvery line. The pectoral fin has a dark spot at the base of the three or four upper rays.

ACANTHOPTERYGII.   *MUGILIDÆ.*

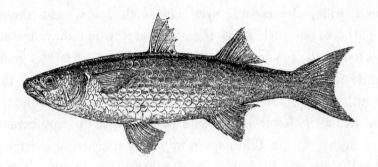

THE THICK-LIPPED GREY MULLET.

Mugil chelo,   Cuvier, Règne An. t. ii. p. 232.
  ,, *labrosus,*   Risso, Hist. Nat. t. iii. p. 389.
  ,, *chelo,*   P. Musignano, Faun. Ital. pt. vi.
  ,,   ,,   Couch, MS.

Mr. Couch appears to be the only naturalist who has noticed the appearance of *Mugil chelo* on the British coast. A decided difference observed in the habits of this Grey Mullet compared with those of *M. capito* led to an examination of its specific characters, and a knowledge of the fact that it was a distinct species, which, though well known to modern Continental Ichthyologists, had not previously been noticed by observers here.

Mr. Couch's communication is as follows :—

" This Grey Mullet is gregarious, frequenting harbours and the mouths of rivers in the winter months in large numbers, all of which are just of one size. I have heard of so many as two tons being taken at one time : but the fish which I shall here describe was taken with about four hundred others as they were left in a pool of our river, forsaken by the tide. This species has, like the other, the habit of escaping from a net by leaping over the head-lines. The length of the specimen was ten inches : the head wide, depressed; eyes one inch apart, and three-eighths of an inch from the angle of the mouth, not connected with any membrane ; nostrils close together, and, while the fish is alive, moveable on each contraction of the mouth : a prominent superior maxillary bone, minutely notched at its lower or posterior edge ; upper lip protuberant and fleshy, with a thin margin minutely notched or ciliated ; the lip appears behind as projecting under the maxillary. Carina of the under jaw prominent and square ; edge of the lower lip fine and simple. Body solid, round over the back : pectoral fins high on the side, pointed, rounded below, the first rays short. The first dorsal fin five inches and three-eighths from the snout, the origin of the first three rays approximate, the first ray the longest ; the first two rays of the anal fin short : tail broad, concave ; scales large. Colour of the head and back greenish ; all besides silvery, with six or seven parallel lines along the sides of the same colour as the back." The number of fin-rays are—

$$D. 4. 9 : P. 14 : V. 1 + 5 : A. 3 + 8 : C. 16.$$

The figure of this fish is taken from the *Fauna Italica* of the Prince of Musignano, who attaches to this species the following specific characters:—

" Head of moderate size, subtruncated in front; upper lip thickened, under lip very slightly margined ; the descending portion of the maxillary bone projecting below the suborbital bone ; the space between the edges of the inter-opercula very narrow ; the rays of the spiny dorsal fin longer than the half of the depth of the body."

The characters of *M. chelo*, as given by Cuvier in the *Règne Animal*, are, that it is distinguished particularly by its very large and fleshy lips, the edges of which are ciliated, and through their thickness the teeth penetrate like so many hairs : the maxillary bone is curved, and shows itself behind the commissure.

These short descriptions of the Thick-lipped Grey Mullet, as given by the Prince of Musignano and Cuvier, have been added here to show, by their general accordance with the account of Mr. Couch, the correctness of that gentleman's views of this species.

The Arun, it has been stated, is proverbially celebrated for its Grey Mullet. The vignette below represents part of this river near its mouth, with Arundel Castle on the right ; taken from a drawing most obligingly made by Mr. Lear for this work.

ACANTHOPTERYGII. *MUGILIDÆ.*

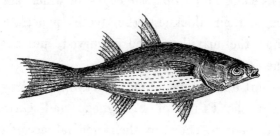

THE SHORT GREY MULLET.

Mugil curtus, YARRELL.

HITHERTO but one species of Grey Mullet has been described and figured as belonging to the British coast; but this is probably owing to the want of close comparative examination of specimens from different localities. Cuvier, in the *Règne Animal,* says the European Mullets have not been well determined; and from the general distribution* of the species of this genus, it is not unlikely that more than those at present known may yet be made out. The Prince of Musignano, as before mentioned, has described and figured five species belonging to the Mediterranean; but the Small Grey Mullet of the present article appears to be as yet unknown, at least as far as I have been able to ascertain by existing descriptions. Its principal distinction, as a species, is in the extreme shortness of the body, which has induced me to adopt for it the specific term *curtus.*

* Species belonging to the genus *Mugil* of authors have been found more or less plentiful at the Cape de Verd Islands, Caspian Sea, Japan, New South Wales, Sandwich Islands, and in the Bay of Mexico, besides the other localities that have been previously named.

The number of its fin-rays are—

D. 4. 1 + 8 : P. 11 : V. 1 + 5 : A. 3 + 8 : C. 14.

The length of the head as compared with that of the body and tail is as one to three, the proportion in the common Grey Mullet being as one to four; the body is also deeper in proportion than in *M. capito*, being equal to the length of the head; the head is wider, the form of it more triangular, and also more pointed anteriorly; the eye larger in proportion; the fin-rays longer, particularly those of the tail; the ventral fins placed nearer the pectorals, and a difference exists in the number of some of the fin-rays: the colours of the two species are nearly alike; and in other respects, except those named, they do not differ materially.

The proportions of the figure of the Grey Mullet in Mr. Donovan's History of British Fishes, plate 15, approach more closely to those of *M. curtus*, than to those of the common Grey Mullet of this country.

Of this Grey Mullet I have only obtained the single specimen that served for the representation given, which is exactly the natural size of the fish. I caught this with the young of the common Grey Mullet, and various other fry, when fishing with a small but very useful net between Brownsey Island and South Haven, at the mouth of Poole Harbour.

The net used is called a keerdrag, and as it is an effective machine, where the ground is smooth, for the collectors of small fishes and various other marine animals, I have made a representation of it the subject of the vignette annexed, and will shortly describe the apparatus and the manner of working it.

The bottom and sides of the oblong mouth of the net are formed of an iron rod about seven feet long, of which about

P 2

fifteen inches at each end are bent once at right angles :
to these ends a straight beam of wood three inches diameter
is fixed, which should be rounded for the convenience of
handling. The wood by its buoyancy, when the net is in
use in the water, tends to preserve the vertical position of
the framework.

To the mouth of the net thus formed by the union of the
iron and wood a piece of netting is to be applied all round,
which should diminish gradually, both in the size of the net
and its mesh, till, at the distance of seven or eight feet from
the framework, it should terminate in a round open mouth
about the size of the top of a stocking. The mesh of the
net for the last three feet should be very small, as it is at
this part the most strenuous efforts to escape will be made ;
particularly by the *Syngnathi*.

The net is to be drawn along the ground by a slight rope,
over the stern of the boat, which should not be rowed fast.
This tow-rope ends towards the net by a three-tie bridle, one
of which is attached to the centre of the wooden beam : of
the other two, one goes to each side, and thus the mouth of
the net is not only kept square to the front, but its vertical
position is also preserved.

The open tail of the net being closed and securely tied,
and the apparatus put overboard from a row-boat, keeping
hold of the tow-rope, and taking care that the mouth of the
net preserves its position, it should be towed leisurely about,
the iron bottom traversing the ground, and the quantity of
contents obtained soon lead to a knowledge of the best
localities. Should the mouth of the net get foul of any opposing substance on the ground, it is only necessary to push
the boat back in the line of its previous course, and the net
comes away clear, being thus pulled upon in the opposite
direction.

When inclined to examine the net, the framework may be raised by the tow-rope high enough to lodge the wooden beam over the edge of the boat's stern,—but higher than that is unnecessary: the tail of the net is to be handed in, untied, and the contents shaken into a tub for examination. The tail of the net being retied, the frame may be lowered and towed about as before; and while the net is at work at the bottom, the collector may be engaged over the contents of his tub at the top.

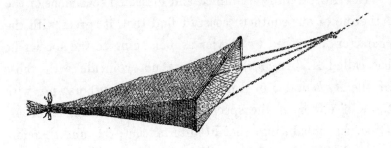

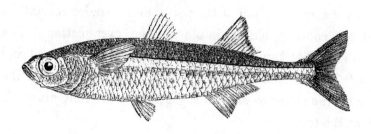

THE ATHERINE, OR SANDSMELT.

Atherina presbyter, Cuvier, Règne An. t. ii. p. 235.
   ,,   *hepsetus,*   *Atherine,* Penn. Brit. Zool. vol. iii. p. 434, pl. 76.
   ,,   ,,   ,,   Don. Brit. Fish. pl. 87.
   ,,   ,,   ,,   Flem. Brit. An. p. 217, sp. 160.

Generic Characters.—Body rather elongated ; two dorsal fins widely separated ; ventral fins placed far behind the pectorals ; sides with a broad longitudinal silver band ; teeth minute ; branchiostegous rays 6.

Having carefully examined and dissected specimens of the Atherine of our southern coast, I find that it agrees with the characters described by Cuvier as belonging to the species he has called *A. presbyter,* and does not coincide with those of the *A. hepsetus* of Linnæus and other authors, nor with those of either of the species described by Brunnich or M. Risso as inhabiting the Mediterranean. I am therefore induced to believe that our British Atherine is the *A. presbyter* of Cuvier, and I have adopted that name accordingly.

Cuvier considers that more than two species of Atherine have been confounded under the name of *hepsetus.*

The following observations are from Colonel Montagu's notes :—

" The Atherine is as plentiful on some parts of the

southern coast of England as the Smelt is on the eastern
coast, and each appears to have its limits, so as not to intrude
upon the other ; at least, as far as our observation has gone,
where one is, the other is not. We have traced the Smelt
along the coast of Lincolnshire, and southward into Kent,
where the Atherine appears to be unknown ; but in Hampshire the Atherine is extremely plentiful, especially about
Southampton, where, for want of knowing the true Smelt,
this is sold under that denomination. On the south coast
of Devon they are caught in great abundance in the creeks
and estuaries, but never in rivers above the flow of the tide ;
and they appear to continue near shore through the months
from autumn to spring, being caught for the table more
or less during the whole of that time; but are greatly superior
in the spring, when the males are as full of milt as the females are of roe. The Atherine is a well-flavoured fish ;
but, in our opinion, not so good as the Smelt: it is more
dry ; but when in season, and fried without being embowelled, the liver and roe make it a delicious fish."

To this account by Colonel Montagu, I may add that
I have not known the Atherine taken east of the county
of Sussex ; and to his view of its non-appearance upon
our north-eastern shores I can only find one exception :
Mr. Neill says, in the Memoirs of the Wernerian Society,
he has repeatedly found the Atherine washed ashore about
Figget Whins, in the Frith of Forth, after easterly winds.
The Atherine is a delicate, and perhaps a tender fish,
averse to cold: Mr. Couch says, that during severe frosts
large quantities are sometimes killed and left by the tide.
It appears to be a very well-known fish, common in most
of the sandy bays of our southern coast. It is taken in
the first bay east of Beachy-head, and probably comes as
far as Rye bay or Dungeness ; but keeping close in shore

in the smooth water, it perhaps very seldom ventures into the increased rapidity of the Channel tide, in its rush through the Straits of Dover.

Mr. Couch says it is found in Cornwall at all seasons, and sometimes in such numbers that three small boat-loads have been enclosed in a sean at once.

The Atherine is a common fish at Brighton, where it is called Sandsmelt. Large quantities are eaten by the inhabitants and visitors during the winter months. They possess a little of the cucumber smell and flavour of the true Smelt; and as they are very pretty in appearance, from the fine broad silver stripe along the side, they look attractive as arranged by the fishmongers in their shops, and obtain a ready sale.

The net used for taking them is made of fine threadlike twine, the mesh of course very small: the net is thirty yards long, and about eighteen feet deep. It is drawn along near the edge of the water, by two parties; one of which in a boat, having the head and ground-line of the seaward end of the net, row gently on; the other party on the shore, at or near the edge of the water, advance in a line with the boat, holding and drawing on their end of the net, and thus sweep the circle of the bays and sandy shores. For those caught for the supply of Brighton market I have seen the fishermen going westward, probably to the sandy shore of Shoreham or Worthing. I have also seen this mode of taking Atherine adopted in the bay close to the sea-houses near Eastbourne.

Another method is practised in Portsmouth harbour. The fishermen use a concave circular net suspended from an iron ring of four feet diameter, kept horizontal by a three-slip bridle. The net is lowered steadily in eight feet water, among the timber floating on the side of the harbour nearest the dock-yard. Pounded crabs sprinkled over the net as bait is the attraction; and the net is occasionally raised gently to the surface. In this way five or six dozen are obtained during the flood-tide.

The Atherine is a handsome small fish, from five to six inches long, but very rarely brought to the London market. It spawns in May or June.

The number of fin-rays are—

1st D. 8 : 2nd D. 1+12 : P. 15 : V. 1+5 : A. 1+14 : C. 17 : Vertebræ, 50.

The length of the head, from the point of the under jaw to the edge of the operculum, compared to the length of the body and tail, is as one to four; the depth of the body not quite equal to the length of the head; a silver-coloured band, half as broad as the space above it, and one-third as broad as the space below it, passes from the upper edge of the operculum and the base of the pectoral fin, to the centre of the base of the tail; four rows of scales above the silver band, and six rows below it; the band itself occupying two rows.

The form of the head rather short; nose blunt; upper jaw capable of considerable protrusion; lower jaw the longest when the mouth is open; one row of minute teeth along the edge of each jaw: the eye large; top of the head flat, with a ridge descending on each side to the nose.

The first dorsal fin commences about the length of the head from the nape; the second dorsal begins at the same distance behind the origin of the first, and ends at the same distance from the base of the caudal fin-rays; the ventrals originate, on a vertical line, with the ends of the pectoral fin-rays, and the ventral and anal fins begin a little in advance of each dorsal fin respectively : the tail deeply forked, the longest rays nearly equal to the length of the head ; the vent a small circular orifice in a line under the tips of the first dorsal fin-rays when folded down.

Colour of the cheeks, irides, gill-covers, base of pectoral fins, and broad side-band, shining silvery white ; the other parts of the body a pale transparent flesh colour ; the upper part of the back and head freckled with small black spots ; the membranes of the fins yellowish white.

Considerable numbers of the Atherine are caught by anglers from projecting points at various localities along the southern coast. Poole Quay is a favourite spot. The fish bite voraciously at any bait that is offered to them, and even at that season when they are heaviest with roe, which is not the case with fishes generally.

ACANTHOPTERYGII.   *GOBIOIDÆ.**

MONTAGU'S BLENNY.

Blennius Montagui,   FLEM. Brit. An. p. 206 & 207, sp. 121.
   ,,   *galerita,*   MONTAGU, Mem. Wern. Soc. vol. i. p. 98, pl. v. fig. 2.
              Diminutive Blenny, PENN. Brit. Zool. vol. iii. p. 277.

Generic Characters.—Head rounded and blunt; body smooth, unctuous, compressed; a single elongated dorsal fin; ventral fins placed before the pectorals, and containing generally but two rays, united at the base; teeth slender, in a single row.

SOME difficulty occurs on endeavouring to reconcile the synonymes of the Blennies of British authors, from the want of correct representations and more detailed descriptions.

But five species, as the genus is now restricted, will be figured in the present work: four of these, having the line of the edge of the dorsal fin interrupted, will be arranged according to the number of appendages on the head, these appendages being considered only as specific characters, beginning with that species which possesses the greatest number: the fifth species, having an uninterrupted dorsal

* The family of the Gobies.

fin, a more elongated slender body, short ventral fins, and longer anal fin, exhibits in these various particulars so many relations to the characters of the genus next in succession, and is therefore placed last.

The fishes of this genus are of little value : they swim in small shoals, feeding on minute crustaceous animals, and some of them are remarkably tenacious of life. They are most frequently found, left by the retiring tide, in small pools on the rocky parts of the coast, are active and vigilant, hiding themselves in small crevices or under sea-weed, and remaining concealed till the return of the tide.

The example of Dr. Fleming has been followed in considering this Blenny, described by Colonel Montagu in the Wernerian Memoirs before referred to under the term *galerita*, as distinct from the *galerita* of Linnæus : the uniformity in the boundary line of the dorsal fin in the true *galerita*, and the interrupted line in the fish figured and described by Montagu, being one of the most obvious characters for distinction. The number of the rays in the dorsal and anal fins in **B. Montagui** are only as thirty to fifty-one in the dorsal, and eighteen to thirty-six in the anal, as compared with the *galerita* of Linnæus, which will be hereafter described.

Not having been so fortunate as to obtain a specimen of this fish, the account given is derived from Colonel Montagu, and the figure is from a drawing by Mr. Couch, who in his MS. briefly refers to this fish as occurring in Cornwall, and as being very active and difficult to catch.

" Body rather more slender than that of the Smooth Blenny. Head much sloped ; eyes high up, approximating, gilded ; the upper lip furnished with a bony plate that projects at the angles of the mouth into a thin lamina

that turns downwards, the ends of which are orange-co-
loured: on the top of the head, between the eyes, is a
transverse, fleshy, fimbriated membrane; the *fimbriæ* of
a purplish brown colour, tipped with white; the nostrils
furnished with a minute bifid appendage: behind the crest
are several minute, erect, filiform *appendiculæ*, between
that and the dorsal fin, placed longitudinally: the lateral
line considerably curved near the head; the pectoral fins
are large and ovate, reaching as far as the vent; the ventral
fins two unconnected rays: the dorsal fin extends from
the head to the tail, and appears like two distinct fins,
by reason of the slope to the thirteenth ray, which is not
above half the length of the anterior ones, and the sudden
elongation of the fourteenth ray: this fin is very broad, and
in one specimen there was an ovate black spot between the
first and second ray, and another obscure one between the
next rays; but this is not a constant character. The anal
fin is equally broad, and extends from the vent to the tail,
the rays margined with black and tipped with white: caudal
fin slightly rounded."

" The colour above is generally olive green, spotted
with pale blue, shaded to white: the belly white, and
the pectoral fins spotted with orange. The number of
fin-rays are—

D. 30 : P. 12 : V. 2 : A. 18 : C. 14.

" Not fewer than eight or ten of this species have
come under my inspection, the greater part of which did
not exceed an inch and a half in length; but two at
present before me measure nearly two inches and a half,
and differ in nothing but the spots on the dorsal fin.
The crest is not capable of being erected,—at least no
voluntary motion could be observed while the fish was

examined alive in sea-water; but this appendage is invariably transverse, and generally conic or angular, but sometimes irregularly truncated, though always fimbriated."

" This is occasionally taken, with others, among the rocks on the south coast of Devon, in the pools left by the retiring tide."

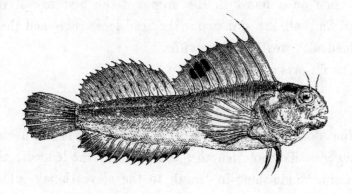

THE OCELLATED BLENNY, OR
BUTTERFLY FISH.

Blennius ocellaris,	Brunnich, p. 25, sp. 35.	
,, ,,	Bloch, pt. v. pl. 167.	
,, ,,	Cuvier, Règne An. t. iii. p. 237.	
,, ,,	*Ocellated Blenny,* Montagu, Mem. Wern. Soc. vol. ii. p. 443, pl. 22.	
,, ,,	,, ,, Flem. Brit. An. p. 206, sp. 119.	

THE OCELLATED BLENNY was first described as a British fish by Colonel Montagu, who obtained three specimens by dredging on the south coast of Devon.

The example from which the present description and figure were taken, was obtained among the rocks of the island of Portland.

The length near three inches; the head rounded and blunt: teeth in a single row, small, elongated, rather uneven at the edge, as if some of them had been broken off; the last tooth on each side, both above and below, considerably longer than the others: the eyes large, irides golden; attached to the anterior edge of the orbits are two large filamentous and fimbriated appendages, three-eighths of an inch in length; a small pedicle of skin behind the nape on each side on a line with the origin of the first ray of the dorsal fin; all the skin about the head loose, here and there studded with small warty *papillæ*.

The fin-rays are—

D. 26 : P. 12 : V. 2 : A. 17 : C. 11.

The dorsal fin begins at the nape, and is connected throughout its whole length; the first ray the longest, the next nine diminishing in length to the eleventh ray, which is the shortest, and marks the place of the interruption to the uniformity of the line, the twelfth ray being as long again as the eleventh : the second portion of the dorsal fin rounded in form, the membrane beyond the last ray being united to the base of the tail. The pectoral fin large, rounded, the middle rays about as long as the body of the fish is deep. The ventrals in this specimen with no more than two rays; the anal fin begins about half way between the nose of the fish and the end of the fleshy portion of the tail, and in a line but little in advance of the depression in the dorsal fin : the tail rounded, the rays about as long as those of the pectoral fin.

The general colour of the body is a pale brown, with occasional patches of darker reddish brown; the pectoral and ventral fins rather darker than the other fins, but the edges of the dorsal and anal fins rather darker than the part of the membrane nearer the body. The rounded spot on the dorsal fin is placed between the sixth and eighth rays : it is of a dark brown colour, with a slight indication of a lighter co-loured circle around it. The irides golden.

Montagu mentions his suspicion that the spot on the dorsal fin is not always present; but the form and elevation of the dorsal fin, would, without the spot, be sufficiently characteristic to mark the species.

This fish is a native of the Mediterranean, described by Rondeletius, Brunnich, and others. M. Risso says it lives much among weeds, feeding on minute crustaceous and molluscous animals, and spawns in the spring.

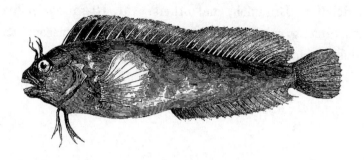

THE GATTORUGINOUS BLENNY.

Blennius gattorugine, Brunnich, p. 27, sp. 37.
    ,,   ,,   Cuvier, Règne An. t. ii. p. 237.
    ,,   ,,   *Gattoruginous Blenny*, Penn. Brit. Zool. vol. iii. p. 278,
                          pl. 39.
    ,,   ,,   ,,   ,,   Don. Brit. Fish. pl. 86.
    ,,   ,,   ,,   ,,   Montagu, Mem. Wern. Soc.
                          vol. ii. p. 447.
    ,,   ,,   ,,   ,,   Flem. Brit. An. p. 206, sp. 120.

The Gattoruginous Blenny appears to be a rare
fish on some parts of our sea-shore. Pennant first recorded
it as British from a specimen taken on the Anglesey coast.
Colonel Montagu obtained two in Devonshire, but considered it rare. Mr. Couch finds it frequently in Cornwall;
and specimens of one inch and a half, two inches and a
half, and five inches and a half, each, are now before me. For
the first of these I am indebted to Mr. Couch, the second
I obtained myself somewhere on our southern coast, but
have neglected to preserve any note of the exact locality;
and the largest example was given me by my friend Mr.
Thomas Bell, who brought it from Poole Harbour. It is
also said to have been taken at Belfast.

Mr. Couch considers it a common species in Cornwall,

that " keeps in the neighbourhood of rocks, in water of four or five fathoms depth. I have heard of its taking the hook, but it is more commonly caught in crab-pots, and consequently occurs in spring and summer, when that fishing is chiefly followed. It is called Tompot by the Cornish boys. At the end of May I have found it large with roe, the grains of which are, some of them of a mulberry, others of a lead colour; I have also seen numerous and minute young ones at the same season. In its stomach I have found various bivalve shells, parts of a star-fish, and of the common jointed corallines, and brown seaweed. Specimens occasionally measure eight or nine inches in length."

Some differences have been noticed in the descriptions and figures of this fish among several of the early, as well as of the more modern authors, and it is probable that a nearly allied species may have been sometimes mistaken for the *gattorugine*. I have, however, ventured to consider the Gattoruginous Blenny of Pennant, Montagu, and Donovan, as the same with that now described. A dried specimen of *gattorugine* from the Mediterranean, now before me, is the same as the English fish.

The forehead slopes considerably: viewed in front, a groove appears between the eyes, which ends in a channel, passing downwards behind each eye, formed by the elevation of the bones of the orbit on each side; from the upper and rather the posterior part of each eyelid arises a branched membrane, the eyelids extend considerably over the cornea all round; the nostrils are circular, in a depression, and above each is a small fimbriated membrane, plainly observable with a lens: the lips are thin and loose, turning up or down to a considerable extent, exposing the teeth; these are placed in a single row in each jaw, are long, slender, and semi-transparent, unequal in length in the front, almost

Q 2

every other one having had a small piece apparently broken off; the teeth on the sides of the mouth more uniform. The gill-cover ends in two angular points directed backwards, the edge of the membrane being continued under the throat to the gill-cover on the other side.

The body is compressed, and deepest on the line of the middle of the pectoral fins, from whence it tapers gradually to the end of the fleshy portion of the tail. The lateral line proceeds straight from the centre of the tail as far as the line of the commencement of the anal fin, and then arches high over the pectorals.

The nape of the neck rises high, upon which the dorsal fin commences on a line with the preoperculum. The first ray is shorter than the second, the next ten nearly equal in length, and about half the height of the body; the thirteenth ray shorter, and the fourteenth nearly one-fourth longer than the thirteenth, forming the interruption to the line of the dorsal fin; the remaining rays are nearer together than those that precede them, each portion of the fin occupying about the same space, with thirteen stiff rays in the first portion, and twenty flexible rays in the second; the membrane beyond the last ray extending to the base of the upper caudal fin-ray.

The pectoral fins are broad and rounded, the central rays the longest, and equal to the length of the head. The ventral fins slender, of two rays each only, about three-fourths of the length of the longest of the pectoral fin-rays. The anal fin is half as long as the head and body of the fish; it commences on a line rather before the depression in the dorsal fin, and immediately behind the vent: the rays of this fin project beyond the edge of the membrane connecting them, the last ray joined by a membrane to the body of the fish but does not quite reach the tail fin. The tail itself is slightly rounded, the rays about equal in length to those of the pectoral fin.

The number of fin-rays are—

D. 33 : P. 14 : V. 2 : A 23 : C. 11.

The prevailing colour of this specimen is a dark reddish purple brown, the lower part of the head, belly, and hinder portion of the body pale brown, all the fins dark brown. The smaller examples previously referred to, differ only in colour, being barred transversely, and clouded with a reddish brown over a light brown surface; the membranes of the fins also of a much lighter brown.

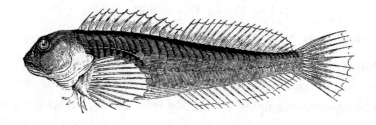

THE SHANNY, OR SMOOTH SHAN.

Blennius pholis, Linnæus.
    ,,   ,,   Cuvier, Règne An. t. ii. p. 238.
    ,,   ,,   *Smooth Blenny,* Penn. Brit. Zool. vol. iii. p. 280, pl. 40.
    ,,   ,,   ,,   ,,   Don. Brit. Fish. pl. 79.
Pholis lævis,   *Smooth Shan,*   Flem. p. 207, sp. 123.

THE Shanny is by no means uncommon at most of the rocky parts of our coast, and is easily distinguished among the Blennies by the want of any appendages on the head; the line of the dorsal fin is also interrupted. The term Smooth Blenny has not been continued here; as this name conveys no specific distinction, all the British Blennies being smooth.

"Destitute of a swimming-bladder, this fish," says Mr. Couch, "is confined to the bottom, where it takes up its residence on a rock or stone, from which it rarely wanders far, and beneath which it seeks shelter from ravenous fishes and birds; for cormorants, with their long and sharp beaks, drag multitudes of them from these retreats, and devour them. When the tide is receding, many of these fishes hide beneath the stones or in pools, but the larger individuals quit the water, and by the use of the pectoral fins creep into convenient holes, rarely more than one in each, and there, with the head outward, they wait for a few hours, until the return of the water restores them to liberty. If discovered or alarmed in these chambers, they retire by a backward motion to the bottom of the cavity. These circumstances show that the Shanny is retentive of life; in confirmation of which I have known it continue lively after a confinement of thirty hours in a dry box, notwithstanding which it soon expires in fresh water."

Furnished with long and firm incisor teeth, the Shanny is able to separate from the rocks, muscles, limpets, &c. on which to feed. The spawn is deposited in summer, and soon comes to life.

The head is rounded over the eyes, descending from thence rapidly to the nose; between the eyes a deep groove; the irides scarlet, no appendages either to the orbit or eyelids; the nostril pierced in a depression, with a small fimbriated membrane above it, a narrow oblong aperture on each side in front of the edge of the orbit; the mouth small, angular, much the widest at the gape, the lips large, broad, the posterior angle on each side free; the teeth small, a single row in each jaw, with occasionally a longer tooth projecting above the rest; the cheeks tumid; the gill-aperture large, the membrane continuing unattached, and extending under the throat to the other side.

The number of fin-rays are as follows—

D. 31 : P. 13 : V. 2. : A. 19 : C. 11.

The dorsal fin commences on a line over the union of the operculum with the body, the first portion consisting of twelve rays, the last of which is the shortest, the thirteenth as long again as the twelfth, forming the interruption; eighteen others succeed, nearly equal in height, the last of

which is united to the upper edge of the fleshy portion of
the tail by a continuation of the membrane connecting the
fin-rays : the ventrals of two rays, which originate before the
pectorals, and immediately behind the edge of the gill-cover :
the pectoral fins are large and rounded, the longest rays,
which are in the middle, reaching as far as the vent : the
anal fin commences immediately behind the vent, and under
the depression in the dorsal fin ; the last ray is attached to
the tail ; all the rays in this fin extend beyond the membrane : the tail is rounded ; the lateral line proceeding forwards, is straight for two-thirds of the distance along the
side, it then curves over the pectoral fin to the upper edge
of the operculum.

" It has justly been observed, that this species is extremely
variable in colour ; out of twenty or more examined at the
same time, not two were to be found alike ; some are prettily mottled with reddish brown, others quite plain, and one
variety is of a uniform dusky colour, even on the under
parts."

" This species of Blenny is remarkably tenacious of life,
and will live out of water for many days in a damp place,
or put in fresh grass or moss moistened with water ; and
probably, with a little attention, might be kept alive in this
way for many weeks.   If put into fresh water, it swims and
does not appear to feel any inconvenience, but does not long
survive the change."—*Montagu's MS*.

It rarely exceeds five inches in length.

ACANTHOPTERYGII. *GOBIOIDÆ.*

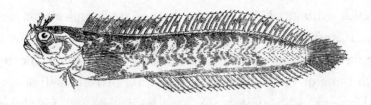

THE CRESTED BLENNY.

Blennius palmicornis, Cuvier, Règne An. t. ii. p. 237.
   ,,   *galerita,*   Strom. Linnæus. Gmelin.
   ,,   ,,   *Crested Blenny,* Penn. Brit. Zool. vol. iii. p. 276, pl. 39.
   ,,   ,,   ,,   ,,   Flem. Brit. An. p. 207, sp. 122.
   ,,   ,,   Nilsson, Ichth. Scand. p. 102.

I AM indebted to the kindness of Dr. George Johnston, of Berwick-upon-Tweed, for the only specimen of this fish I have ever seen ; and it proves to be a valuable acquisition, by affording an opportunity of giving a detailed description of the species, which, from the evidence to be quoted, I am induced to believe to have been first confounded by Strom, and afterwards by Linnæus, with the true *galerita* of Rondeletius, the *alauda cristata sive galerita* of Willughby and Ray.

Linnæus, in the tenth edition of his *Syst. Nat.* 1758, quotes Artedi only for his *Blennius galerita,* but without giving any number of fin-rays ; the account of Artedi, taken from Rondeletius, not including that part of the subject. In 1762, Strom published his account of the Fishes of the extreme North-Western portion of the coast of Norway and its Islands, which, under the name, and with a reference to the *galerita* of Artedi, contains a Blenny with an enumeration of fin-rays, which appears then for the first time, and was probably obtained from a specimen. Linnæus, in his twelfth edition, 1766, quotes both Artedi and Strom for his *Blennius galerita*, adding the number of fin-rays from Strom; thus coupling the characters of the northern Blenny with those of the Mediterranean *galerita* of Rondeletius.

Pennant, who appears to have been the first to obtain on our shore a specimen of the northern Blenny of Strom, referred it to the *galerita* of Linnæus. Gmelin in his work followed Linnæus and Pennant.

The error of Gmelin was first pointed out by Bloch, Schneider, page 169, note, with a reference also to Linnæus and Strom, but it was reserved for Cuvier to call more marked attention to Pennant's Crested Blenny, and to give to it the specific name which it will hereafter bear. Cuvier considers the *galerita* of Rondeletius to be the same with the *B. pavo* of Risso's Hist. t. iii. p. 235, sp. 124; a fish having only thirty-six rays in the dorsal fin, and but twenty-four rays in the anal fin.

A comparison of the figure at the head of this article with that of the crested Blenny of the British Zoology, will leave but little or no doubt that they are intended to represent the same species; yet the Crested Blenny of Pennant, with its two pair of appendages on the head, was referred by Strom, Linnæus, Pennant, and Gmelin, to the *galerita* of Rondeletius; a Mediterranean species, furnished with only a single pair of very short and scarcely remarkable appendages over the eyes, and having, in addition, on the back part of the head a transverse fold of skin, which at a certain period becomes enlarged. Linnæus's acquaintance with the true *B. galerita* appears to have been founded solely on the works of Artedi. The reference by Gmelin of the Crested Blenny of Pennant to *B. galerita*, has led many Northern zoologists to give that specific name to Pennant's fish; and not only the later edition of Pennant's work, but the works also of Dr. Fleming and Professor Nilsson have so recorded it.

Dr. Fleming, in his History of British Animals, has described, under the term *B. galerita*, a species of Blenny, obtained in Loch Broome, which differs but little from the specimen obtained by Dr Johnston in Berwick bay; and Professor Nilsson, in his Prodromus of the fish of Scandinavia, which has been frequently quoted, has described also as the *B. galerita* of Linnæus a fish occurring among seaweed on the coast of Norway, and living on crustaceous and molluscous animals. His description,* though short, bears evidence of having been taken from the fish; it contains a reference to some peculiarities mentioned by Dr. Fleming, but claims for it three rays in the ventral fins.

The number of fin-rays in the fish termed *B. galerita* by Strom, Dr. Fleming, and Professor Nilsson, as described in their works, and that found to exist in the specimen received from Dr. Johnston, are here added, to show by their general agreement the great probability that all four sets of numbers refer to the same fish.

---

* As this useful little book may not be in the possession of many, I here add the description referred to :—" Bl. tentaculis duobus supraciliaribus ramosis ; radiis pinnæ ventralis tribus ; capite superius barbato ; corpore rutilo, maculis 10—12 dilutioribus rotundis ad latera dorsi. Obs. Alia specimina furviora sunt et maculis dorsi dilutioribus carent. In aliis exemplis spinæ 3—4, dorsales anteriores ceteres sunt longiores et appendicibus crassis ramosis ornatæ ; in aliis hæ spinæ breviores sunt & appendicibus simplicibus, gracilibus terminantur."

Strom.

D. 50 : P. 10 : V. 2 : A. 36 : C. 16.

Dr. Fleming.

50   14   2   39   16.

Professor Nilsson.

51   14   3   39   14.

Dr. George Johnston's fish.

51   14   3   36   14.

By a reference to the four species already described in this work, which, with the present fish, constitute all that have been at present ascertained as belonging to our coast, it will be seen that no one of the Blennies of our seas at all approaches the present in the number of its dorsal or anal fin-rays, and the Crested Blenny cannot therefore he confounded with either of them. It is much more nearly allied to two species belonging to the Mediterranean.

The description which now follows, taken from the fish caught in Berwick bay, will be found to contain most if not all the characters embraced in the four descriptions of Linnæus, Pennant, Dr. Fleming, and Professor Nilsson.

The whole length of the specimen was three inches and three-eighths ; depth of the body alone, seven-sixteenths of an inch ; including the dorsal and anal fins, three-quarters of an inch. The body is much compressed ; the head more oval, the profile rounded ; the outline of the mouth, when viewed from above, forms a half circle ; viewed laterally the the angle of the mouth is depressed, the mouth in front appears wide ; the lips capable of extensive motion ; the teeth smaller and shorter than those of any other British Blenny.

At the superior anterior margin of the eye on each side is a small fimbriated appendage, which is connected with that on the opposite side of the head by a fold of skin form-

ing a transverse union, passing in its passage over the forehead, which is prominent; behind these two small appendages
are two other tentacula, one on each side, about twice the
length of the anterior pair, and also fimbriated.   On the
nape of the neck, and for some distance towards the commencement of the dorsal fin, the skin is smooth, with the
exception of various small papillæ, as noticed by Dr. Fleming; the eyes lateral, large for the size of the head, but not
so large by comparison as those of the other Blennies.

The dorsal fin commences three-eighths of an inch behind
the last pair of tentacula; it is uniform in height throughout,
and reaches to the tail; the first ray a little shorter than the
second; the first three rays with membranous filaments, as
described by Dr. Fleming.   The membrane connecting the
first four rays darker in colour than the other parts of the
fin; the points of all the rays projecting beyond the edge
of the connecting membrane; the last dorsal fin-ray united
to the tail by an intervening membrane; all the rays
simple.

The ventral fins, only three-sixteenths of an inch in length,
are placed rather before the pectorals, and are supported
by three rays, ascertained by carefully dissecting off the
investing membrane on one side.   The pectoral fins are
rounded when spread, pointed when closed, the middle
rays being the longest, and extending over two-thirds of
the space between the edge of the operculum and the commencement of the anal fin.   The vent is placed immediately in advance of the anal fin, which in length is equal to
half the length of the whole fish; the first anal fin-ray
shorter than the second, the others are as long as those of the
dorsal fin, but the ends project further, the connecting membrane not being so deep.

The tail is rounded, or rather slightly lanceolate, the

central rays being the longest.   The lateral line proceeds from the tail in a direction straight to the upper edge of the operculum, about its junction with which there are several open mucous pores ; the membrane connecting both opercula is continuous under the throat.

The general colour of the body and fins is a pale brown, mottled on the sides with darker brown ; the head, the anterior part of the body, the ventral and pectoral fins, being darker than the other parts.

ACANTHOPTERYGII.   *GOBIOIDÆ.*

THE SPOTTED GUNNEL, OR BUTTERFISH.

SWORDICK. *Orkney.*

Murænoides guttata,	Lacepede.
Blennius gunnellus,	Linnæus.
,, ,,	Bloch, pt. ii. pl. 71, fig. 1.
,, ,,	*Spotted Blenny,* Penn. Brit. Zool. vol. iii. p. 282, pl. 60.
,, ,,	,, ,, Don. Brit. Fish. pl. 27.
Gunnellus vulgaris,	*Common Gunnel,* Flem. Brit. An. p. 207, sp. 124.

Generic Characters.—Head small, muzzle obtuse; body elongated, smooth, scales minute, covered with a mucous secretion; dorsal fin extending the whole length of the back, the rays simple; ventral fins very small; teeth small, pointed, detached.

THE SPOTTED GUNNEL, or Butterfish, as it is fre-quently called, from the consistence and quantity of mucous secretion with which its sides are covered, is sufficiently distinguished from the true Blennies by its dorsal fin, but little elevated above the line of the back, and by its elon-gated, slender, and compressed body, from which it has obtained in the Orknies, and in some of the countries of the North of Europe, the names of Swordick and Svardfisk, *Norway,* from a supposed resemblance in shape to the blade of a sword.

It is a common small fish on our sea-shores, where it may be frequently found in pools left by the tide, and occasionally under stones or seaweed. In such situations as those last named, from its great tenacity of life, it appears to suffer little or no inconvenience, though left for several hours ; moistened, externally only, by contact with the wet seaweed or damp rocks. When found in a pool of water it is observed to swim rapidly, and is difficult to catch, shifting its situation with great quickness, and creeping into very small apertures ; it is not easy to retain even when in hand, from the abundance as well as the nature of the slimy secretion aiding its muscular endeavours to escape. Its food is marine insects, the spawn of other fishes and their fry. It occurs generally on the rocky parts of the southern coast, sometimes under stones in soft mud, and is found in Cornwall, Devonshire, and from thence eastward to the mouth of the Thames. It has been taken in Berwick bay, the Frith of Forth, in Orkney, and Zetland. Linnæus includes this species in his *Fauna Suecica,* and other Northern Naturalists have found it on the coast of Norway, as well as on various parts of the shores of the Baltic.

In Greenland the flesh of this fish, though hard, is dried and eaten ; in this country it is seldom if ever made use of, except as bait for sea-lines. It is said to attain the length of ten inches : the more frequent size on our shores is from five to seven inches.

The length of the head is equal to the depth of the body, and is, when compared with the whole length of the head and body of the fish, without including the tail fin, as one to eight.

The head is small ; the line of the mouth directed obliquely upward, the angle depressed, the lower jaw rather the longer ; the teeth placed in a single row in each jaw, small, short, pointed, and sharp, each tooth separated from the next by a space equal to the breadth of the tooth itself; the eyes lateral, moderate in size, the irides dark blue; the cheeks tumid, from the size of the muscles, which enable it to bite hard. The membrane connecting the opercula continuous under the throat; a row of mucous pores descending from the nape to the upper edge of each operculum.

The number of fin-rays are—

D. 76 : P. 11 : V. 1 + 1 : A. 2 + 43 : C. 15.

In long fins of numerous rays, the number, it should be remarked, is at all times liable to variation, and it is not, it may be added, always alike even in those species with short fins.

The dorsal fin commences a little behind the line of the origin of the pectoral fin, and extends the whole length of the back, joining the tail: both the rays and the membranes of this fin are short, or but little elevated, but all the rays project their sharp points beyond the edge of the membrane. The pectoral fin, small and oval in shape, arises immediately behind the free edge of the operculum; the ventral fins are very small, near each other, on the under part of the throat, and appear each like a single sharp spine projecting through a small fleshy tubercle partly supported by one soft ray. The vent is situated under the thirty-fourth ray of the dorsal fin, at about an equal distance between each extremity of the fish; the anal fin commences immediately behind the vent, and extends to the tail, to which it is united: the rays as well as the membranes in this fin are longer and deeper than those of the dorsal fin; the first two rays are spinous, but the others, which are branched and articulated, project further beyond the edge of the membranes. The tail fin is moderate in size, and slightly rounded. The lateral line proceeds straight from the centre of the tail, rather below the middle of the fish, forming, with the upper and lower boundaries of the body, three nearly parallel lines.

The general colour of the body is a mixture of purple brown and yellow brown, sometimes dappled, occasionally assuming a waved or banded appearance. Along the line of the base of the dorsal fin are from nine to twelve conspicuous dark spots with a narrow but well-defined white stripe before and behind, and sometimes encircling each of them : the under surface of the head, the pectoral fins, and belly to the vent, are of a more uniform pale brown ; from the eye a dark brown stripe descends, behind the angle of the mouth, to the lower jaw. The spots described as dark along the back are occasionally not very conspicuous, and specimens sometimes occur in which they are entirely wanting.

A specimen of a spotted Gunnel from America, for which I am indebted to the kindness of Mr. Audubon, proves on comparison to be in every respect so similar to the British Gunnel, that there is little doubt it is the same species. The American specimen measures seven and a quarter inches; the largest British example I have measures only five and three quarters.

ACANTHOPTERYGII. *GOBIOIDÆ.*

THE VIVIPAROUS BLENNY.

EELPOUT, GUFFER, AND GREENBONE. *Scotland.*

Zoarcus viviparus,	Cuvier, Règne An. t. ii. p. 240.			
Blennius	,,	Bloch, pt. ii. pl. 72.		
,,	,,	*Viviparous Blenny,*	Penn. Brit. Zool. vol. iii. p. 283, pl. 61.	
,,	,,	,,	,,	Don. Brit. Fish. pl. 34.
Gunnellus	,,	,,	*Gunnel,*	Flem. Brit. An. p. 207, sp. 125.

Generic Characters.—Body elongated, covered with a mucous secretion ; head smooth, muzzle blunt ; ventral fins situated before the pectorals ; dorsal, anal, and caudal fins united ; all the fins very thick ; vent anterior to the middle of the body, its situation marked by a tubercle ; teeth conical, placed in a single row ; branchiostegous rays 6.

THE VIVIPAROUS BLENNY differs from the other British Blennies in the circumstance to which its name refers—that of bringing forth its young alive, which seem perfectly able to provide for themselves from the moment they are excluded. Mr. Low, in his *Fauna Orcadensis,* says, when he first observed this, he put a number of the small fishes into a tumbler-glass of sea-water, and kept them alive for many days, changing the water every tide. They grew a good deal bigger, and continued very lively, till in a hot

R 2

day, forgetting to refresh them with clean water, they died
to the last fish.

While they were very young and transparent, they made
excellent objects for the microscope, for viewing the circula-
tion of the blood.

The females of this species appear to produce their young
more or less grown according to their own size.

Mr. Neill says, " though not a delicate morsel, this fish
is often brought to the Edinburgh market." In the month
of February 1807, this gentleman saw a female fifteen inches
long in the fish-market, from which several dozens of young
escaped alive : these fry were from four to five inches long.
In a female of seven inches, obtained by myself on the
Kentish coast, full of young, these, when excluded, were only
one inch and a half long ; but such was the perfection of the
internal organization of this female, that after the specimen
had been kept for months in diluted spirit of wine, on
making slight pressure upon the abdomen, the young were
excluded one after another, and invariably with the head
first.

This viviparous species appears to be more common on
our east and north-east coast than in the south. Montagu
considered it a scarce fish in Devonshire, only obtaining
a single specimen in several years. As a species its earliest
describer was Schonevelde, whose name and discoveries have
been previously referred to. Sir Robert Sibbald first noticed it in Scotland. It occurs on the Norfolk and Yorkshire coasts, in Berwick bay, in the Forth, and on the coasts
of Norway and Sweden, where, hiding itself, as it does on
our own shore, under sea-weed, which is called tang, it has
acquired the name of Tanglake.

The whole length of the specimen described was seven
inches ; the length of the head, as compared with the whole

length of head, body, and tail, is as one to six : the head more elongated than in the last species, the muzzle more protruded and sharper ; the upper jaw the longest ; the teeth short, conical, sharp, with a second row round the front only of the lower jaw ; the lips fleshy ; the eyes small, lateral, irides blue ; the nostrils half-way between the inferior edge of the upper lip and the edge of the orbit, each nostril with a small membranous tubercle ; numerous mucous pores above the lips ; cheeks flat ; the membranous free edge of the operculum ending in an angle directed backwards : the pectoral fins large, broad, rounded, nearly as long as the head, and reaching half-way from the operculum to the commencement of the anal fin ; the membrane of one operculum not continuous under the jaw to the other as in the true Blennies : the ventrals small, narrow and pointed, about one-third the length of the pectorals, and placed in advance of them ; the investing membrane being dissected off, exposes three branched rays.

The dorsal fin commences at the nape, over the angle of the operculum ; the membrane investing and connecting its rays is too dense to admit of their number being ascertained with certainty or facility. The edge of the dorsal fin is straight till within a short distance from the tail, where a slope or emargination takes place. The form of the tail is lanceolate, but not distinguished by any separation from the dorsal or anal fin.

The anal fin in continuation underneath, in this specimen of seven inches, is four inches long ; the vent immediately in advance of its commencement.

The number of fin-rays are in the dorsal, anal, and caudal fin, as united,

About 148 : P. 18 : V. 3.

The general form of the body is lanceolate, tapering gra-

dually both in thickness as well as depth from the shoulder
to the end of the tail. The colour is pale brown; the
dorsal fin, upper surface and sides, mottled, and banded
with darker brown; the under part of the head, pectoral
fins, belly, and anal fin, uniform pale brown. The lateral
line traverses the centre of the body, slightly elevated only
as it approaches the anterior third of the fish. The surface
of the body appears, under a lens, to be studded with circular depressions.

ACANTHOPTERYGII. *GOBIOIDÆ.*

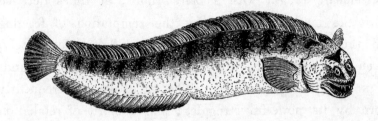

THE WOLF-FISH.

SEA-WOLF, SEA-CAT. *Scotland.* —SWINE-FISH. *Orkney.*

Anarrhichas lupus, LINNÆUS.
  ,, ,, BLOCH, pt. iii. pl. 74.
  ,, ,, CUVIER, Règne An. t. ii. p. 240.
  ,, ,, *Wolf-fish,* PENN. vol. iii. p. 201, pl. 27.
  ,, ,, *Striped Sea-wolf,* DON. pl. 24.
  ,, ,, *Wolf-fish,* FLEM. Brit. An. p. 208, sp. 127.

Generic Characters.—Head smooth, rounded in form, muzzle obtuse ; body elongated, covered with minute scales ; dorsal and anal fins long, distinct from the caudal ; no ventral fins : teeth of two kinds ; those in front elongated, curved, pointed ; the others on the vomer, as also on the jaws, truncated, or slightly rounded : branchiostegous rays 6.

CUVIER considers the species of *Anarrhichas* as Blennies destitute of ventral fins. One of them, the Wolf-fish of the British coast, is almost exclusively a northern fish, and has been seldom observed on our southern shore. It is taken off the coasts of Norfolk and Yorkshire, in the Frith of Forth, and among the Orkneys ; it is well known also on the shores of the North of Europe, in Greenland and Iceland.

The appearance of this fish is not prepossessing. Independently of a ferocious-looking cat-like head, with an

exceedingly thick, coarse skin, covered with slime, it possesses most formidable teeth, and neither wants the will
nor the power to attack others or defend itself. It is
occasionally caught with a baited hook, at times decoyed
into the meshes of a net by the temptation of feasting
on the fishes already entangled; but fights desperately,
even when out of its own element, inflicting severe wounds
if not cautiously avoided. The nets also are frequently
torn by its powerful struggles; and a spirit of retaliation
for the labour thereby occasioned, or for personal injury
inflicted by it, brings a speedy death to the unfortunate
fish. Handspikes and spars of wood are articles always at
hand in fishing-boats, and the savage Sea-cat is speedily
rendered incapable of doing further harm by heavy well-
aimed blows upon the head.

According to Mr. Neill, specimens of small size, about
two feet in length, are frequently brought to the Edinburgh
market; and those who are able to overcome the prejudice
excited by its appearance find it good food. Mr. Hoy

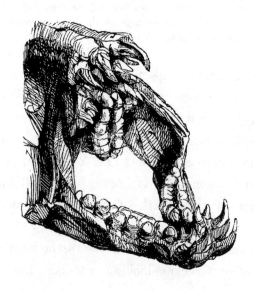

and Mr. Low have borne their testimony to the excellence of its flesh, and Mr. Donovan states that it is delicious. It may be observed here, that this is the general character of the flesh of those fishes that feed on crustaceous animals. It is eaten by the Norwegians and Greenlanders, as well as by most of the inhabitants of the northern parts of Europe, the head and skin being first taken off. The skin is converted into very durable bags and pockets.

The food of the Wolf-fish consists of crustaceous and testaceous animals, which its powerful jaws and rounded molar teeth enable it to break down sufficiently for its purpose. The vignette at the bottom of the preceding page, being a representation of the jaw-bones and teeth of a Wolf-fish, shows the formidable nature of the weapons with which it is furnished; while its German and Danish names have reference to a supposed power of crushing even stones in its mouth. It swims rapidly, with a lateral undulating motion; and has acquired the name of Sea-wolf from its voracity. It is called Swine-fish in the Orkneys, from a particular motion of the nose. It approaches the shore to deposit its spawn in the months of May or June; and the young, of a green colour, are occasionally found among sea-weed.

The number of fin-rays are—

D. 74 : P. 20 : A. 46 : C. 16.

The head is slightly flattened on the top; the nose rounded and blunt, nostrils small; eyes near the end of the nose, irides pale yellow; mouth large; lips fleshy; the form and arrangement of the teeth are shown in the vignette; mucous pores abundant about the eye, the gill-cover, and lower jaw on each side. Body elongated, compressed towards the tail; the dorsal fin extends from the nape of the neck almost to the tail, but is not joined to it; pectoral fins broad and rather long; ventral fins wanting; the anal fin extends the length of the posterior half of the body; the tail rounded.

The upper part of the head, the sides, back, and fins, are of a brownish grey; the body crossed by vertical bands, and varied with spots of darker brown, some of which extend over portions of the dorsal fin; the belly and under surface generally are white.

This fish attains the length of six or seven feet, and in the colder and more extreme northern seas is said to become still larger.

ACANTHOPTERYGII. *GOBIOIDÆ.*

THE BLACK GOBY, ROCK-FISH.

Gobius Niger, LINNÆUS.
    ,,   ,,   CUVIER, Règne An. t. ii. p. 243.
    ,,   ,,   *Black Goby,* PENN. Brit. Zool. vol. iii. p. 288, pl. 42.
    ,,   ,,   MONTAGU, MS.

Generic Characters.—Head depressed, with pores between the eyes ; dorsal fins two, distinct, rays of both flexible ; ventral fins united at the edges, forming a circle ; anal aperture with a tubercle ; body covered with scales, the free edges ciliated ; teeth small, numerous ; branchiostegous rays 5.

THE species of this genus are easily recognised by the peculiar form of the ventral fins; the short anterior rays, and the long posterior ones, on each side, being united together, making a circle, with which they have been supposed to possess the power of attaching themselves to rocks, by forming a vacuum. The Gobies are of little value, except as supplying food to other fishes. Of this genus the Black Goby is the most rare on our shores.

This species appears to be chiefly an inhabitant of the rocky parts of our coast, and on that account is not so frequently taken by the net : it is, however, sometimes captured in that manner on the coast of South Devon, particularly in the estuary of Kingsbridge, from whence, says Colonel Montagu, we have obtained several specimens of tolerable size, the largest about five inches.

" The head is large, the cheeks inflated, and the lips very thick : the mouth is wide, and furnished with numerous small and very short teeth in several indistinct rows in both jaws ; the under jaw is roughened by them like a rasp : the eyes are high up on the head, and approximate ; the upper part of them dusky, partaking of the colour of the head, the lower part of the irides golden: between the eyes are two small pores, the anterior one more than double the size of the other, but not distinguishable without the assistance of a lens : the nostrils are placed before the eyes, on the outside of each of which is a small fleshy appendage, rather elevated. The cheeks and opercula of the gills are furnished . with lines of very minute papillæ, which appear like spines : most of these lines are transverse, · but some run longitudinally, observable only with the aid of a glass. On the top of the head a longitudinal sulcus runs as far as the commencement of the first dorsal fin. The colour is uniformly dusky in the more matured fishes, except from the chin to the vent, which is whitish, with some deep purplish black between the gills beneath ; the ventral fins usually more or less black. It is, when fresh, covered with a thick mucous secretion ; but after having been in spirits, the fish becomes extremely rough to the touch if rubbed the reverse way. This roughness is occasioned by the scales, which are large in proportion, being ciliated at their free edges."

" The ventral fins, which supply the great generic character, are connected, forming a funnel-shaped appendage of twelve branched rays; and the anus is furnished with an elongated tubercle. We never could discover that the Black Goby ventured into fresh water, and with us certainly spawns in the sea. With respect to the union of the ventral fins, it would seem to be for the purpose of forming an instrument of adhesion; but in no instance have we observed that they adhered either to rocks, or to the bottom of the glass vessel in which they have been kept alive for several days."—*Montagu's MS*.

The number of fin-rays are—

D. 6. 17. : P. 17 : V. 12 : A. 12 : C. 15.

The lower jaw is the longest, with fine carding-like teeth in several rows; the tongue square at the end; gill-apertures small; behind the vent a small conical tubercle. The adult fish are from five to six inches in length. They spawn in May or June, depositing the ova on stones. The young are to be seen abundant in summer; and are lighter in colour, particularly on their under surface. They are to be found on various parts of the coast from Cornwall to the Orkneys. Mr. Couch has observed a peculiarity in the habits of the Black Goby, in which it resembles the Shanny,—that of carrying off its prey in its mouth to a resting-place, and there struggling with it. The *Gobius niger* of Mr. Donovan and Dr. Fleming appears to be distinct from the *G. niger* of authors, and identical with the *G. bipunctatus* of this work, the species next to be described.

The Black Goby inhabiting the rocky parts of our coast is called Rock Goby and Rock-fish, to distinguish it from the other British species of this genus, which frequent sandy bays. The vignette annexed represents the under surface of the Black Goby, showing the ventral fins forming a circle by their double union. The use of the anal tubercle is only conjectured.

ACANTHOPTERYGII. GOBIOIDÆ.

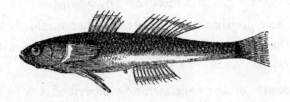

THE DOUBLY-SPOTTED GOBY.

Gobius bipunctatus, YARRELL.
   ,,   *niger,*   *Black Goby,*   DON. Brit. Fish. pl. 104.
   ,,   ,,   ,,   ,,   FLEM. Brit. An. p. 206, sp. 117.

THE little Goby which forms the subject of the
present article has been considered identical with *Gobius
niger* by some authors, and by others has apparently not
been distinguished from *Gobius minutus.* As this species
is not uncommon, and is widely distributed, it has been
obtained by various collectors ; and few that have had the
opportunity of comparing it side by side with *G. niger* and
minutus, but have been convinced that it was distinct from
both. This, at least, was the opinion of several to whom
specimens were shown. As I have been unable to reconcile any of the existing descriptions with this fish, I have
ventured to propose for it the name which its two conspicuous and constant spots on each side have suggested.

I have received specimens from Belfast, by the kindness
of William Thompson, Esq. who considered it distinct,
and had publicly noticed the differences existing between

it and the two Gobies that have hitherto been considered
as our only indigenous species ; I have also received it from
Holyhead and Cornwall : I have taken it myself on the
coast of Dorsetshire, and have had specimens sent me from
Berwick bay by the kindness of Dr. Johnston. I am not
aware of any peculiarity in its habits that would distinguish
it from the other Gobies, but I have never met with it in
fresh water.

The length of the specimen now described was two inches
and one-eighth ; the upper part of the head and nape flattened ; the eyes large, placed laterally ; the mouth large,
the line of the gape slanting obliquely upwards, the angle
depressed, the lower jaw much the longest when the mouth
is opened ; both jaws furnished with numerous slender,
sharp teeth, curving inwards.

The number of fin-rays are as follows :—

D. 7. 12 : P. 15 : V. 12 : A. 12 : C. 11.

The first dorsal fin commences a little in advance, on
a vertical line, of a conspicuous dark spot on the side just
behind the origin of the pectoral fin ; the second dorsal fin
commences in a line over the vent : all the rays of both dorsal
fins are slender and flexible. The pectoral fin large, and when
spread covers, but from the transparency of the fin-mem-
brane does not entirely conceal, the dark spot on the side
before referred to : the ventral fins, arising a little behind
the origin of the pectorals, are united, the longest rays
extending considerably beyond those of the pectoral fins.
The vent with its tubercle are in a line under the commencement of the second dorsal fin ; the rays of the anal fin possess
the same slender, flexible character as those of the dorsal ;
the tail nearly square, with a conspicuous dark spot at the
base of the caudal rays.

The prevailing colour of the head and upper parts of the body is a nutmeg brown, produced by a double series of diagonal lines taking opposite directions; the under part of the head, body, pectoral and ventral fins, very pale brown, almost white.

The two spots on each side, and the darker brown colour, distinguish this species from *G. minutus* which is next to be noticed.

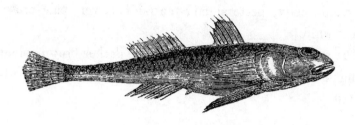

THE SPOTTED GOBY.

POLEWIG.　*Thames Fishermen.*

Gobius minutus,	LINNÆUS.		
,,	,,	*Boulereau blanc,*	CUVIER, Règne An. t. ii. p. 243.
,,	,,	*Spotted Goby,*	PENN. Brit. Zool. vol. iii. p. 290, pl. 41.
,,	,,	,,　　,,	DON. Brit. Fish. pl. 38.
,,	,,	,,　　,,	FLEM. Brit. An. p. 206, sp. 118.

THE SPOTTED GOBY of Pennant and others, which
might rather be termed the Freckled Goby, is not only
common on all our sandy shores, where it is constantly to
be obtained of the shrimpers in whose nets it is taken, but
is also most plentiful in the Thames, where it is known
to the fishermen by the names of Polewig, or Pollybait;
the larger sized specimens being at times taken to sea by
the line fishermen to be placed on their hooks for bait.

The length of the adult is usually about three inches;
the head is large; the eyes near the upper surface closely
approximating, the irides blue; the mouth wide, furnished
with numerous small pointed teeth in several rows, curving
inwards: the dorsal fins distinct, the rays slender and flexible, the anterior rays of the second dorsal fin rather longer
than the posterior ones; pectoral and ventral fins large; the

vent placed just half-way between the nose and the end of the fleshy portion of the tail ; anal fin ending nearly on the same plane as the second dorsal ; tail nearly square, or but very slightly rounded.

The prevailing colour of the body is a pale yellowish white, minutely freckled over with pale ferruginous, occasionally with a row of larger spots along the lateral line ; the tail slightly barred by lines formed of minute spots.

The number of fin-rays are—

D. 6. 12 : P. 20 : V. 12 : A. 13 : C. 12.

s 2

THE SLENDER GOBY.

Gobius gracilis, Jenyns.

I have been favoured by the Rev. L. Jenyns with the following particulars of a fourth species of Goby.

"Dorsals remote, the second with the posterior rays longest : eyes closely approximating : length three inches two lines."

"Form closely resembling that of *G. minutus,* but more elongated and slender throughout; greatest depth barely one-seventh of the whole length ; snout rather longer ; opercle approaching more to triangular, the lower angle being more cut away, and the ascending margin more oblique ; a larger space between it and the pectorals ; the two dorsals further asunder : rays of the second dorsal longer ; these rays also gradually increasing in length, instead of decreasing, as in *minutus,* the posterior ones being the longest in the fin, and rather more than equalling the whole depth : rays of the anal fin in like manner longer than in *G. minutus.* The fin-rays are—

D. 6. 12 : P. 21 : V. 12 : A. 12 : C. 13.

In all other respects similar."

"Apparently a new species, though probably of not less frequent occurrence than the Spotted Goby, with which it may be easily confounded.

"My specimens were obtained from Colchester, and were supposed to have been taken somewhere off the Essex coast." —*Jenyns' MS.*

ACANTHOPTERYGII. *GOBIOIDÆ.*

THE GEMMEOUS DRAGONET.

YELLOW SKULPIN. *Cornwall.* — GOWDIE. *Scotland.*

Callionymus lyra, LINNÆUS.
   ,,   ,,   CUVIER, Règne An. t. ii. p. 247.
   ,,   ,,   *Gemmeous Dragonet,* PENN. Brit. Zool. vol. iii. p. 221,
                          pl. 31.
   ,,   ,,   ,,   ,,   DON. Brit. Fish. pl. 9.
   ,,   ,,   ,,   ,,   FLEM. Brit. An. p. 248, sp. 126.

Generic Characters.—Head depressed ; eyes on the upper surface, approximated ; body smooth, without scales ; two dorsal fins, distinct ; ventral fins separated under the throat, larger than the pectoral fins ; mouth capable of great protrusion ; teeth small, numerous, on the bones of the jaws only ; the males, and probably the males only, with a postanal tubercle, and with the first ray of the first dorsal fin elongated, reaching to the tail ; branchiostegous rays 6 ; preoperculum ending with three spines ; gill-aperture very small, at the upper edge of the operculum.

THE GEMMEOUS DRAGONET, so called from its brilliant gem-like colours, was first described as a British fish by Dr. Tyson, in the twenty-fourth volume of the Philosophical Transactions. The second term, that of Dragonet, was deduced by Pennant from the trivial name, attached to the second British species, *dracunculus* ; that name, in its turn, having probably been given with a double reference to its speckled appearance, and also to its large wing-like ventral and pectoral fins ; which induced Bellon, Seba, and others, to consider these species as allied to the flying fishes.

When examining different books to obtain local and provincial names for the various fishes to be included in this work, I find that I made the mistake of transferring the name of Gowdie to the Great Weever, instead of to the Gemmeous Dragonet. The prevailing colour of this last-named fish is a golden yellow ; it is called Yellow Skulpin in Cornwall,—and the Northern term Gowdie, from *gowd*, gold, means yellow or golden : the gowan is a golden flower ; hence gowd and Gowdie are probably derived from the French word *gaude*, which is also a yellow flower.

The species of the genus *Callionymus* have two very strongly marked characters. The branchial aperture on each side is only a small orifice near the nape of the neck, while the ventral fins, widely separated, and situated under the throat, are even larger than, as well as placed in advance of, the pectoral fins. The Gemmeous Dragonet is a handsome fish, with a smooth skin, and having the head singularly spotted and striped with blue on a yellowish ground. When fresh from the water, these colours are vivid, and the appearance of the fish attractive. Linnæus indulged his fancy by attaching the term *Callionymus*, which signifies literally, beautiful name, to a prettily marked species; and the word *lyra* was doubtless suggested by the resemblance of its elongated dorsal filament and fin-rays to the strings of a musical instrument.

The Gemmeous Dragonet is not a common fish on our coast, and, according to my own observation, is much more rare than the Sordid Dragonet. It has been taken on the coast of Cumberland and at Belfast. Mr. Couch has met with it occasionally in Cornwall, where it frequents deep water, generally keeping close to the bottom. Colonel Montagu considered it rare, and only obtained one specimen, about nine inches long, which was taken off the bar at Salcombe in Devonshire, in the autumn of 1809 : it has also been obtained at Weymouth and Hastings. On the eastern coast, it has been noticed at Harwich, Yarmouth, and Scarborough. Mr. Neill records it among the fishes of the Forth ; and Mr. Low, in his *Fauna Orcadensis*. It is included by Nilsson among the fishes of the coast of Norway, and is mentioned by most of the Northern Ichthyologists. Brunnich, M. Risso, and the Prince of Musignano, also record this species as belonging to the Mediterranean.

The Gemmeous Dragonet occasionally takes a bait, but is more frequently caught in a net,—sometimes, when of small size, by the shrimpers in sandy bays. Young specimens only six inches in length possess the elongated dorsal filament. Its food is testaceous animals, which are swallowed whole, molluscous animals, and worms. The flesh is said to be white, firm, and of good flavour. It is very frequently the prey of other fishes.

The length of the specimen described was ten inches ; the length of the head, compared to the whole length of the fish, as one to four : the form of the head oblong, ovate, measuring two and a half inches in length, and but one inch and a half in breadth ; the anterior half of the length is before the eyes, the orbits occupy one-third, while the space behind is equal to the breadth of the orbits. The branchial apertures are small orifices, one on each side the nape of the neck, at the upper edge of the operculum. The upper part of the head is flat; the profile of the nose convex; the under surface of the head flat; inferior angle of the preoperculum ending in three spines, directed upwards; the free edge of the operculum hid by the continuation of the common covering of the body.

The mouth is deeply divided, measuring seven-eighths of an inch from the angle of the gape to the point of the upper jaw; the teeth occupy a broad surface in the front, which becomes narrower as the band proceeds backwards; the point of the lower jaw with a single row of teeth, longer and more curved, anterior to the others; the inside of the mouth furnished with two transverse folds of lining membrane to admit the extension of the moveable portion of the upper jaw.

The first dorsal fin, of four rays, commences in a line with the origin of the pectorals: the first ray very much elongated, reaching to the base of the tail; the second ray two-thirds of the length of the first ray; the third ray half the length of the second; the fourth ray short, about one inch in length. The number of fin-rays are—

D. 4. 9 : P. 20 ; V. 5 : A. 9 : C. 10.

The second dorsal fin has eight rays of equal length, and about as long as the body of the fish is deep; the ninth ray double, and nearly as long again as any of the preceding rays of that fin; the origin of the last dorsal fin-ray being in a line over the last ray but two of the anal fin. The ventral fins are large, all the rays branched, supporting a dense and strong membrane; the last ray attached by a membrane to the body of the fish, and to the base of the pectoral fin. The pectoral fin triangular in shape, the central rays the longest; all the rays slender and branched, the connecting membrane delicate and transparent. The vent and postanal tubercle are in a line under the second ray of the second dorsal fin; the elongated tubercle is perforated, and admits a fine probe which passes to the urinary organs. The anal fin commences under the third ray of the second dorsal fin; the last anal fin-ray is as long again as the preceding ray, and reaches to the end of the fleshy portion of the tail. The caudal rays are elongated, articulated, and branched.

The body of the fish is much narrower than the head, and rounded in form: it is nearly cylindrical, but tapers gradually to the tail. The lateral line is a well-marked elevated ridge.

The prevailing colour of the body is yellow, of various shades in different parts, striped and spotted on the head and sides with sapphirine; the irides orange, pupils blue. The membranes of the dorsal fins pale brown, varied with darker longitudinal bands; the ventral, anal, and caudal fins, bluish black; under surface of the head and body white, with a dark patch exterior to the root of the tongue.

The Prince of Musignano, in his *Fauna Italica*, has figured the female of the same colour as the male, but without the elongation of the fin-rays

THE SORDID DRAGONET.

FOX. *Kentish Coast.* — SKULPIN. *Cornwall.*

Callionymus dracunculus, LINNÆUS.
    ,,   ,,   CUVIER, Règne An. t. ii. p. 247.
    ,,   ,,   *Sordid Dragonet,* PENN. Brit. Zool. vol. iii. p. 224,
                                pl. 32.
    ,,   ,,   ,,   ,,   DON. Brit. Fish. pl. 84.

THE SORDID DRAGONET, so called probably from the
dingy hue of its colours as compared with those of its generic
companion, is the most common species of the two on various parts of the coast, but generally occurs of small size. It
is frequently taken at the mouth of the Thames, where, on
account of its reddish appearance, it is called the Fox.

The general accordance in the situation of the fins and
the number of fin-rays in the two British examples of Dragonets, has induced a suspicion, first entertained by Gmelin,
that the two fishes are but males and females of the same
species. Mr. Neill, in the Wernerian Memoirs, vol. i. p. 529,
supports this opinion ; having found that the specimens of

C. lyra examined by him were all males, while those considered as *C. dracunculus* were all females. Dr. George Johnston, of Berwick, has, on the other hand, recorded in the third volume of the Zoological Journal, page 336, (note,) that he had found a Sordid Dragonet with a milt, or soft roe. The differences between the two fishes are on some points so great and so obvious, that I have considered them distinct. Mr. Couch has observed a certain difference in their habits : " The Yellow Skulpin prefers deeper water ; whereas the other will often approach the margin of the tide, where I have watched its actions with great interest. They keep at the bottom, among sand or stones, and never rise but to pass from one station to another, which is done with great suddenness and rapidity. They possess great quickness of sight, and dart with swiftness when alarmed, though not to a great distance ; and I have seen the Sordid Skulpin repeatedly mount after prey, and invariably return to the same spot again. This motion is chiefly performed by the ventral fins ; and the eye is well adapted to the habit, the muscles of that organ being fitted to direct the sight upward, but not downward. They sometimes take the hook, though rarely ; and they are much devoured by the larger fish, in the stomachs of which they are often found. They feed on shell-fish, worms, and molluscous animals."

The whole length of the specimen described was nine inches ; the length of the head compared to the whole length of the fish as one to five ; the head triangular, as wide as it is long ; both head and body much more depressed than those of *C. lyra ;* the eyes removed only one diameter of the orbit from the nose ; the mouth measured but half an inch from the angle of the gape to the point of the upper jaw ; the preoperculum armed with three spines ; the fins similar to those of *C. lyra* in situation and in the number of fin-rays, but the rays of the first dorsal fin are shorter than those of the second dorsal fin, and the rays of the second dorsal fin are of uniform length. The number of fin-rays are—

D. 4. 9 : P. 20 : V. 5 : A. 9 : C. 10.

The prevailing colour is a reddish brown, which in young specimens is varied by a few dark spots on the sides of the body; the dorsal fins are pale brown without stripes; all the under surface of the body, pectoral, ventral fins, and head, uniformly white; anal fin even whiter than the belly. The intestines in the Dragonets are so transparent that their contents may frequently be ascertained without further exposure. They have no swimming-bladder.

In proof of the distinction of the species, it may be stated that the colours of the body and fins are decidedly different; that in *C. lyra*, the head is to the whole length as one to four; the eyes removed two diameters from the end of the nose; the head elongated and elevated; the distance from the point of the nose to the posterior edge of the orbit, and thence to the origin of the first dorsal fin-ray, equal; the mouth large; the lateral line prominent. In *C. dracunculus*, the head is to the whole fish as one to five; the eyes but one diameter above the snout; the head depressed, strictly triangular; the distance from the eye to the first dorsal fin-ray double that of the distance from the point of the nose to the eye; the lateral line much less distinct, and the mouth only half as deeply divided. The vignettes show the comparative capacity of the mouths in two specimens of nearly equal size.

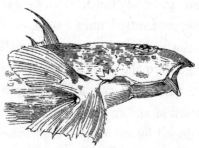

ACANTHOPTERYGII.   *PECTORAL FINS FEET-LIKE.*

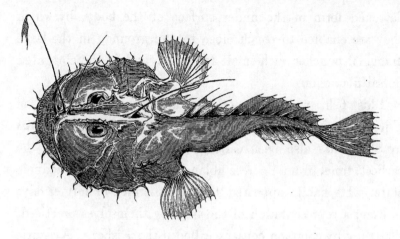

THE FISHING FROG. ANGLER.

SEA DEVIL. WIDE GAB. *Scotland.*

Lophius piscatorius, LINNÆUS. BLOCH, pt. iii. pl. 87.
,,   ,,   CUVIER, Règne An. t. ii. p. 251.
,,   ,,   *Common Angler,* PENN. Brit. Zool. vol. iii. p. 159, pl. 21.
,,   ,,   ,,   ,,   DON. Brit. Fish. pl. 101.
,,   ,,   *Angler,*   FLEM. Brit. An. p. 214, sp. 147.

Generic Characters.—Head very large, depressed; body slender, smooth,
without scales: two dorsal fins separated; pectoral fins broad and thick, somewhat resembling feet; ventral fins small, placed considerably before the pectorals: teeth differing in size, numerous, conical, sharp, curving backwards;
tongue broad; branchial cavities large, with only a small opening behind the
pectoral fins; branchiostegous rays 6.

THE ANGLER, as this fish is called for reasons that
will be given hereafter, belongs to a small and singular
group of fishes, which Cuvier has designated *Pectorales
Pédiculées,* from the peculiar conformation of the pectoral
fins, by which some of the species can creep on land almost
like little quadrupeds. The ventral fins, palmate in form,
are placed very far forward on the body; and the pectorals,
from their position, perform the office of hinder feet. Some

relation to the species of the genus *Callionymus*, last described, particularly *C. lyra*, will be manifest in the very flattened form of the under surface of the body, by which they are enabled to couch close to the ground, in the large branchial pouches with small apertures, and in the elongated dorsal filaments.

This fish, which is not uncommon in all the seas of Europe, and was in consequence called *Lophius Europæus* by Shaw, has also been called Frog and Frogfish from the earliest time, from its resemblance to a frog in the tadpole state. Its habits appeared to the fishermen of former days so exact a representation of the art they themselves practised, that they by common consent called it the Fisher. Aristotle calls it a sort of Frog, which he says is also called a Fisher ; and he adds, that this fish owes its name to the tact and industry it exercises to procure food.

This fish has been taken on the coast of Londonderry, Antrim, Dublin, Waterford, and Cork, in Ireland; in England, on the coasts of Cornwall, Devonshire, Norfolk, and Yorkshire; in Scotland, in the Forth and among the Northern Islands. It is also named by authors as common on the shores of the Baltic and Norway.

In its appetite this fish is most voracious ; and as it is not a rapid swimmer, possessing but little power in its pectoral fins, it is supposed to be obliged to have recourse to art in order to satisfy its appetite. Upon the head, as will be seen in the figure, are two slender elongated appendages, the first of them broad and flattened towards the end, and having at this dilated part a shining silvery appearance. These elongated filaments are curiously articulated at the base with the upper surface of the head. They have great freedom of motion in any direction, the first filament more especially, produced by numerous muscles, amounting, according to M. Bailly, to twenty-two. The figure on the left side of the vignette beneath shows the manner in which these two elongated appendages are attached, as well as the kind of motion of which, by the action of various muscles, they are capable. The first is articulated by a process resembling two links of a chain, by which universal motion is obtained; the second is more limited in its action, and appears, except as far as flexibility may assist it, to be only capable of being brought forward or backward.*

These elongated shafts are formed of bone covered by the common skin; and as the soft parts are abundantly supplied with nerves, they may also serve the Angler as delicate organs of touch. The uses to which they are applied are singular. While couching close to the ground, the fish, by the action of its ventral and pectoral fins, stirs up the sand or mud : hidden by the obscurity thus produced, it elevates these appendages, moves them in various directions by way of attraction as a bait, and the small fishes

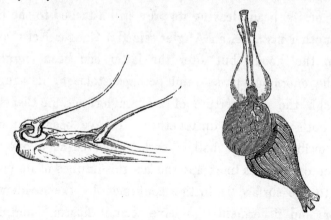

* The three-part figure on the right hand of the vignette above represents the heart of the Angler, from the Transactions of the Royal Society. The lower portion is the auricle, before entering which the large veins unite. The auricle opens into the side of the middle portion, which is the ventricle. The upper part is the branchial artery, dilated, forming the *bulbus arteriosus*. Above the bulb the branchial artery divides to form three, and further division takes place before passing to the branchial arches.

approaching either to examine or to seize them, immediately become the prey of the Fisher.

Numerous are the writers who have borne their testimony to this habit, and some have extolled it as raising the intellectual character of this fish beyond that of most of its class. Half the animal world seem destined to destroy each other, some by open violence, others by stratagem ; and this design in the Angler, though singular, is not more wonderful than that of spiders among insects, who spin and repair their widely-spread webs to catch other insects upon which they subsist.

The Angler has been known to measure five feet in length, but the most common size is about three feet.   Mr. Couch says, " It makes but little difference what the prey is, either in respect of size or quality. A fisherman had hooked a Cod-fish, and while drawing it up he felt a heavier weight attach itself to his line : this proved to be an Angler of large size, which he compelled to quit its hold by a heavy blow on its head, leaving its prey still attached to the hook. In another instance, an Angler seized a Conger Eel that had taken the hook ; but after the latter had been engulphed in the enormous jaws — and perhaps stomach, it struggled through the gill-aperture of the Angler, and in that situation both were drawn up together.   I have been told of its swallowing the large ball of cork employed as a buoy to a bulter, or deep-sea line ; and the fact this implies of its mounting to the surface is further confirmed by the evidence of sailors and fishermen, who have seen it floating, and taken it with a line at mid-water.   These fishes sometimes abound, and a fisherman who informed me of the circumstance found seven of them at one time on the deck of a trawl-boat : on expressing his surprise at the number, he was told that it was not uncommon to take a dozen at once."—*Couch's MS.*

" When this fish is taken in a net, its captivity does not destroy its rapacious appetite, but it generally devours some of its fellow-prisoners, which have been taken from the stomach alive, especially Flounders. It is not so much sought after for its own flesh, as for the fish generally to be found in its stomach : thus, though the fishermen reject the fish itself, they do not reject those that the fish has collected."

" A female examined measured three feet three inches, the breadth across the body at the pectoral fins fifteen inches. Within the teeth, on the lower jaw, is a loose skin of a brown colour, like the back of the fish, forming a sort of bag, which probably assists in preventing the escape of its smaller prey. A male examined was three feet five inches long. When this fish was suspended by the head, the contents of its stomach were readily seen, and I perceived several Cuttle-fish. The sexes are distinctly marked by external appendages, as in some species of *Raia*."—*Montagu's MS*.

The number of fin-rays are—

D. III. 12 : P. 20 : V. 5 : A. 8 : C. 8.

The head is wide, depressed ; the mouth nearly as wide as the head ; lower jaw the longest, bearded or fringed all round the edge ; both jaws armed with numerous teeth of different lengths, conical, sharp, and curving inwards ; teeth also on the palatine bones and tongue ; three elongated unconnected filaments on the upper part of the head, two near the upper lip, one at the nape, all three situated in a depression on the middle line ; eyes large, irides brown, pupil black : pectoral fins broad and rounded at the edge, wide at the base ; branchial pouches in part supported by the six branchiostegous rays. Body narrow compared with the breadth of the head, and tapering gradually to the tail ; vent about the middle of the body ; the whole fish covered with a loose skin.

VOL. I.   T

Colour of the whole upper surface of the body uniform brown ; fin membranes darker ; under surface of the body, ventral, and pectoral fins, white ; tail dark brown, almost black.

The figure at the head of this article represents this fish as seen from above. To give a better notion of its capacity to contain food, an outline vignette of a side view is added.

Mr. Couch informs me by letter that he has reason to believe he saw a specimen of *Lophius parvipinnis* of Cuvier, and regrets that circumstances prevented his taking a minute description. It was of small size, scarcely exceeding fifteen inches in length, thicker in form than the common Angler, with somewhat of a different structure of the pectorals, and regularly and even beautifully mottled with black patches.

Two short notices of another species of *Lophius* are supposed to refer to mutilated examples of *L. piscatorius*, or to specimens deformed in drying.

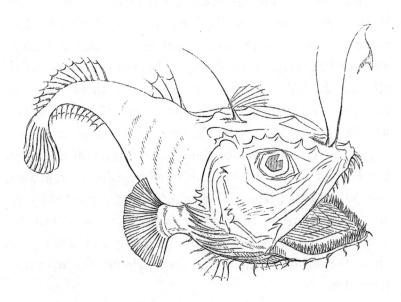

ACANTHOPTERYGII. LABRIDÆ.*

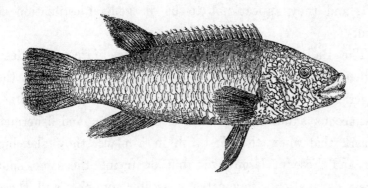

THE BALLAN WRASSE.

Labrus maculatus, BLOCH, pt. ix. pl. 294.
,, ,, Ballan Wrasse, PENN. Brit. Zool. vol. iii. p. 334, pl. 55.
,, tinca, Ancient Wrasse, or Old Wife, DON. Brit. Fish. pl. 83.
,, Balanus, Ballan Wrasse, FLEM. Brit. An. p. 209, sp. 130.
,, ,, CUVIER, Règne An. t. ii. p. 255.

Generic Characters.—Body elongated, covered with large thin scales : a single dorsal fin, extending nearly the whole length of the back, part of the rays spinous, the others flexible ; behind the point of each spinous ray a short membranous filament : lips large and fleshy ; teeth conspicuous, conical, sharp ; cheek and operculum covered with scales ; preoperculum and operculum without serrations or spines.

THE species of this family are numerous even on our own shores alone, and from the variations in colour to which they seem liable, they have not been very well determined : the more permanent characters of the proportions of the head and body, the form of the different parts of the gill-covers, the structure and relative position of the fins and fin-rays, afford the best points for specific distinction. The Wrasses, or Rock-fish, as they are also called,

* The family of the Wrasses, or Rock-fishes.

T 2

inhabit principally the rocky parts round our coast, spawning in spring or summer, just previously to which their colours are in the highest perfection. The flesh is said to be soft, and they appear not to be in general estimation as food.

The Ballan Wrasse, Mr. Couch says, "frequents deep gullies among rocks, where it shelters itself among the larger kinds of sea-weeds, and feeds on crabs and other crustaceous animals. It takes a bait freely, and fishermen remark that when they first fish in a place, they take but few, and those of large size; but on trying the same spot a few days after, they catch a greater number, and those smaller; from whence they conclude that the large fish assume the dominion of a district, and keep the younger at a distance. The spawn is shed in April, and the young, scarcely more than an inch in length, are seen about the margin of the rocks in shallow water through the summer."

A fine specimen, eighteen inches long, and weighing three pounds seven ounces, was taken in January 1831, in Swansea bay, of which a notice and short description was furnished me by L. W. Dillwyn, Esq. The colour was red, becoming pale orange on the belly; the body ornamented with bluish green oval spots; the fins and tail green, with a few red spots; the dorsal fin had spots along the base only.

About the same time a specimen of nearly the same size was obtained in the London market, and presented to the Zoological Society by Sir Anthony Carlisle. Both these specimens are referred to in the first volume of the Proceedings published by the Society, pages 17 and 34.

The deep blue colours of the latter were removed with astonishing rapidity when the specimen was placed in spirits.

I have also seen several specimens of large size that were

taken on the coasts of Down and Antrim. These fish, with many others, taken on various parts of the Irish coast, were exhibited at the Zoological Society by William Thompson, Esq. of Belfast.

This species occurs along our eastern coast, as well as at various places on the shores of Dorsetshire, Devonshire, and Cornwall. A fine specimen sent to me by Dr. Johnston from Berwick afforded the following measurements, and in its colours had more of the orange red than the London specimen before mentioned. The length of the head, compared to the whole length of head, body, and tail, as one to four; the depth of the body alone equal to the length of the head : the lower edge of the scaly portion of the cheek rounded, the scales only half as large as those on the operculum; preoperculum without scales, the horizontal and vertical edges forming an angle somewhat obtuse, the ascending line being oblique, the margin entire ; operculum broad, covered with large scales, and ending in a membranous projection over the upper part of the origin of the pectoral fin. The pectoral fin broad and rounded ; the membranes connecting the rays of all the fins spotted with verditer, rather inclining to blue than green ; the fin-rays reddish orange, with six or seven scales in succession between each ray of the caudal fin. Back and sides bluish green, paler on the belly ; all the scales margined with orange red, the margins varying in breadth in different specimens, and thus producing the prevalence of the blue or orange colour ; six rows of scales between the lateral line and the middle portion of the dorsal fin. Head and cheeks bluish green, reticulated with orange red lines ; lips flesh colour ; about eighteen teeth in each jaw, conical, those in front the longest ; the tail slightly rounded at the upper and under corner, the tip dusky.

The number of fin-rays are—

D. 20 + 11 : P. 15 : V. 1 + 5 : A. 3 + 9 : C. 13.

This species possesses teeth on the pharyngeal bones, as figured on Pennant's plate of the Ballan Wrasse. The vent is in a line under the eighteenth spinous ray of the dorsal fin.

Bloch's figure was probably taken from a preserved specimen, the colours of which had faded.

ACANTHOPTERYGII. *LABRIDÆ.*

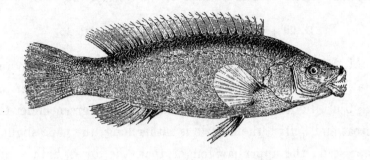

THE GREEN-STREAKED WRASSE.

Labrus lineatus, Streaked Wrasse, Don. Brit. Fish. pl. 74.
  ,,   ,,   ,,   ,,   Flem. Brit. An. p. 209, sp. 132.
  ,, *psittacus,* Risso, Hist. t. iii. p. 304, sp. 203.

This species is here given on the authority of Mr. Donovan ; and as there is reason to believe it is identical with the *Labrus psittacus* of Risso, I have added a reference to that author. I have never seen either an English or Mediterranean specimen.

"This fish," Mr. Donovan says, "is an occasional visitor, as he was informed, to the coast of Cornwall, where it is provincially known by the name of Green-fish : it usually appears in the summer, and is esteemed the rarest species of its tribe by the fishermen in those parts. The specimen now in our possession, and from which the figure in the plate is delineated, was taken on this coast a few years ago by Captain Bray. This specimen is seven inches long ; and having been carefully divested of the flesh while perfectly fresh, and the skin well prepared, the natural colours of the fish are admirably well retained. Besides this genuine British specimen, we possess another in excellent preservation from the Mediterranean Sea, that differs only in being smaller, and having the head, back, and sides of the body of a brighter green." The number of the fin-rays are—

D. 20 + 10 : P. 14 : V. 8 : A. 3 + 8 : C. 15.

M. Risso's description represents this species as having the body elongated, of a fine meadow-green, darker on the back, lighter on the sides, and yellowish green under the throat and belly : the muzzle is rather long, the nape slightly depressed ; the upper jaw longer than the lower, both armed with teeth, the longest of which are in front ; the operculum angular ; the lateral line curved on its approach to the tail ; the fins green. The female is of a uniform green colour above, silvery on the belly.

The fin-rays, according to M. Risso, are—

D. 18 + 12 : P. 14 : V. 1 + 5 : A. 3 + 9 : C. 14.

According to his MS. notes, Colonel Montagu, who possessed a copy of Mr. Donovan's History of British Fishes, had taken this species on the Devonshire coast.

ACANTHOPTERYGII. *LABRIDÆ.*

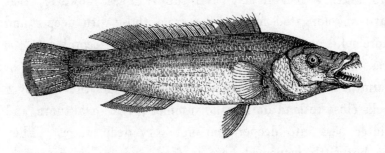

THE BLUE-STRIPED WRASSE.

Labrus variegatus, Gmel. Linn.
    ,,   ,,   *La vieille rayée,* Cuvier, Règne An. t. ii. p. 255.
    ,,   ,,   *Striped Wrasse,* Penn. Brit. Zool. vol. iii. p. 337, pl. 57.
    ,, *coquus,* *The Cook Wrasse,* Penn. ,, ,, ,, iii. p. 340.
    ,, *variegatus,* *Striped Wrasse,* Don. Brit. Fish. pl. 21.
    ,, *coquus,* *The Cook Wrasse,* Flem. Brit. An. p. 209, sp. 133.
Sparus formosus, Shaw, Nat. Misc. pl. 31.
Labrus pavo, Risso, Hist. t. iii. p. 299, sp. 196.

This species, first described as British by Pennant, is one of the most beautiful of the *Labridæ,* and appears to be much more frequently met with on the southern than upon the eastern shores of England : those from the southern shores are also most conspicuous for the beauty of their appearance. It is liable to some variation occasionally in its colours and markings, as the various figures of this fish will show, but the general form of body and fins are permanent. I have obtained two specimens in the London fish-market.

This species is not unfrequent in Cornwall, and has been taken at Cork, Dublin, Down, and Antrim.

" In its habits," Mr. Couch says, " this species keeps, like the others, among rocks, seeking shelter among the larger sea-weeds, where it feeds chiefly on crustaceous animals, and takes a bait freely. All the Wrasses, however, that have an elongated form, differ from those with deeper and more solid bodies, in changing their quarters according to the season, and that too without much reference to the cold or warmth. They enter harbours and frequent the shallower rocks close to land during the summer; but in autumn and winter pass into deeper, but not very deep water. They are but little esteemed here as food, and are chiefly sought after as bait for other fish."

The whole length of the specimen described is twelve inches; the length of the head, compared with the length of the head and body without the tail-rays, is as one to three; the distance from the teeth in front to the edge of the preoperculum equal to the depth of the body alone; the depth of the body and dorsal fin included is equal to the whole length of the head: lips thin, flexible, capable of considerable extension; a single row of teeth in each jaw, those of the lower jaw smaller than those of the upper jaw, and of these last those in front are the largest. The dorsal fin commences in a line over the origin of the pectoral fin; the pectoral fins short and rounded; the vent situated under the sixteenth spiny ray of the dorsal fin; the anal fin ending rather before the end of the soft portion of the dorsal fin; the fleshy portion of the tail and its rays elongated, the latter slightly rounded: the body deepest at the origin of the pectoral fin; in its whole form elongated; the scales of moderate size, with six rows, following the diagonal line of their succession, between the base of the dorsal fin and the lateral line, and twenty rows between the lateral line and the bottom of the belly.

The general colour of the body and head varies in tint from orange yellow to orange red, darkest on the back, lightest on the belly ; sides of the head and body striped with fine blue ; the irides of two colours, orange and blue ; the membrane connecting the first twelve rays of the dorsal fin blue, the upper edge orange, the remaining portion orange with blue spots ; pectoral, ventral, anal, and caudal fins, orange tipped with blue. The number of fin-rays are—

D. 17 + 13 : P. 15 : V. 1 + 5 : A. 3 + 10 : C. 11.

Pennant's specimen of this fish was taken off the Skerry Isles, on the coast of Anglesey ; and Cuvier, in his *Règne Animal*, tom. ii. page 256, (note,) has borne testimony to the excellence of Pennant's representation of this species, by stating that he did not know any good figure of this fish except that in the British Zoology. Some specimens are darker on the upper part of the back than others ; and the sides of Pennant's fish are described as having been marked with four parallel lines of greenish olive, and the same number of most elegant blue.

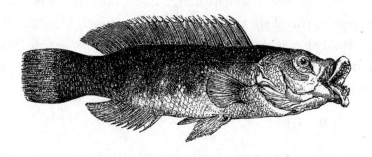

THE SEA WIFE.

Labrus vetula, BLOCH, pt. ix. pl. 293.

A SPECIMEN of this fish, measuring nine and a half inches in length, I obtained in the London market, and had a drawing of it made at the time, from which the figure above, reduced in size, was taken. In the proportions of the body and its different parts, it is intermediate between *L. maculatus* and *L. variegatus*. It is not so short a fish in proportion to its depth as the first, yet shorter and deeper than the second, and distinct, apparently, from both in colour. Though approaching some specimens of *L. variegatus*, it is distinguished by the ascending line of the preoperculum being much more oblique, forming with the inferior margin a more obtuse angle ; by the six spinous rays at the commencement of the anal fin : the teeth are also smaller and more numerous, in the upper jaw especially. The length of the head, compared to the whole length of the fish, is as two to seven ; the depth of the body is to the whole length of the body, without the caudal fin, as one to four : the vent in a vertical line under the thirteenth spinous ray of the dorsal fin. The number of fin-rays,

$$D. \ 16 + 13 : P. \ 15 : V. \ 1 + 5 : A. \ 6 + 8 : C. \ 12.$$

The whole of the upper part of the back, neck, and sides, very dark purple black, becoming lighter on the belly ; lips, and anterior part of the head, flesh colour tinged with purple ; irides blue ; the teeth had all the characters of those of a *Labrus*, and the branchiostegous rays were five in number. All the fins blue ; the ventrals tipped with black.

I could not ascertain from what part of our coast this fish had been brought.

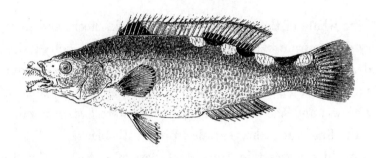

THE RED WRASSE. THREE-SPOTTED WRASSE.

DOUBLY-SPOTTED WRASSE.

Labrus carneus,	Bloch, pt. ix. pl. 289.	
,,　　　,,	Cuvier, Règne An. t. ii. p. 256.	
,,　　*trimaculatus,*	*Trimaculated Wrasse,* Penn. Brit. Zool. v. iii. p. 336, pl. 56.	
,,　　　,,	,,　　　,,　Don. Brit. Fish. pl. 49.	
,,　　　,,	*Three-spotted* and *Doubly-spotted Wrasse* of Mr. Couch's MS.	
,,　*quadrimaculatus,* Risso, Hist. t. iii. p. 302. sp. 199.		

The Red Wrasse is a well-marked species, first described by Ascanius, and appears to be occasionally met with at various places on our coast. Pennant's fish was taken on the coast of Anglesey. Mr. Donovan, Colonel Montagu, and Mr. Couch, have each described from specimens obtained in Devonshire and Cornwall. Mr. Neill has recorded his notice of several taken in the Frith of Forth; and naturalists situated still farther north include it among the fishes of the Baltic and coast of Norway. Two specimens in my own collection, one of six inches in length, prepared dry, the second of nine inches, preserved in diluted spirits of wine, supply materials for the present article.

With less variation in its colours than either the *Lab. maculatus* or *variegatus*, the Red Wrasse is still a beautiful fish. In its habits it resembles the last of these two named species ; it feeds on crustaceous and testaceous animals ; approaches the shore to deposit its spawn in March or April : and Muller says that its flesh is good food. Risso states that in the Mediterranean the females are found full of ova twice in the year.

The length of the head, measuring from the teeth to the backward projecting angle of the operculum, is to the head and body, without including the caudal rays, as one to three ; the depth of the body and dorsal fin equal to the length of the head ; the depth of the body alone, in a line with the origin of the ventral fins, is to the whole length of the fish as one to four : the scales small. The number of fin-rays,

D. $17 + 13$: P. 15 : V. $1 + 5$: A. $3 + 11$: C. 14.

The prevailing colours are a fine red orange over all the upper parts of the body, becoming lighter as it descends the sides, and ending in pale orange yellow on the belly; all the fins rich orange, with a tinge of darker colour at the edges of the membranes ; part of the anterior spinous portion of the dorsal fin is of rich purple, with two spots at the base of the hinder soft-rayed part of the same fin, and one still farther back, at the upper part of the fleshy portion of the tail, of the same deep purple colour.

Alternating with the last three dark spots are four light-coloured ones, of a delicate rose colour or fleshy tint, which appear to have given origin to the name of Doubly-Spotted Wrasse. There are occasionally but two dark spots at the hinder part of the body. Risso includes the dark blotch on

the anterior part of the dorsal fin in his enumeration of the
spots, and calls the species four-spotted : he also adds, that
those specimens which frequent the more rocky districts of
the Mediterranean are observed to be most inclined to red
in colour.   Mr. Couch's coloured drawings of Cornish specimens, which are remarkably red, are in accordance with
Risso's remark.   The lips and fleshy portions of the under
jaw, not covered with scales, are of a delicate flesh colour.

ACANTHOPTERYGII. *LABRIDÆ.*

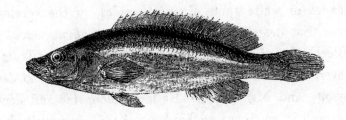

THE COMBER WRASSE.

Labrus comber, GMELIN. LINN.
    ,,   ,,   *Comber Wrasse,* PENN. Brit. Zool. vol. iii. p. 342, pl. 58.
    ,,   ,,   *Comber,*   FLEM. Brit. An. p. 209, sp. 131.

PENNANT says he received this species from Cornwall, and supposed it to be the Comber of Mr. Jago.

It was of a slender form : the colour of the back, fins, and tail, red ; the belly yellow ; beneath and parallel to the lateral line ran a smooth even stripe from gill to tail, of a silvery colour. The number of fin-rays were—

D. 20 + 11 : P. 14 : V. 5 : A. 3 + 7 : C. number not given.

Mr. Couch has also met with this species, and the account in his MS. is as follows :—" Compared with the Common Wrasse, the Comber is smaller, more slender, and has its jaws more elongated. The two upper front teeth are very long : a white line passes along the side from head to tail, unconnected with the lateral line. It has distinct blunt teeth in the jaws and palate : the ventral fins are somewhat shorter than in others of the genus." " Such is the note I made," says Mr. Couch, " on inspecting one of this

VOL. I.   U

species several years since ; but I have not lately had an opportunity of a re-examination ; it is consequently scarce."

Cuvier, in the *Règne Animal*, tom. ii. p. 255, note, considers the Comber of Pennant only a red variety, with a succession of white spots along the sides, of the species of *Labrus* first described in this work, the *Labrus maculatus* of Bloch, *La vieille tachetée* of French authors. It may, however, be stated, that this fish has occurred to Jago, Pennant, and Mr. Couch. Its more elongated and slender form, both in reference to head and body, as described and figured by Pennant, and, in further confirmation, again so described by Mr. Couch, who had made notes from a specimen, and who has great opportunities from his locality of examining the various species of this family, has induced me to give it a place here as a species, inviting the investigation of ichthyologists to the subject. In the elongation of its form, and the lengthened light-coloured band along the side, not made up by a series of spots, this fish leads very naturally to Cuvier's next division of this family, of which the Rainbow Wrasse of Pennant is a beautifully coloured example.

ACANTHOPTERYGII.　　　　　　　　　　*LABRIDÆ.*

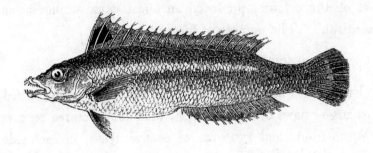

THE RAINBOW WRASSE.

INDENTED STRIPED WRASSE.

Julis Mediterranea, Risso, Hist. t. iii. p. 309, sp. 209.
Labrus Julis,　　Linnæus. Bloch, pt. viii. pl. 287, fig. 1.
Julis,　　　　*La Girelle,*　　　　Cuvier, Règne An. t. ii. p. 257.
Labrus Julis,　　*Indented Striped Wrasse,* Don. Brit. Fish. pl. 96.
　　,,　　　,,　　*Rainbow Wrasse,*　　Penn. Brit. Zool. vol. iii. p. 343.
Julis vulgaris,　Flem. Brit. An. p. 210, sp. 134.

Generic Characters.—Head smooth; cheeks and gill-covers without scales; the lateral line bent suddenly downwards when opposite the end of the dorsal fin; in other respects the generic characters are similar to those of the genus *Labrus.*

THIS very beautiful species, which appears to have been known to the oldest ichthyologists as an inhabitant of the Mediterranean, was first made known as a British fish by Mr. Donovan, in his Natural History of British Fishes, in which he figured and described a specimen taken on the coast of Cornwall in the summer of 1802.

"This specimen," Mr. Donovan observes, "rather exceeded the length of seven inches; it was of a slender, or elongated form, and remarkable for the elegant distribution of its colours, which were changeable in various directions of

U 2

light ; but the most striking peculiarity was the broad den-tated stripe, extending along each side, from the head nearly to the tail, the colour of which was fulvous, and, with the rest of the colours, produced an effect equally singular and beautiful." The number of fin-rays were—

D. 9 + 13 : P. 12 : V. 1 + 5 : A. 2 + 13 : C. 13.

From other sources we learn that the head is compressed ; lips large ; jaws of equal length, with four pointed recurved teeth in front, and two rows of conical teeth on each side ; palate and tongue smooth ; nostrils pierced near the eyes with four apertures, those anterior round, the others oval ; irides orange, the pupil black ; body elongated and narrow ; back and belly round ; lateral line elevated ; scales of the body adherent, small, and thin.

Risso says an assemblage of beautiful colours pervades the body of this species ; the back is greenish blue ; the longitu-dinal band is orange, beneath that are lilac-coloured bands on a silvery ground ; the head varied with brown, yellow, blue, and silver ; the dorsal fin orange, with a purple spot on the membrane connecting the first three spinous rays, which are elongated beyond the others.

Although this species is reported not to be uncommon in the Mediterranean, and Risso states that it frequents all the rocky shores of that sea, but little appears to be known of its habits. A translator of Oppian says :

> " On some thick beds of mossy verdure grow,
> Sea grass, and spreading wrack are seen : below,
> Gay Rainbow-fish, and sable Wrass resort."

The food of this fish is small fry and crustaceous animals. Elian says the flesh is poisonous : Galen and Bloch consider it wholesome.

ACANTHOPTERYGII. *LABRIDÆ.*

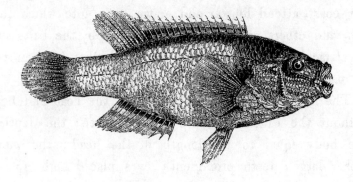

THE GILT-HEAD. CONNOR.

GOLDEN MAID.

Crenilabrus tinca, RISSO, Hist. t. iii. p. 315, sp. 215.
Labrus ,, LINNÆUS.
Crenilabrus ,, FLEM. Brit. An. p. 208, sp. 128, but not all the synonymes.

Generic Characters.—The *Crenilabri* have all the external and internal characters of the true *Labri*, but are distinguished from them by the denticulation
of the edge of the preoperculum. They are generally a little deeper also in the
body compared with their length.

THIS is the Ancient Wrasse and Common Wrasse of
those authors who describe a blue and yellow species with
a denticulated preoperculum. The Ancient Wrasse of Pennant and others, described as having the margin of the
preoperculum entire, is the same with Pennant's Ballan
Wrasse, and the *Labrus maculatus* of Bloch.

This species is not uncommon, and I have obtained it
on the coasts of Sussex, Hampshire, and Dorsetshire; it
occurs also on the coast of South Wales; and has been
taken at Londonderry, Dublin, and Belfast bay. In its
habits, as far as I am acquainted with them, it resembles

the Goldfinny ; and, like that fish, the specimens I have obtained, or seen in the possession of others, have been procured from those fishermen who, 'on the rocky parts of our coast, attend lobster and prawn pots, into which these fish, and others occasionally, are enticed by the baits used to attract the *crustacea*, and for which they in their turn are usually cut up into bait.

The head alone is to the length of the head and body, without the caudal rays, as one to three: the depth of the body equal to the length of the head : the mouth rather large ; teeth prominent ; eyes placed high up near the line of the profile, irides orange ; the horizontal and ascending lines of the preoperculum forming nearly a right angle, the margin strongly denticulated. The dorsal and pectoral fins commence on the same plane ; the spinous rays of the dorsal fin rather short, the flexible rays elongated ; the pectoral fins large and rounded ; the ventral fins with an elongated scale between them, extending over a portion of the inner and shorter rays of each ventral fin ; the longest ventral fin-rays reaching to the anal aperture ; the aperture itself in a line under the thirteenth spiny ray of the dorsal fin ; the anal fin with three spiny rays, the first the shortest, the third but little shorter than the flexible rays, which are elongated ; the soft portions of the dorsal and anal fins terminate on the same plane ; the fleshy portion of the tail longer than the caudal rays, with the upper and under edges parallel ; the rays rounded. The number of the various fin-rays are—

D. 16 + 9 : P. 14 : V. 1 + 5 : A. 3 + 10 : C. 13.

The head is blue, striped and spotted on the cheeks and gill-covers with reddish orange ; the general colour of the body is red, varied with green ; all the fins greenish blue ;

the membranes of the dorsal and anal fins with a longitudinal stripe or two of darker blue. Specimens from shallow water are said to be the finest in colour.

This fish spawns in April. The example here described measured six inches in length, and I have seldom seen any specimens much larger.

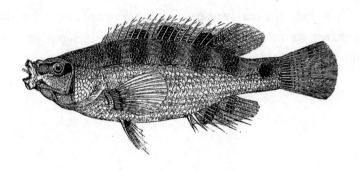

THE GOLDFINNY, OR GOLDSINNY.

Crenilabrus Cornubicus, Risso, Hist. t. iii. p. 325, sp. 233.
Labrus,   ,,   *Goldsinny,* Penn. Brit. Zool. vol. iii. p. 339, pl. 58.
    ,,   *Cornubius,* *Goldfinny,* Don. Brit. Fish. pl. 72.
    ,,   *Cornubicus,*   ,,   Montagu's MS.
    ,,   ,,   *Corkwing,* Couch, Loudon Mag. Nat. Hist. vol. v.
                                        p. 17, fig. 4.

THE GOLDFINNY has been called *Cornubius, Cornubicus,* and *Cornubiensis,* and though supposed originally, as its name would seem to imply, to be exclusively Cornish, it is not confined to that western part of England. Montagu and others have taken it frequently in Devonshire; and it has been obtained as far east as Beachy-head. The mouth is small; the teeth regular; the eye of moderate size; the serrations on the edge of the preoperculum very strongly marked; "the lateral line is straight till it reaches the posterior part of the dorsal fin, where it deflects," says Colonel Montagu, "almost at right angles, and again turns at a similar angle to go to the tail fin, dividing that part equally."

"This fish varies in colour, but the shape is like that figured in the British Zoology. It is generally more or less green or yellowish ; darkest on the back ; the sides generally marked with longitudinal lines of a darker colour, mostly green, but sometimes not very conspicuous. The dark spot at the base of the caudal fin, on the lateral line, appears to be a constant specific character. One fish, about three inches long, has the dorsal and anal fins mottled with purplish brown."

The fin-rays are—

$$D. 16 + 8 : P. 14 : V. 1 + 5 : A. 3 + 10 : C. 14.$$

Mr. Couch says " this species rarely takes a bait, and as its haunts are among rocks, where nets are rarely cast, it has only been caught within my knowledge in the wicker-vessels set to take lobsters and crabs, on which account also I have only seen it in summer. It is less abundant than most of the other species."

My own specimens of this fish are of various sizes, measuring from one inch and a half to four inches ; and, as far as my own observations have gone, the dark spot on the side of the fleshy portion of the tail, at the end of the lateral line, close to the base of the caudal rays, is a constant character. I have quoted Risso in the present instance, as also in that of the fish last described, *Crenilabrus tinca*, because in both cases the descriptions agree with our specimens as closely as fishes from such distant localities can be expected to coincide, and in both instances also Risso refers to Pennant.

I have seen a specimen of the Goldfinny from the coast of Ireland with twenty-three rays in the dorsal fin, of which the first thirteen only were spinous, the others soft.

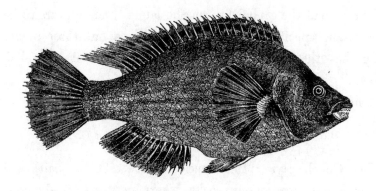

THE GIBBOUS WRASSE.

Crenilabrus gibbus, Flem. Brit. An. p. 209, sp. 129.
　　　,,　　　　,,　　　Cuvier, Règne An. t. ii. p. 259, note.
Labrus　　　　,,　　　Gmelin. Linn.
　　,,　　　　,,　　*Gibbous Wrasse,* Penn. Brit. Zool. vol. iii. p. 338, pl. 56.

The Gibbous Wrasse is inserted on the authority of Pennant, whose figure and description from the last edition of the British Zoology are here repeated.

" This species was taken off Anglesey : its length was eight inches ; the greatest depth three inches : it was of a very deep and elevated form, the back being vastly arched, and very sharp or ridged. From the beginning of the head to the nose, was a steep declivity ; the teeth like those of the others ; the eyes of a middling size ; above each a dusky semilunar spot ; the nearest cover of the gills finely serrated." The fin-rays were—

D. 16 + 9 : P. 13 : V. 1 + 5 : A. 3 + 11 : C. 14.

" The first sixteen rays of the back fin strong and spiny, the other nine soft and branched ; the pectoral fins consisted of thirteen ; the ventral of six rays ; the first ray of the ventral fin was strong and sharp ; the anal fin consisted of fourteen rays, of which the first three were strongly acuminated. The tail was large, rounded at the end, and the rays branched ; the ends of the rays extending beyond the webs : the lateral line was incurvated towards the tail : the gill-covers and body covered with large scales ; the first were most elegantly spotted, and striped with blue and orange, and the sides spotted in the same manner ; but nearest the back the orange was disposed in stripes ; the back fin and anal fin were of a sea-green, spotted with black ; the ventral fins and tail a fine pea-green ; the pectoral fins yellow, marked at their base with transverse stripes of red."

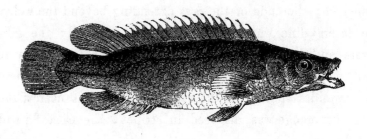

THE SCALE-RAYED WRASSE.

Crenilabrus luscus, The Scale-rayed Wrasse of Couch's MS.
Labrus,   Linnæus.
     ,,   ,,   Couch, Loudon Mag. Nat. Hist. vol. v. p. 18. and
                     p. 741 & 2, fig. 121.

" The specimen was twenty-two inches long; the greatest
depth exclusive of the fins two and a quarter inches; the
body plump and rounded.   Head elongated; lips membranous; teeth numerous, in several rows, those in front
larger and more prominent, rather incurved.   Eye moderately large; anterior gill-plate serrate; six gill-rays; body
and gill-covers with large scales; lateral line nearer the back,
descending with a sweep opposite the termination of the
dorsal fin, thence backward straight.   Dorsal fin twenty-one
firm rays, eight soft rays; the fin connected with the latter
expanded, reaching to the base of the tail: pectorals round,
fourteen rays; ventrals six rays, the outermost simple, stout,
firm, tipped; between these fins a large scale.   Anal fin,
six firm, eight soft rays, the soft portion of the last expanded.
Tail round, fifteen rays; between each ray of the dorsal,

anal, and caudal fins, is a process formed of firm, elongated, imbricated scales. Colour a uniform light brown, lighter on the belly; upper eye-lid black; at the upper edge of the base of the caudal fin a dark-brown spot. Pectorals yellow; all the other fins bordered with yellow." The number of fin-rays are here repeated for the sake of uniformity.

$$D. 21+8 : P. 14 : V. 1+5 : A. 6+8 : C. 15.$$

" I have seen only one specimen of this Wrass, of the size here mentioned, which was taken in the month of February 1830, at the end of a very cold season; but I have met with specimens about three inches long that resemble it too nearly to permit me to consider them as different species."

Mr. Couch very kindly sent me one of the specimens here referred to, which has been engraved as an appropriate vignette to this article. This fish exhibits an elongated scale, attached by the base only, at the origin of the ventral fins, partly covering the inner edges of both, and extending nearly half-way along the fins; but it is not peculiar to this fish : I observe such an appendage in both the first and second British species of *Crenilabrus* described in this work.

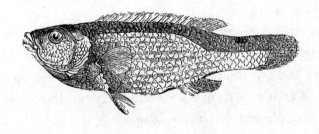

THE TRUMPET-FISH.   SEA-SNIPE.

BELLOWS-FISH.   *Cornwall.*

Centriscus scolopax, Linnæus.
,,   ,,   Bloch, pt. iv. pl. 123, fig. 1.
,,   ,,   *Snipe-nosed Trumpet-fish,* Penn. vol. iii. p. 190.
,,   ,,   *Snipe-fish,*   Don. Brit. Fish. pl. 63.
,,   ,,   *Trumpet-fish,* Flem. Brit. An. p. 220, sp. 170.
,,   ,,   Cuvier, Règne An. t. ii. p. 268.

Generic Characters.—Snout produced ; mouth small ; teeth wanting : body compressed, oval, covered with scales ; two dorsal fins, the rays of the first spinous ; ventral fins united ; abdomen carinated.

The Trumpet-fish is, in a British series, the last of the *Acanthopterygian* fishes, or those having some of their fin-rays spinous.

One specimen of this singular-looking fish, recorded in the eighth volume of the Transactions of the Linnean Society to have been taken in England, was thrown ashore at St. Austle's bay, in Cornwall, early in the year 1804. This example was five inches long.

* The family of tube-mouthed fishes with spiny fin-rays.

Mr. Donovan, in his History of British Fishes, refers to two instances of its capture within his knowledge, and appears to have had two specimens in his collection.

The Trumpet-fish was first described and figured by Rondeletius, and is not uncommon in some parts of the Mediterranean. Risso says it prefers a muddy bottom in moderately deep water, and that it spawns in spring. The young are seen near the shore in autumn, shining with a brilliant silvery lustre, not having then acquired the golden red of the adult fish ; they are not very numerous, but they do not wander far from the locality in which they are bred. Their food is not mentioned by authors, though it probably consists of minute crustaceous animals ; and in reference to their tubular mouths, it is probable that by dilating their throat they can draw their food up their cylindrical beak, as water is drawn up the pipe of a syringe. The beak-like mouth is also well adapted for detaching minute animals from among the various sorts of sea-weed. The flesh of the Trumpet-fish is considered good.

The snout is elongated, the jaw-bones forming a tube extending an inch and a half before the eyes ; the mouth at the extremity small, without teeth ; the eyes large, irides silvery, streaked with red, the pupils black. The back in the specimen now before me, and from which the figure was taken, is elevated, forming a slight ridge, and ending in a short spine just in advance of the long and strong denticulated spine of the first dorsal fin. The first dorsal fin, in my specimen of this fish, has but three spinous rays, as shown in the figure ; but authors generally state them to be four. The first spine is three times as long, and also much stronger, than the others, pointed, moveable, and toothed like a saw on the under part, constituting a formidable weapon of defence ; the other spines are short, but their points project beyond the membrane by which they are united ; the rays of the second dorsal fin are soft. The shape of the body is oval, compressed ; the pectoral fin small ; the ventral fins also small, with a depression behind in which they can be lodged. The anal fin is elongated, the rays short. The whole number of fin-rays are—

D. 4. 12 : P. 17 : V. 4 : A. 18 : C. 16.

The colour of the back is red, the sides rather lighter ; sides of the head and belly silvery, tinged with gold colour ; the scales on the body hard, rough, minutely ciliated at the free edge, and the surface granulated. All the fins greyish white.

*ABDOMINAL
MALACOPTERYGII.*   CYPRINIDÆ.†*

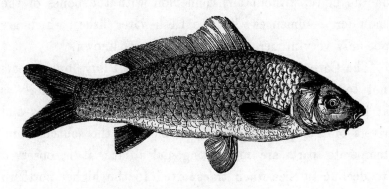

THE COMMON CARP.

Cyprinus carpio, Linnæus. Bloch, pt. i. pl. 16.
    ,,   ,,   Carp, Penn. Brit. Zool. vol. iii. p. 467, pl. 81.
    ,,   ,,   ,,   Don. Brit. Fish. pl. 110.
    ,,   ,,   ,,   Flem. Brit. An. p. 185.
    ,,   ,,   Cuvier, Règne An. t. ii. p. 271.

Generic Characters.—Body covered with large scales; a single elongated dorsal fin; lips fleshy; mouth small; teeth in the pharynx, but none on the jaws; branchiostegous rays 3.

Having concluded Baron Cuvier's first order, the *Acanthopterygian* fishes, or those bearing numerous spines, which support the whole or part of the membranes of some of the fins, and of which various examples have been given, the species forming the next great division are called *Malacopterygian* fishes, or those having their fin-membranes supported by flexible rays, which are either jointed or branched, or both. These are again divided into three orders, according to the position of the ventral fins, or in wanting the

---

* Soft-finned fishes; the fin-rays almost universally flexible.
   † The family of the Carp.

ventral fins altogether. The first order of this second division, the second order in the whole subject, called *Abdominal Malacopterygii*, or those having the ventral fins situated on the belly, without any connexion with the bones of the shoulder, commences with the fresh-water fishes, which are not only very numerous, but also the best known.

The Carp is noticed by Aristotle and Pliny, but appears not to have been held in the same estimation formerly as at the present day. It is found in most of the lakes and rivers of Europe generally, but those of the southern and temperate parts are most congenial to it; it is observed to decline in size when transported to the higher northern districts, and is said to be even now unknown in Russia.

Their growth is, however, particularly cultivated in Austria and Prussia, and considerable traffic in Carp prevails in various parts of the European continent, where an acre of water will let for as much yearly rent as an acre of land, and where fresh-water fishes, as articles of food, are held in higher estimation than in this country. Carp are said to live to a great age, even to one hundred and fifty, or two hundred years; but they lose their rich colour,—their scales, like the productions of the cuticle in some other animals, becoming grey and white with age.

Neither the exact period when, nor the particular country from which, Carp were first brought to England, appear to be distinctly known. Leonard Mascall takes credit to himself for having introduced the Carp, as well as the Pippin; but notices of the existence of the Carp in England occur prior to Mascall's time, 1600. In the celebrated Boke of St. Albans, by Dame Juliana Barnes, or Berners, the Prioress of Sopewell Nunnery, printed at Westminster by Wynkyn de Worde in 1496, Carp is mentioned as a " deyntous fisshe;" and in the Privy Purse expenses of

King Henry the Eighth in 1532, various entries are made of rewards to persons for bringing " Carpes to the King."*

The old couplet is certainly erroneous, which says,—

" Turkies, Carps, Hops, Pickerell, and Beer,
Came into England *all* in one year."

Pike or Pickerell were the subject of legal regulations in the reign of Edward the First. Carp are mentioned in the Boke of St. Albans, printed in 1496; Turkies and Hops were unknown till 1524, previous to which *wormwood* and other *bitter plants* were used to preserve beer, and the Parliament in 1528 petitioned against *hops*, as a *wicked weed*. Beer was licensed for exportation by Henry the Seventh in 1492, and an excise on beer existed as early as 1284, also in the reign of Edward the First.

In this country the Carp inhabits ponds, lakes, and rivers; preferring in the latter those parts where the current is not too strong, and thriving best on soft marly or muddy bottoms. They are very prolific, breeding much more freely in lakes and ponds than in rivers. Bloch found six hundred thousand ova in the roe of a female of nine pounds weight, and Schneider seven hundred thousand in a fish of ten pounds weight. They spawn towards the end of May or the beginning of June, depending on the temperature of the water and the season; and the ova are deposited upon weeds, among which the female is followed by two or three males, and the fecundation of a large proportion of the ova is by this provision of nature effectually secured; but they both breed and grow much more freely in some waters than in others, without any apparent or accountable cause. But few Carp exist even in preserved waters in Scotland, and these breed but slowly, and in some instances not at all.

* Pickering's splendid edition of Walton, page 207, note.

x 2

Carp are said to have been introduced into Ireland in the reign of James the First, and are preserved in the counties of Cork, Dublin, Kilkenny, and probably several others.

The larvæ of insects, worms, and the softer parts of aquatic plants, are the food of the Carp. They probably eat little or nothing during winter, and are supposed to bury themselves in mud. White, in his Natural History of Selbourne, says, " in the garden of the Black Bear Inn in the town of Reading, is a stream or canal, running under the stables, and out into the fields on the other side of the road : in this water are many Carps, which lie rolling about in sight, being fed by travellers, who amuse themselves by tossing them bread : but as soon as the weather grows at all severe these fishes are no longer seen, because they retire under the stables, where they remain till the return of spring."

They feed well in stews ; and Mr. Jesse says of some Carp and Tench, so retained by him," that they were soon reconciled to their situation, and eat boiled potatoes in considerable quantities ; and the former seemed to have lost their original shyness, eating in my presence without any scruple."

The Carp is exceedingly tenacious of life out of water. Several authors have stated that it is a common practice in Holland to keep them alive for three weeks or a month, by hanging them in a cool place, with wet moss in a net, and feeding them with bread steeped in milk ; taking care to refresh the animal now and then by throwing fresh water over the net in which it is suspended.

Though not so rapid in their growth as some fish, Carp have attained three pounds weight by their sixth year, and six pounds weight before their tenth year. The largest I can refer to are thus noticed in Daniel's Rural Sports :

—" Mr. Ladbroke, from his park at Gatton, presented Lord Egremont with a brace that weighed thirty-five pounds, as specimens to ascertain whether the Surrey could not vie with the Sussex Carp." In 1793, at the fishing of the large piece of water at Stourhead, where a thousand brace of killing Carp were taken, the largest was thirty inches long, upwards of twenty-two broad, and weighed eighteen pounds.

At Weston Hall, Staffordshire, the seat of the Earl of Bradford, the painting of a Carp is preserved which weighed nineteen and a half pounds. This fish was caught in a lake of twenty-six acres, called the White Sitch, the largest of three pieces of water which ornament this fine estate.

Carp are difficult to take by angling, or rather very uncertain,—great success one day, and little or none another, happening to the same angler at the same water. Carp manage equally to avoid a net, burying themselves in the mud, and allowing a heavily-loaded ground-line to pass over them without their moving; but if disturbed from their hiding-places, frequently endeavouring, like the Grey Mullet, to escape over the corked head-line. Carp are in season for the table from October to April, and are greatly indebted to cooks for the estimation in which they are held.

The mouth is small; no apparent teeth; a barbule or cirrus at the upper part of each corner of the mouth, with a second smaller one above it on each side; the nostrils large, pierced at the second third of the distance between the lip and the eye; the eye small; operculum marked with striæ radiating from the anterior edge; nape and back rising suddenly. The fin-rays are—

D. 22 : P. 17 : V. 9 : A. 8. : C. 19. Vertebræ 36.

First dorsal fin-ray short and bony; the second also

bony, strongly serrated on the posterior surface ; the third
ray flexible, and the longest ray in the fin ; all the other
rays flexible, the last ray double : the dorsal and anal fins
ending on the same plane. The pectoral fin arises immediately behind the free edge of the operculum, its origin
semicircular, concave forwards, upper ray the longest, all
the rays flexible. Ventral fin commences, in a vertical line,
under the third ray of the dorsal fin ; first anal fin-ray bony
and strong, serrated posteriorly ; the other rays flexible, the
last double. The tail forked, the longest rays as long again
as those of the centre. The caudal rays of the two halves
of the tail always unequal in number in the *Cyprinidæ*.
The body covered with large scales, about twelve rows
between the ventral and dorsal fins : the general colour
golden olive brown, head darkest ; irides golden ; belly
yellowish white ; lateral line interrupted, straight ; the fins
dark brown.

ABDOMINAL
MALACOPTERYGII.   *CYPRINIDÆ.*

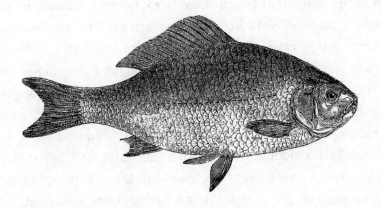

THE CRUCIAN CARP. PRUSSIAN CARP.

CROWGER. *Warwickshire.*

Cyprinus gibelio,	Bloch, pt. i. pl. 12.		
,,	,,	*Gibele Carp,*	Penn. Brit. Zool. vol. iii. p. 480, pl. 83.
,,	,,	*Gibel,*	Flem. Brit. An.
,,	,,	Cuvier, Règne An. t. ii. p. 271.	

PENNANT considers the Crucian Carp as a naturalised species in this country, into which it is said to have been introduced from Germany ; but, as in the case of the Common Carp, the country from which the Crucian was brought, and the year in which it was introduced, are both enveloped in obscurity.

The Crucian Carp is found in some of the ponds about London, particularly in the vicinity of the Thames, a few miles upwards to the west of the metropolis ; and the floods that occasionally happen are supposed to be the means by which this species has been carried into the Thames, from whence individuals of large size are sometimes obtained. It is also known to be very common in several counties of

England, and probably exists in most of them. This fish spawns about the end of April, or early in May: it is very prolific, and the roe, as might be expected, is in great quantity. Though known to be very numerous in some situations, but little success attends the angler who endeavours to catch them, as they seldom bite freely. They attain considerable size, sometimes weighing a pound, or a pound and a half; in one instance a specimen brought to me from the Thames, in October 1829, weighed two pounds eleven ounces; but the most common size is about half a pound. The flesh is white and agreeable. This fish is exceedingly tenacious of life. I have known them recover and survive after having been kept out of water thirty hours. The number of fin-rays are—

D. 18 : P. 14 : V. 9 : A. 8 : C. 19. Vertebræ 30.

The length of the head, compared to the length of the head and body without the tail, is as one to four; including the tail, as one to five; the caudal rays being as long as the head, and one-fifth of the whole length. The body is deepest on the line of the commencement of the dorsal and ventral fins; and the depth, compared to the whole length, including the tail, is as one to three.

The form of the head obtuse; the mouth and eyes small; the body rather short and thick: the scales large; seven scales in an oblique line between the base of the first dorsal fin-ray and the tubular scale of the lateral line, and six scales below between that and the origin of the ventral fin; thirty-four or thirty-five scales along the lateral line; this line descending by a gentle curve from the upper free angle of the operculum below the middle of the body, thence straight to the tail: the pectoral fin commences in a line under the posterior point of union of the operculum with the suboperculum ; the dorsal and ventral fins commence on the same vertical plane ; the length of the base of the dorsal fin nearly equal to the depth of the body ; anal fin small, placed in a vertical line half before and half behind the origin of the last ray of the dorsal fin ; the stronger bony ray of the dorsal and anal fin finely serrated, compared with the serrations in the rays of these fins in the Common Carp. The tail forked, divided into two nearly equally rounded halves, the longest rays about one-third longer than the short rays of the middle portion ; the upper part with ten rays, the lower portion with nine.

The top of the head and back olive brown ; the sides lighter in colour ; the belly almost white ; the whole fish shining with a brilliant golden metallic lustre : irides golden ; cheeks and gill-covers brilliant golden yellow ; the dorsal fin and upper part of the tail, brown tinged with orange ; pectoral, ventral, and anal fins, orange red ; lower part of the tail tinged with the same colour.

This fish has been considered here as the *C. gibelio* of Bloch ; but in one circumstance it does not agree with his fish. Bloch says the *C. gibelio* has but twenty-seven vertebræ ; those of *C. carassius* are stated by Bloch to be thirty, but the specimens examined for the description now given had also thirty vertebræ. The fin-rays agree in number with those of Bloch's *gibelio,* and do not coincide with the number considered to be present in *carassius.*

From some measurements, and other particulars in my note-book, I have great reason to believe that specimens of *C. carassius* have been more than once brought to me from the Thames ; but not at that time contemplating the want of them for the present purpose, the specimens were not preserved. In order, however, that other inquirers may identify this species should it occur, a reduced figure of

C. carassius, from Bloch, pl. 11, is here supplied as a vignette. The fin-rays are—

D. 21 : P. 13 : V. 9 : A. 10 : C. 21. Vertebræ 30.

The length of the head is to the depth of the body as one to two, and to the whole of head, body, and tail, as one to five; the depth of the body compared to the whole length, as two to five : the tail nearly square at the end.

Several authors have wholly omitted the *gibelio* in their systematic works ; others have considered it as a distinct species ; and some have supposed it only a variety of *carassius*. The most obvious points of distinction between these two fishes are, in *C. gibelio*, depth of body compared to the whole length as one to three ; depth of body not equal to twice the length of the head ; tail forked. In *C. carassius*, depth of body compared to the whole length as two to five ; depth of body equal to twice the length of the head ; tail nearly square. The length of the head, compared to the whole length of the fish, is the same in both, viz. as one to five. Dr. Turton, in his British Fauna, has described the *carassius*, p. 108, sp. 119. Mr. Pennant and Dr. Fleming have described the *gibelio*.

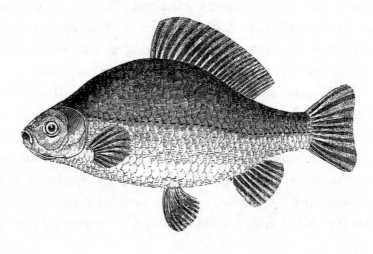

ABDOMINAL
MALACOPTERYGII.   *CYPRINIDÆ.*

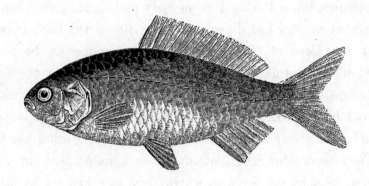

THE GOLD CARP.

Cyprinus auratus, Linnæus.
    ,,   ,,   Bloch, pt. iii. pl. 93, 94.
    ,,   ,,   *Gold Carp*, Penn. Brit. Zool. vol. iii. p. 490.
    ,,   ,,   *Golden Carp*, Flem. Brit. An. p. 185, sp. 3.
    ,,   ,,   Cuvier, Règne An. t. ii. p. 272.

The date of the first introduction of the Golden Carp, or Gold and Silver Fish, as they are more frequently called, is differently stated by authors : 1611, 1691, and 1728, are each recorded as the particular year in which they were first brought over. The earliest seen in France were sent there for Madame Pompadour.

Pennant says, " In China the most beautiful kinds are taken in a small lake in the province of Che-Kyang. Every person of fashion keeps them for amusement, either in porcelain vessels, or in the small basins that decorate the courts of the Chinese houses. The beauty of their colours, and their lively motions, give great entertainment, especially to the ladies, whose pleasures, from the policy of that country, are extremely limited." The Chinese call their fish with a whistle to receive their food.*

A correspondent in Loudon's Magazine of Natural History, vol. iii. page 478, considers " that they were probably introduced into Portugal at an early period, after the people of that country had discovered the route to the East Indies by the Cape of Good Hope, as they appear to be now completely naturalised there, and abound in many of their streams, whence they are brought to us by trading vessels from Lisbon, St. Ubes, &c. in large earthen jars, and may be had at a very easy rate before they get into other hands. They have also been introduced and naturalised in the Mauritius by the French, where they now abound in fish-ponds and streams, and are served up at table as agreeable food, with the other fresh-water fishes, to the brood of which they are thought to be very inimical, by destroying their spawn and young fry. The extreme elegance of the form of the Golden Carp, the splendour of their scaly covering, the ease and agility of their movements, and the facility with which they are kept alive in very small vessels, place them amongst the most pleasing and desirable of our pets."

" They even recommend themselves by another agreeable quality—that of appearing to entertain an affection for each other. A person who kept two together in a glass, gave one of them away ; the other refused to eat, and showed evident symptoms of unhappiness till his companion was restored to him."—*Jesse's Gleanings.*

This fish breeds freely in small ponds and even in tanks in this country ; but particularly so if, by any means, the

---

* I think it was Sir Joseph Banks who used to collect his fish by sounding a bell, and Carew, the Cornish historian, brought his Grey Mullet together to be fed by a noise made with two sticks, see page 202.

temperature of the water can be maintained at an elevation above the ordinary mean.

" It is well known that in manufacturing districts, where there is an inadequate supply of cold water for the condensation of the steam employed in the engines, recourse is had to what are called engine-dams or ponds, into which the water from the steam-engine is thrown for the purpose of being cooled : in these dams, the average temperature of which is about eighty degrees, it is common to keep Gold-fish ; and it is a notorious fact, that they multiply in these situations much more rapidly than in ponds of lower temperature, exposed to the variations of the climate. Three pair of this species were put into one of these dams, where they increased so rapidly, that at the end of three years their progeny, which were accidentally poisoned by verdigris mixed with the refuse tallow from the engine, were taken out by wheelbarrows-full. Gold-fish are by no means useless inhabitants of these dams : they consume the refuse grease, which would otherwise impede the cooling of the water by accumulating on its surface."

A few authentic notices of the power of fishes in bearing extremes of high and low temperature may not improperly be introduced here.

" Desfontaines found a *Sparus* of Lacépède, the *Chromis* of Cuvier, in the hot waters of Cafsa in Barbary, in which Reaumur's thermometer rose to thirty degrees, equal to eighty-six of Fahrenheit. Shaw saw small fishes of the Mullet and Perch kind in these springs."—*Travels in Barbary*, folio edit. Oxford : 1738, p. 231.

Saussure, speaking of the hot springs of Aise in Savoy, says : " I have frequently examined the temperature of these waters at different seasons, and have always found it very nearly alike (about 113 Fahr.). Notwithstanding the heat of these waters, living animals are found in the basins which receive them. I saw in them eels, rotifera, and infusoria, in 1790."

" At Feriana, the ancient Thala," says Bruce, " are baths of warm water without the town : in these were a number of fish, about four inches in length, not unlike Gudgeons. Upon trying the heat by the thermometer, I remember to have been much surprised that they could have existed, or even not been boiled, by continuing so long in the heat of this medium."

" The facts mentioned by Sonnerat and other travellers induced Broussonnet to make some experiments on the degree of heat which river fish are capable of enduring. The details of the degrees of heat are not stated, but many species lived for several days in water which was so hot that the hand could not be retained in it for a single minute."

The five preceding notices are from Dr. Hodgkin's additions to the translation of Dr. W. F. Edwards's French work " On the Influence of Physical Agents on Life."

" In the thermal springs of Bahia in Brazil, many small fishes were seen swimming in a rivulet which raises the thermometer eleven and a half degrees above the temperature of the air."

" Humboldt and Bonpland, when travelling in South America, perceived fishes thrown up alive, and apparently in health, from the bottom of a volcano, in the course of its explosions, along with water and heated vapour that raised the thermometer to two hundred and ten degrees, being but two degrees below the boiling point."

The power of fishes to sustain a low temperature is equally extraordinary; " for that these," says John Hunter, in his Animal Œconomy, " after being frozen, still retain so much of life as when thawed to resume their vital actions, is a fact so well attested that we are bound to believe it."

" Perch have been frozen, and in this condition transported for miles. If, when in this state, fishes are placed in water near a fire, they soon begin to exhibit symptoms of reanimation ; the fins quiver, the gills open, the fish gradually turns itself on its belly, and moves slowly round the vessel, till at length, completely revived, it swims briskly about."*

But to return to the fish before us : I need not occupy space by attempting to describe a species so well known, and of which the variations in colour, fin-rays, and even in the fins, are so numerous, as to appear to bear some proportion to the degree and extent of the domestication. M. de Sauvigny, in his *Histoire Naturelle des Dorades de la Chine*, published at Paris in 1780, has given coloured representations of eighty-nine varieties of this Carp, exhibiting almost every possible shade or combination of silver, brilliant orange, and purple. I have referred to variations in the fins themselves. These fishes are sometimes seen with double anal fins, and others with triple tails : when this occurs, it is generally at the expense of the whole or part of some other fin : thus the specimens with triple tails are frequently without any portion of the dorsal fin, and such specimens have been figured by Bloch and others. Among two dozen Gold-fish for sale in London, were some with dorsal fins extending more than half the length of the back ; some, on the contrary, had dorsal fins of five or six rays only, and one specimen without any dorsal fin whatever ; yet this fish appeared to preserve its perpendicular position with the same ease as any of the others. This induced me to make an experiment, in order to ascertain

* T. S. Bushnan's Introduction to the Study of Nature.

whether the sudden privation of the dorsal fin would produce any more apparent inconvenience than was observable in the specimen just referred to.

For this purpose I attended at the Zoological Society's Garden a short time before the hour at which the Otter was fed daily with his accustomed meal of living fish. Nine or ten Roach and Dace were placed with plenty of water in a large tub of three feet diameter. Five or six of these fish I took from the tub one after another, and with a pair of scissors cut off the whole of the dorsal fin close to the back, returning each fish to the water. They were but little or scarcely at all affected, and each fish appeared to preserve its perpendicular position, or to ascend or descend in the water with the same ease and certainty as before the privation ; the mutilated, swimming among the unmutilated, seemed to possess the same powers. I did not carry the experiment beyond ascertaining this point, and in a few minutes the fish were consigned to the Otter.

When Gold-fish breed in ponds or tanks under favourable circumstances, the young attain the length of five inches in the first twelve months, but their growth afterwards is much less rapid. I have not seen any specimen that exceeded ten inches in length. The young are dark-coloured at first, almost black, changing more or less rapidly according to constitutional power.

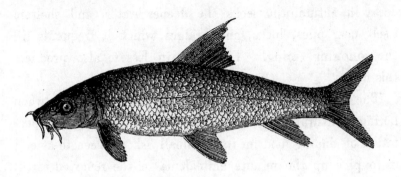

THE BARBEL.

Barbus vulgaris,	FLEM. Brit. An. p. 185, sp. 58.
,, ,,	CUVIER, Règne An. t. ii. p. 272.
Cyprinus barbus,	LINNÆUS. BLOCH, pt. i. pl. 18.
,, ,,	*Barbel,* PENN. Brit. Zool. vol. iii. p. 472, pl. 82.
,, ,,	,, DON. Brit. Fish. pl. 29.

Generic Characters.—Distinguished from *Cyprinus* in having the dorsal and anal fins short ; a strong, serrated, bony ray at the dorsal fin ; mouth furnished with four barbules, two near the point of the nose, and one at the angle of the mouth, on each side.

THE BARBEL is said to have been so called from the barbs or wattles attached about its mouth. It is readily distinguished by these appendages, in conjunction with the great extension of the upper jaw beyond the lower.

This fish was well known to the older ichthyologists. The warm and temperate parts of Europe appear to be its natural locality, and it is abundant in the Rhine, the Elbe, and the Weser.

Near London, the Thames, from Putney upwards, produces Barbel in great quantities, and of large size ; but they are held in little estimation, except as affording sport to the angler. During summer this fish in shoals frequents the weedy parts of the river, but as soon as the weeds begin to decay in autumn, it seeks the deeper water, and shelters itself near piles, locks, and bridges, which it frequents till the following spring. The Lea, in Essex, also produces this fish.

The Barbel feeds on slugs, worms, and small fish: when boring and turning up the loose soil at the bottom, in expectation of finding food for itself, small fish are seen to attend it to pick up the minute animalcula in the removed earth. The Barbel spawns in May or June: the ova, amounting to seven or eight thousand in a full-sized female, are deposited on the gravel, and covered by the parent fishes. These are vivified in a warm season between the ninth and fifteenth day.

Mr. Jesse, when describing the habits of the different sorts of fishes kept in a vivarium, says, "the Barbel were the shyest, and seemed most impatient of observation; although in the spring, when they could not perceive any one watching them, they would roll about and rub themselves against the brick-work, and show considerable playfulness. There were some large stones, round which they would wind their spawn in considerable quantities."

So numerous are the Barbel about Shepperton and Walton, that one hundred and fifty pounds' weight have been taken in five hours; and on one occasion it is said that two hundred and eighty pounds' weight of large-sized Barbel were taken in one day. The largest fish I can find recorded weighed fifteen and a half pounds. Mr. Jesse, and other anglers, have occasionally caught Barbel when trolling or spinning with Bleak, Gudgeon, or Minnow, for large Thames Trout.

" Barbel appear to be in a torpid state in very cold weather, so much so that they may be taken up by the hand.
The fishermen provide themselves with a net fastened to an
iron hoop, having a handle to it, which they place near the
fish, and with a pole push it into the net, so perfectly inanimate are they at this season. Shoals of them also congregate
under the lee of a sunken boat, lying one upon the other,
and are often taken by letting a hook down amongst them,
and then pulling it up."

The length of the head is, to the whole length of the
fish, as one to five; the head the same length as the longest
of the caudal rays : depth of the body not equal to the length
of the head, and compared to the whole length of the fish
as one to five and a half. The head elongated, wedge-
shaped; nose produced; upper jaw much the longest; under
jaw very short; upper lip fleshy, forming three-fourths of a
circle round the under jaw; opening of the mouth horizontal,
admirably adapted to feeding on the ground; one pair of
cirri or barbules at the front of the nose, and a single one at
each end of the upper lip, near the angle of the mouth;
nostrils about one-third nearer the eye than the end of the
nose : form of the body elongated : dorsal fin commencing
half-way between the point of the nose and the end of the
fleshy portion of the tail; the base of the fin shorter than the
longest ray, the third ray the longest as well as the strongest,
denticulated on its hinder surface; pectoral fin half as long
as the distance between its origin and the origin of the ventral fin; the ventral fin commencing in a vertical line under
the fourth ray of the dorsal fin; anal fin commencing half-
way between the origin of the ventral fin and the end of the
fleshy portion of the tail; the base of the fin half as long as
the longest ray : the tail deeply forked, the longest rays
three times as long as the middle short ones.

Y 2

The general colour of the upper part of the head and body is greenish brown, becoming yellowish green on the sides ; cheeks, gill-covers, and scales tinged with bronze ; belly white ; irides golden yellow ; lips pale flesh colour ; dorsal and caudal fins brown, tinged with red ; pectoral, ventral, and anal fins flesh red ; the lateral line nearly straight throughout its whole length.

The number of fin-rays are—

D. 11 : P. 16 : V. 9 : A. 7 : C. 19 : upper half 10.

The Barbel, in the coat of Bar, forms one of the quarterings of the arms of Margaret of Anjou, queen of Henry the Sixth, and founder of Queen's College, Cambridge. She was daughter of René, Duke d'Anjou, titular King of Jerusalem. These arms are very beautifully painted in glass in the windows of a curious old manor-house at Ockwells, in Berkshire, near Bray, on the banks of the Thames ; well known to antiquaries from the engravings in Lysons' history of that county.

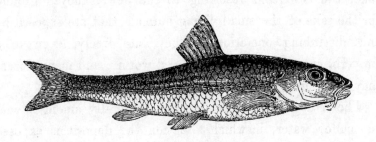

THE GUDGEON.

Gobio fluviatilis,	Willughby, p. 264, Q. 8, fig. 4.
,, ,,	*Gudgeon,* Flem. Brit. An. p. 186, sp. 60.
,, ,,	Cuvier, Règne An. t. ii. p. 273.
Cyprinus gobio,	Linnæus. Bloch, pt. i. pl. 8, fig. 2.
,, ,,	*Gudgeon,* Penn. Brit. Zool. vol. iii. p. 476.
,, ,, ,,	Don. Brit. Fish. pl. 71.

Generic Characters.—The species of this subgenus have, like those of the last, the dorsal and anal fins short ; are furnished with barbules or cirri about the mouth, but have no strong, bony, serrated ray at the commencement of either the dorsal or anal fins. In other respects like *Cyprinus.*

THE GUGDEON is found in many streams that in their course flow over gravelly soils : it appears to delight in slow rivers that have shallow scours over which the current of the water is increased. The Thames, Mersey, Colne, Kennet, and the Avon, produce abundance of the finest Gudgeons. Daniel, in his Rural Sports, says they thrive well in ponds that are supplied with fresh water from brooks running into them. Gudgeons swim together in shoals, feeding on worms, aquatic insects and their larvæ, small molluscous animals, ova, and fry, affording excellent amusement to those anglers who are satisfied with numbers rather than weight ; the

Gudgeon rarely exceeding eight inches in length, and being seldom so large. When angling for them, it is usual to scratch the gravel at the bottom of the water with an iron rake, the Gudgeons resorting to the newly moved ground for the sake of the small living animals that are exposed by this disturbing operation. They bite freely, even to a proverb, at a small portion of a red worm ; and many dozens may be caught, in some situations, in a few hours.

The Gudgeon spawns in May, generally among stones in shallow water, in which situation the deposit is exposed to the vivifying influence of the sun's rays ; the fry are about an inch long by the beginning of August. The Thames fishermen take them in shallow water with a casting net, keeping them in their well-boats till wanted. The London fishmongers are also able to keep Gudgeons alive several weeks in leaden or stone tanks, which are constantly supplied with fresh cold water ; and Colonel Montagu says that a very considerable quantity of these fishes are taken with the casting net in the Avon near Bath, long famous for its Gudgeons, which are exposed for sale alive in shallow tubs of water, and are thus obtained in the highest perfection for invalids, being considered easy of digestion.

The length of the head is, to the whole length of the head, body, and tail, as one to five ; the depth of the body, which is greatest at the commencement of the dorsal fin, not equal to the length of the head ; the lower jaw broad, shorter than the upper ; the mouth wide, with a barbule at the angle on each side ; the nostrils in a circular depression ; the eye placed high up on the side of the head, and about half-way between the point of the nose and the free edge of the operculum ; the dorsal fin commencing on a vertical line rather before the ventrals, the rays slender, the connecting membrane thin and transparent, the base of the fin one-third shorter than its longest ray ; the distance from the point of the nose to the origin of the pectoral fin, from thence to the origin of the ventral fin, again to the anal fin, and from thence to the end of the fleshy portion of the tail, are four very nearly equal distances ; the tail deeply forked, the outer rays nearly as long again as those of the centre ; all the fins rather long, the rays slender, the connecting membrane thin and transparent ; the lateral line straight from the middle of the base of the tail forward till near the operculum, then suddenly rising to its upper edge ; the scales of the body moderate in size, about ten rows, completing the oblique line of their arrangement between the base of the dorsal and the origin of the ventral fins. The fin-rays in number are—

D. 9 : P. 15 : V. 8 : A. 8 : C. 19. Upper half 10.

The colour of the upper part of the head, back, and sides, olive brown spotted with black ; irides orange red, pupil large and dark ; gill-covers greenish white ; all the under surface of the body white ; pectoral, ventral, and anal fins nearly white, tinged with brown ; dorsal fin and tail pale brown spotted with darker brown.

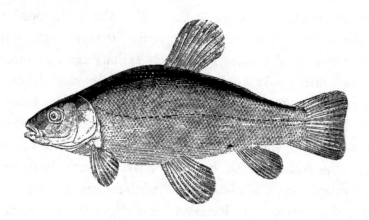

THE TENCH.

Tinca vulgaris, Cuvier, Règne An. t. ii. p. 273.
   ,,   ,,   *Tench,* Flem. Brit. An. p. 186, sp. 61.
   ,,   ,,   *Tinca,* Willughby, p. 251, Q. 5, fig. 1.
Cyprinus tinca, Bloch, pt. i. pl. 14.
   ,,   ,,   *Tench,* Penn. Brit. Zool. vol. iii. p. 474.
   ,,   ,,   ,,   Don. Brit. Fish. pl. 113.

    Generic Characters.—To those common to the Gudgeons (*Gobio*), may be added, that the scales are very small, the mucous secretion on the surface of the body abundant, the barbules or cirri very small.

    The Tench was known to the older writers, but was not held in much estimation. In the present day it inhabits most of the lakes of the European continent. In this country, though frequent in ornamental waters and ponds, it is but sparingly found in the generality of our rivers. There is some doubt whether, like the Carp, its origin be not foreign, and whether those rivers that can now boast of it are not indebted for it to the accidental escape of fish from the preserved waters of neighbouring gentlemen. In rivers it is mostly in those which are slow and deep, that this fish is found, and in such situations it does not appear to be so prolific as in ponds. In deep pits, from which clay for bricks has been dug out, Tench are often abundant; broad shallow waters on muddy bottoms frequently produce great quantities; some very extensive tracts of water a few miles north of Yarmouth in Norfolk, not far inland from a point called Winterton Ness, abound with Tench, which, when removed to stews, feed and thrive on a mixture of greaves and meal till fit for table : their flesh is nutritious and of good flavour.

The Tench appears to decline in numbers in proportion as we proceed northward. In a communication from Carlisle on the subject of fish, obligingly supplied to me by J. C. Heysham, Esq. that gentleman states that the Tench is only now and then taken in the Eden; and occasionally he has known of one being caught in the Solway Frith. A few Tench exist in preserved waters in the neighbourhood of Edinburgh, but they are not very prolific. In a paper by Mr. Whyte, land-surveyor at Mintlaw, which obtained one of the Highland Society's prizes, it is stated, that in some ponds belonging to Mr. Fergusson of Pitfour, in Aberdeenshire, the Tench thrives well; and the Carp, although not very prolific, breeds. This is owing, it is said, to a particular softness in the quality of the water where these fish exist; in fact, it is allowed by Mr. Whyte, in allusion to the Carp-ponds, that they are wholly kept up by rain-water, a very different fluid from that produced by the hard springs of the country.*

In Ireland the Tench is noticed as existing in ponds in the counties of Cork, Dublin, and Kilkenny.

Tench are exceedingly tenacious of life; and experiments have shown that a Tench is able to breathe when the quantity of oxygen is reduced to the five-thousandth part of the bulk of the water : ordinary river water generally containing

---

* The Art of Angling as practised in Scotland, p. 99.

one per cent. of oxygen. The fact, however, as observed by Dr. Roget, shows the admirable perfection of the organs of this fish, which can extract so minute a quantity of air from water, to which that air adheres with great tenacity. This power is strongly illustrated in the instance about to be quoted, which is selected on account of its reference to other points in the history of the Tench.

" A piece of water which had been ordered to be filled up, and into which wood and rubbish had been thrown for years, was directed to be cleared out. Persons were accordingly employed; and almost choked up by weeds and mud, so little water remained, that no person expected to see any fish, except a few Eels, yet nearly two hundred brace of Tench of all sizes, and as many Perch, were found. After the pond was thought to be quite free, under some roots there seemed to be an animal which was conjectured to be an otter; the place was surrounded, and on opening an entrance among the roots, a Tench was found of most singular form, having literally assumed the shape of the hole, in which he had of course for many years been confined. His length, from eye to fork, was thirty-three inches; his circumference, almost to the tail, was twenty-seven inches; his weight eleven pounds nine ounces and a quarter; the colour was also singular, his belly being that of a char, or vermilion. This extraordinary fish, after having been inspected by many gentlemen, was carefully put into a pond, and at the time the account was written, twelve months afterwards, was alive and well."*

" Tench are said to love foul and weedy, more than clear, water; but situation does not always influence their taste. Tench taken out of Munden Hall Fleet, in Essex, belonging to Mr. Western, which was so thick with weeds that

* Daniel's Rural Sports.

the flew-nets could hardly be sunk through them, and where the mud was intolerably fœtid, and had dyed the fish of its own colour, which was that of ink, yet no Tench could be better grown, or of a sweeter flavour; many were taken that weighed nine, and some ten pounds the brace. In a pond at Leigh's Priory, a quantity of Tench were caught, about three pounds' weight each, of a colour the most clear and beautiful, but when some of them were dressed and brought to table, they smelt and tasted so rankly of a particular weed, that no one could eat them. Some that were conveyed alive and put into other water, soon recovered themselves from this obnoxious taint: an experiment that will always answer in this kind of fish, where it is suspected that there is a necessity for cleansing them; and the circumstance is recited to show that no decisive judgment can be formed from the external appearance of the Tench, however prepossessing it may appear."

As the Tench is one of our most useful fresh-water fishes, from the ease with which it may be preserved and its increase promoted, the facility of transportation from its great tenacity of life, and the goodness of its flesh,—which is not, however, generally held in the estimation which I think it deserves,—as the Tench is also, like the Carp, one of those species first selected as stock for ornamental waters, I venture to recommend that large and fine fish be chosen as breeders, as the most certain mode of obtaining sizeable fish for table in the shortest space of time. Two males to one female, or not less than three to two, should be the proportion of the sexes; and from the pond, which is found by experiment favourable for breeding, the small fish should be in part withdrawn from time to time, and deposited elsewhere to afford more space for all. The male of the Tench is recognised by the large size of the ventral fins, which reach far enough to cover the vent, and are deeply concave internally : in the females the ventral fins are smaller, shorter, and less powerful.

In other fishes, besides occasional external sexual distinctions in particular species, it may be stated as a general law, that in the males the head is sharper and longer ; the latter effect being produced by a greater backward dilatation of the operculum, and the body less deep in proportion to its length than in the females, the abdominal line nearly straight, in accordance with the general law, that in males the respiratory cavity, and in females the abdominal cavity, has the greater proportional size. In measuring the length of the head with reference to the length of the body, the sex causes little or no difference, the female obtaining in depth what is wanting in length; her shorter body and head afford the same comparative proportions.

The Tench spawns about the middle of June, with some variation depending on the season. Willughby says it happens when wheat is in blossom. Such coincident circumstances in the seasonal progress of animals and vegetables particularly deserve to be studied, recorded, and remembered : they may be made subservient to many useful purposes ; one, which has a direct reference to fishing, will serve as an illustration. Some London friends, who are enthusiastic fly-fishers, know exactly when to leave home and find the Mayfly on the water in different counties of England by the flowering of certain shrubs and plants in the neighbourhood of London.

The female Tench, when ready to spawn, is usually attended by two males, who follow her from one bunch of weeds* to another, upon which the ova are deposited ; and

* The broad-leaved pondweed, *Potamogeton natans,* is in some counties called Tench-weed.

so engrossed are they at this time in the fulfilment of the Divine command, that I have frequently dipped out all three fish by a sudden plunge of a landing-net. The ova are very numerous, Bloch says near three hundred thousand in a fish of four pounds' weight. The food of the Tench consists of the various soft-bodied animals which inhabit fresh-water, with some vegetable matter, as the contents of the intestines seem to indicate ; and the best bait for them is the dark red meadow-worm, which they take very readily early in the morning throughout the summer. They are said to bury themselves in soft mud during winter, and certainly move very little in the colder months of the year.

The length of the head, compared to the length of the head and body without including the caudal rays, is as two to seven ; the depth of the body compared to the length of the head and body as one to three ; the head rather large and blunt ; the mouth small, with a very small barbule at each corner ; the tongue short ; the lips flesh colour ; the eyes small, the irides golden yellow ; a row of mucous pores down the preoperculum, and thence taking a direction towards the mouth. The body covered with small scales, about forty-eight, in an oblique row between the base of the dorsal fin and the origin of the ventral. The fin-rays in number are—

D. 10 : P. 17 : V. 10 : A. 9 : C. 19.

The dorsal fin commences about the middle of the body, the first ray half as long as the second, which is one-third longer than the whole base of the fin, and more than half as long as the body is deep ; the front line of the fin straight, the upper and hinder edges rounded ; the pectoral fins large and rounded ; the ventral fins arise in a vertical line before the commencement of the dorsal, and exhibit the sexual indication already noticed ; the anal fin commences half-way

between the origin of the ventral fin and the end of the
fleshy portion of the tail ; the caudal rays not so long as the
head ; the posterior edge of the tail in young specimens
concave, afterwards straight, and finally convex.

The general colour of the body greenish olive gold, lightest along the whole line of the under surface ; the fins darker
brown ; the lateral line elevated, distinct, descending by a
curve from the top of the operculum to the middle of the
body, then passing straight to the centre of the tail.

ABDOMINAL MALACOPTERYGII.   *CYPRINIDÆ.*

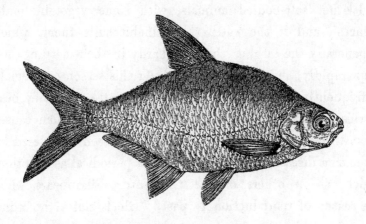

THE BREAM. THE CARP-BREAM.

Abramis brama,   Cuvier, Règne An. t. ii. p. 274.
　　,,　　,,　　Flem. Brit. An. p. 187, sp. 62.
　　,,　　,,　　Jenyns, Syst. Cat. p. 26, sp. 85.
Cyprinus ,,　　Bloch, pt. i. pl. 13.
　　,,　　,,　　Bream, Penn. Brit. Zool. vol. iii. p. 478, pl. 81.
　　,,　　,,　　,,　Don. Brit. Fish. pl. 93.

Generic Characters.—Have not like the Carp or Barbel either strong bony rays or barbules; body deep, compressed; dorsal and abdominal line very convex; the base of the dorsal fin short, placed behind the line of the ventrals; base of the anal fin very long.

The Bream is an inhabitant of many of the lakes and rivers of the continent of Europe generally, even as far north as Norway and Sweden. In this country it appears also to thrive best in large pieces of water, or in the deep and most quiet parts of rivers that run slowly, being found in many counties, and particularly in some of those that contain lakes and canals of considerable extent. The lakes of Cumberland, and some of the most extensive lakes in Ireland, produce large quantities of Bream of great size.

Of the rivers near London producing Bream, the Mole and the Medway are the most noted; it also occurs in the Regent's Canal. Bream swim in shoals, feeding on worms, and other soft-bodied animals, with some vegetable substances; and if the water they inhabit suits them, which is generally the case, as they are hardy in their nature, they grow rapidly, and spawn in May. At this season one female is frequently followed by three or four males, and they bear at this time a whitish tubercle on their scales, which causes them to feel rough to the hand: this has been considered by some as a disease, but is in fact only a periodical assumption, which, as in others of the *Cyprinidæ*, disappears when the season of reproduction is past. Bloch states the number of ova in the female roe at one hundred and thirty thousand.

The flesh of the Bream being generally considered insipid and bony, they are not in great estimation for table, though the breeding of them is cultivated, or rather permitted, as useful to feed Pike, and other voracious fishes. They afford great amusement to the angler; and the more the ground is baited to collect them at a particular spot, the greater the sport. The flesh is in more request on the Continent than in this country, if we may credit the French proverb quoted by Isaac Walton, which says, " he that hath Bream in his pond is able to bid his friend welcome."

It may also be inferred, from a couplet in Chaucer's Prologue to the Canterbury Tales, that the feeding and eating of Bream was more in fashion in the days of Edward the Third than at the present time.

> " Full many a fair partrich hadde he in mewe,
>   And many a Breme and many a Luce* in stewe."

* Luce, a Pike.

Daniel, in his Rural Sports, refers to a pleasant day of Bream-fishing at New Hall Pond in Essex. " The weather was cloudy, and the wind brisk : there were seven rods used by the party, and very frequently there were bites at them all at the same time. When a fish was hooked, and played on the top or near the surface of the water, numbers were seen to follow him, and so soon as the hooks were fresh baited, were alike greedily taken. Some few Perch and Tench were caught, but principally Bream, which averaged at least two pounds a fish ; and of these, from six in the morning till dark in the evening, some hundred weight were taken. The bait used was the large red worm, and the spot had been baited on the morning and evening previous to the day of fishing : the ground-bait used was boiled wheat and tallow-melters' greaves mixed together."

In some of the lakes of Ireland great quantities of Bream are taken, many of them of very large size, sometimes weighing as much as twelve or even fourteen pounds each. A place conveniently situated for the fishing is baited with grains or other coarse food for ten days or a fortnight regularly, after which great sport is usually obtained. The party frequently catch several hundred weight, which are distributed among the poor of the vicinity, who split and dry them with great care to eat with their potatoes. The Bream, as food, is best in season in spring and autumn.

As the fish next to be described after the present Bream is a species of Bream new to the British catalogue, I shall follow the example of Cuvier, in describing the first closely, and when describing the second, to point out more particularly the differential characters.

The whole length of the fish was five inches; a small one was chosen in order to contrast it the better with the specimen of the White Bream which follows : the length

of the head compared with the length of the body was as one
to three ; the head small, the nape depressed ; the diameter
of the eye, compared to the length of the head, as two to
seven, or considerably less than one-third, leaving the space
between the eye and the edge of the preoperculum broad ;
the irides yellow ; the body deep and flat, the dorsal and
abdominal lines very convex ; the head and the fleshy portion of the tail being small and acuminated, produce an
appearance of neatness in shape : the scales of the body
small ; of the two representations of scales forming the
vignette, that on the right hand belongs to this species ; the
number of scales forming in succession the lateral line, about
fifty-six ; the lateral line itself low down on the side, two-
thirds of the space below the dorsal line ; the number of
scales in an upright direction nineteen, of which one punctured scale is on the line itself, with twelve above it, and
six below it ; the first ray of the dorsal fin arises at half the
distance between the point of the nose and the end of the
short central rays of the tail ; the first ray shorter than the
second, the second frequently the longest in the fin, both simple,—that is, not divided or branched,—the third ray nearly
or quite as long as the second, and about twice as long as
the whole base of the fin ; this and all the eight other rays of
this fin branched.　The fin-rays in number are—

D. 11 : P. 17 : V. 9 : A. 29 : C. 19.

The first ray of the pectoral fin the longest and simple,
all the others branched ; the ventral fins placed in a vertical
line in advance of the dorsal fin, the first ray simple, the
others branched ; the anal fin begins on the line of the origin
of the last ray of the dorsal fin, the first ray short, the second
longer, half as long as the third, which is the longest ; these
three rays simple, all the other rays branched, diminishing

in length ; the form of the fin falcate ; the tail long and deeply forked.

The general colour yellowish white, becoming yellowish brown by age, and called Carp-Bream from its colour resembling that of the Carp ; the irides golden yellow ; cheeks and gill-covers silvery white ; the fins light-coloured, the pectoral and ventral fins tinged with red ; the dorsal, anal, and caudal fins tinged with brown.

z 2

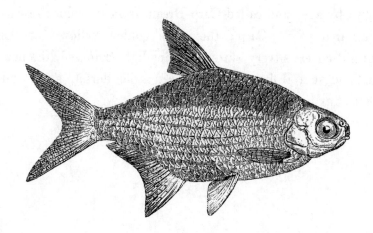

THE WHITE BREAM, OR BREAMFLAT.

Abramis blicca, Cuvier, Règne An. t. ii. p. 274.
   ,,   *latus ?* Gmelin. Jenyns, Syst. Cat. p. 26, sp. 86.
Cyprinus blicca, Bloch, pt. i. pl. 10.

In November 1824, the Rev. Revett Sheppard made a communication to the Linnean Society, of which the following is an extract from the Transactions of that Society, vol. xiv. page 587 :—" There exist in the river Trent, in the neighbourhood of Newark, two species or varieties of Bream. The common one, *Cyprinus brama,* is known there by the name of Carp-Bream, from its yellow colour, and has been taken of nearly eight pounds' weight. The other species or variety, which I believe to be nondescript, never exceeds one pound in weight : it is of a silvery hue, and goes by the name of White Bream."

While investigating the natural history of the county

of Cambridge, the Rev. L. Jenyns has discovered that this second species of Bream inhabits the Cam; and I am indebted to that gentleman for a specimen of the fish, with some further remarks. Referring to the communication of the Rev. Mr. Sheppard, Mr. Jenyns says : " This second species of Bream is very abundant in some parts of the river Cam, where it is called by the fishermen the Bream-flat : it does not attain the size of the Carp-Bream, rarely exceeding ten or twelve inches. It is probably the *Cyprinus blicca* of Bloch ; though I have never seen the fins so red as they are represented in his figure."

I have been very recently favoured by the Rev. Richard Lubbock, of Tombland, near Norwich, with a communication that this fish is also occasionally met with in some of the broads and rivers in that part of the county of Norfolk ; but that it is limited both as to its location and numbers. Its mode of biting, when angled for, is singular: it appears more prone to rise than to descend ; and the float, consequently, instead of being drawn under water, is laid horizontally on the surface, by the attack of the fish on the bait. A specimen of each of the two Bream described in this work, obligingly sent me by Mr. Lubbock from the same locality, differed much less in colour than those previously received from Cambridgeshire. The example of *A. blicca* had two rays less in the anal fin than the fish from the Cam.

That this fish is distinct from the well-known Carp-Bream, the description will evince. Bloch says of this species, that it is very common on the Continent, being found generally in lakes and slow rivers; that it is tenacious of life and breeds fast, but is not in any esteem for table, though encouraged as supplying food for Pike and other voracious fishes.

The length of the head compared to that of the body alone
is as one to three; the head larger, and the fleshy portion
of the tail deeper, than in a Carp-Bream of the same size,
and the fish in consequence has a more bulky and less neat
appearance: the diameter of the eye compared to the length
of the head is as two to six, occupying a full third; it is
therefore larger than in the other Bream, leaving the space
between the eye and the edge of the preoperculum narrower
in comparison, and affecting the proportions of the various
parts of the head : the irides are silvery white tinged with
pink, in the other fish yellow: the nape without any depression, the curve of the back uniform : the scales of the
body larger ; the vignette shows the comparative differences
both in structure and size in two scales, one from the lateral
line of each of two fish of very nearly the same length ; the
colour of the scales bluish white : the lateral line not quite
so low down on the body as in the Carp-Bream ; the number
of scales forming the lateral line about fifty; the number
up the side fifteen, of which nine are above the punctured
scale forming the line itself, and five below it : the character
and relative position of the fins not very dissimilar, except
that the dorsal fin in the White Bream, the subject of the
present article, begins rather nearer the head ; but the number of the rays in the pectoral and anal fins differ considerably, the pectoral fin of the White Bream having three
rays, and the anal fin five rays, less in number than the
same fins in the Carp-Bream. The fin-rays in number
are—

D. 10 : P. 14 : V. 9 : A. 22 : C. 19.

The general colour of the sides is silvery bluish white,
without any of the yellow golden lustre observable in the
last species; the irides silvery white, tinged with pink,

as before stated, which in the Carp-Bream are yellow; the fins in my specimen, which has been some time in spirit, are dusky blue, particularly the dorsal, anal, and caudal fins; the pectoral and ventral fins tinged with red.

The vignette below represents a view of the Thames, looking eastward, from Windmill Hill, near Gravesend.

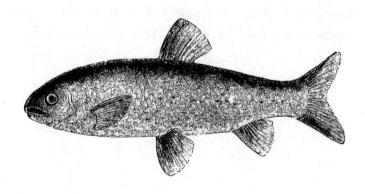

THE IDE.

Leuciscus idus,   Cuvier, Règne An. t. ii. p. 275.
   ,,   ,,   Jenyns, Syst. Cat. p. 26.
Cyprinus ,,   Linnæus. Bloch, pt. i. pl. 36.

Generic Characters.—The anal fin short, as well as the dorsal, but without strong rays at the commencement of either ; no barbules at the mouth.

The species of the first section of this sub-genus, instituted by Klein, have the dorsal fin over the ventral fins.

According to Mr. Stewart, this species was taken at the mouth of the Nith by the late Dr. Walker ; and a figure from Meidinger's plate, which Cuvier considers a better representation than that in Bloch's work, is introduced here, with a short description to assist in identifying this species at any future time.

Bloch says this fish is distinguished by the bulky character of the body, and by the anal fin having thirteen rays. It was first described by Gesner, and is found in Pomerania, Westphalia, Sweden, Norway, Denmark, and Russia. It is an inhabitant of large rocky lakes, from which it seeks its way into rivers in the months of April or May, to deposit its spawn in running water among stones. It feeds on worms and herbage, like the other species of the genus, but grows slowly. The flesh is said to be white, tender, and of good flavour.

The head is large, and appears somewhat truncated ; the muzzle blunt ; the mouth small, without teeth, as is the case also with the other species of this extensive family ; the upper jaw rather the longest ; the eye of moderate size ; the dorsal line convex ; abdominal line almost straight ; the scales of the body large ; the lateral line curved in its descent from the upper edge of the operculum to the centre of the body : the fin-rays in number are—

D. 10 : P. 17 : V. 11 : A. 13 : C. 19 : Vertebræ 41.

In colour the irides are straw yellow, the pupils black ; forehead, nape, and back, very dark bluish black ; the sides bluish grey ; the belly white : pectoral fin orange ; ventrals immediately under the dorsal fin, red in the middle, the first and last rays white ; base of the anal fin white, the other part red ; dorsal fin and tail grey, all the rays branched.

THE DOBULE ROACH.

Leuciscus dobula,   Cuvier, Règne An. t. ii. p. 275.
    ,,   ,,   Yarrell, Linn. Trans. vol. xvii. pt. 1, p. 9.
    ,,   ,,   Jenyns, Syst. Cat. p. 26, sp. 88.
Cyprinus   ,,   Linnæus. Bloch, pt. i. pl. 5.

WHILE fishing in the month of August 1831, in the
Thames below Woolwich, with the mouth of a White-bait
net open against a strong flood-tide, I caught a single specimen of the fish above-named, but have not been fortunate
enough to obtain any more since.

This species inhabits the Oder, the Elbe, the Weser,
and the Rhine, as well as the smaller streams that fall into
them. Like the *Idus,* last noticed, it inhabits also large
lakes, and seeks to enter rivers from March till May for the
purpose of spawning. The larger and older fish spawn earlier than the younger ones, which exhibit small dark spots
on the body and fins at this particular season. The flesh
of the Dobule is white, but full of bones, and little esteemed
for table : they die soon when taken out of the water, and
their food is worms, snails, and small dark-coloured leeches.

The specimen taken was a young male fish of six and a half inches long; the body slender in proportion to its length; the head compared to the length of the head and body alone, without the caudal rays, is as two to nine; the depth of the body equal to the length of the head; the diameter of the eye compared to the length of the head is as two to seven; the nose rather rounded; the upper jaw the longest; the ascending line of the nape and back more convex than any other portion of the dorsal or abdominal line: the first ray of the dorsal fin arising half-way between the anterior edge of the orbit of the eye, and the edge of the fleshy portion of the tail; the first ray half as long as the second, which is the longest, and is as long again as the last ray of this fin, the length of the last ray being equal to the length of the base of the fin: the pectoral fin rather long and narrow; the ventral fins arise just in advance of the line of the origin of the first ray of the dorsal fin; the distance from the origin of the ventrals to the origin of the anal fin, and from the origin of the last ray of the anal fin to the end of the fleshy portion of the tail, are equal; the first ray of the anal fin nearly as long again as the last; the tail considerably forked, the external rays being as long again as those in the centre. The scales of the body moderate in size, fifty forming the lateral line, with an oblique row of seven scales above it under the dorsal fin, and four below it; the lateral line itself concave to the dorsal line throughout its whole length. The fin-rays are in this specimen—

D. 9 : P. 16 : V. 9 : A. 10 : C. 19.

The colour of the top of the head, nape, and back, dusky blue, becoming brighter on the sides, and passing into silvery white on the belly; dorsal and caudal fins dusky brown; pectoral, ventral, and anal fins, pale orange red; the irides orange; the cheeks and operculum silvery white.

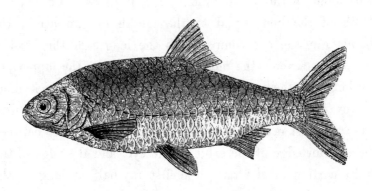

THE ROACH.

Leuciscus rutilus,	Cuvier, Règne An. t. ii. p. 275.	
,, ,,	Flem. Brit. An. p. 188, sp. 65.	
Cyprinus ,,	Linnæus. Bloch, pt. i. pl. 2.	
,, ,,	*Roach,* Penn. Brit. Zool. vol. iii. p. 482.	
,, ,,	,, Don. Brit. Fish. pl. 67.	

THE ROACH is said to be abundant in almost all the rivers throughout the temperate parts of Europe, and in this country appears to be a very common fish, inhabiting most of our rivers, but preferring those that are slow in their course, frequenting the deepest parts by day, and by night feeding on the shallows. A specimen sent to me from Scotland by Sir William Jardine, Bart. was rather shorter and deeper than the Roach of the South. The Rev. David Ure, in a statistical account, when describing the Roach in the parish of Killearn, says, " Vast shoals come up from Loch Lomond, and by nets are caught by thousands : their emigrations from the loch, however, are only for the space of three or four days about the end of May." Mr. Donovan, in his History of the British Fishes, says, " In the river Thames the finest

Roach are caught about the middle of May or early in June, when those fish come up in shoals from the sea to deposit their spawn in the higher parts of the river :" but the Roach in this instance come from the direction only in which the sea lies,—not, I apprehend, from the sea itself : the attempt to gain a higher station in the river, where the oxygen is in greater quantity, is analogous to the movement previously noticed as occurring in Loch Lomond, and also in the allied species, *L. idus* and *dobula*, previously described ; but I have never known a Roach to be taken in the sea into which the fish had entered voluntarily. Montagu, in his MS. referring to Mr. Donovan's statement of this migration from the sea, states his belief that Mr. Donovan was mistaken, and expresses also his belief that the Roach could not exist in sea-water at all ; quoting the following fact which came under his own observation :—In a small river that runs into a large piece of water of nearly two miles in extent, close to the sea, on the south coast of Devon, there is no outlet but by means of percolation through the shingle that forms the barrier between it and the sea : in this situation Roach thrive and multiply beyond all example. About eight or nine years ago the sea broke its boundary, and flowed copiously into the lake at every tide for a considerable time, by which every species of fish were destroyed.

The fish of Lough Neagh, in Ireland, called a Roach, is in reality the Rudd, or Red-eye, *Cyprinus erythropthalmus* of authors, to be hereafter described—a fish belonging to the second division of the genus *Leuciscus* of Klein, which has the dorsal fin over the space between the ventral and anal fins : the Roach has the dorsal fin more forward on the body, and over the ventral fin, not over the space behind it. I may here mention, that the representation of the fish at the bottom of the title-page of the third volume of Pennant's

British Zoology is that of a Rudd or Red-eye, and not that
of a Roach, as stated; which the position of the dorsal fin,
and comparison with the figure of the Rudd, plate 83, page
479, will sufficiently testify.

Roach are gregarious, swimming constantly in large shoals,
and feeding on worms and herbs.   Pennant refers to a
Roach of five pounds' weight.   Mr. Jesse says the largest
he has known to be caught in the Thames weighed three
pounds; and Walton considered a Roach of two pounds
worthy of particular mention.   Mr. Jesse says of the Roach
detained in his vivarium, that he has seen a Carp swim
among a shoal of them without producing the least disturb-
ance; but if a Pike went near them, they made off rapidly
in all directions.   The Roach spawns at the end of May or
the beginning of June, and the scales are then rough to the
touch.   It is in little estimation generally for table, but is
best as food, as well as finest in colour, in October,—a state
produced, probably, by the variety as well as quantity of
nutriment obtained during a long summer; it is in this
month that it is most sought after by the Thames anglers.

"As sound as a Roach," is a proverb that does not carry with
it the degree of conviction that usually attaches to a popular
apophthegm.   It must, however, be remembered, that in the
older ichthyological works this fish was called *Roche*—a term
probably derived from the French.   The meaning stands
confessed, if we admit the pun upon the word; and we ought
then to read, "as sound as a rock."

The French connect the same idea of haleness with the
Ide, a fish previously described, which is known to them by
the name of *Gardon.*   The English say also, "as sound as
a Trout;" and the Italians connect the idea of health with
fish generally, *è sano como il pésce.*   The Roach was first
described by Rondelctius.

The length of the head compared with the whole length of the head, body, and tail, including the rays, is as one to five ; the depth of the body at the commencement of the dorsal fin is to the whole length of the body alone, without the head or caudal fin-rays, as two to five ; the muzzle rather sharp ; the mouth small ; the nostrils double, both pierced in a circular depression, but little in advance of the anterior superior edge of the orbit ; the diameter of the eye equal to one-fourth of the whole length of the head, and occupying the second fourth portion ; the nape and back rising suddenly ; the dorsal line much more convex than that of the abdomen : scales rather large, marked with concentric and radiating lines ; the number of punctured scales forming the lateral line forty-three ; the oblique line from the base of the dorsal fin down to the scale on the lateral line contains seven scales ; below the lateral line to the origin of the ventral fin, three scales ; the lateral line falls by a curve from the upper part of the operculum below the middle of the body, and from thence nearly straight to the tail.

The first ray of the dorsal fin arises exactly half-way between the point of the nose and the end of the fleshy portion of the tail ; the first ray short, the second the longest in the fin ; both rays simple, all the others diminishing in length and branched ; the sixth ray as long as the base of the fin : the upper ray of the pectoral fin the longest and simple, all the others branched ; the length of the fin equal to the distance from the front of the eye to the free edge of the operculum : the ventral fins arise, on a vertical line, directly under the first ray of the dorsal fin ; the upper ray the longest and simple, the others branched : the anal fin commences on a line with the ends of the rays of the dorsal fin when folded down, the first ray short, the second ray the longest, both simple, the rest branched ; the tail deeply forked, the central rays scarcely half as long as the outer rays. The fin-rays in number are—

D. 12 : P. 17. : V. 9. : A. 13 : C. 19.

The colour of the upper part of the head and back dusky green with blue reflections, becoming lighter on the sides, and passing into silvery white on the belly; the irides yellow; cheeks and gill-covers silvery white; dorsal and caudal fins pale brown tinged with red; pectoral fins orange red; ventrals and anal fins bright red.

ABDOMINAL
MALACOPTERYGII.

CYPRINIDÆ.

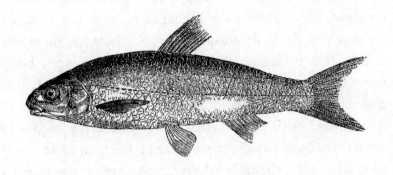

THE DACE, DARE, OR DART.

Leuciscus vulgaris, Cuvier, Règne An. t. ii. p. 275.
    ,,   ,,   *Dace,* Flem. Brit. An. p. 187, sp. 63.
Cyprinus leuciscus, Linnæus. Bloch, pt. iii. pl. 97.
    ,,   ,,   *Dace,* Penn. Brit. Zool. vol. iii. p. 483.
    ,,   ,,   ,,   Don. Brit. Fish. pl. 77.

THE DACE and the Roach are somewhat allied in their habits, and a little so in their appearance; but the former is not so plentiful as the latter, nor is it so generally dispersed, being comparatively more local. The Dace inhabits Italy, France and Germany, and in this country is found in the deep and clear water of quiet streams. Its food is worms and other soft substances; but, like the Trout, it will occasionally rise at an artificial fly, and it is frequently taken by fly-fishers while whipping for that fish. The Dace is gregarious, swimming in shoals, and spawning in June. The flesh is considered preferable to that of the Roach, but is not generally in much estimation: it seldom exceeds nine or ten inches in length. The Dace is frequently used as bait for Pike in trolling, on account of its

VOL. I.

2 A

silvery brightness ; but where live bait are required, as for night hooks, Roach are preferable, on account of their being more tenacious of life.

The length of the head, compared with the length of the head and body, without the caudal rays, is as two to nine ; the depth of the body compared to the whole length, as one to five ; the muzzle pointed ; the mouth rather large, being more deeply cut than in a Roach of the same size ; the nostrils very similar ; the eye not so large, the diameter of it compared with the length of the head being as two to seven ; the back but slightly elevated ; the form of the body elongated and elegantly shaped ; the scales considerably smaller than those of the Roach when the two fishes are of the same size ; the number of scales composing the lateral line fifty-two, on an oblique line above it eight, and descending from the line to the origin of the ventral fin four.

The dorsal fin commences rather farther back than in the Roach, the first ray being behind the middle of the body ; the ventral fins rather in advance of the line of the commencement of the dorsal fin : in other respects the fins in these two species are very similar. The rays in number are—

D. 9 : P. 16 : V. 9 : A. 10 : C. 19.

The colour of the upper part of the head and back dusky blue, becoming paler on the sides and white on the belly ; the irides straw yellow ; cheek and gill-covers silvery white ; dorsal and caudal fins pale brown ; pectoral, ventral, and anal fins almost white, tinged with pale red.

ABDOMINAL
MALACOPTERYGII.   *CYPRINIDÆ.*

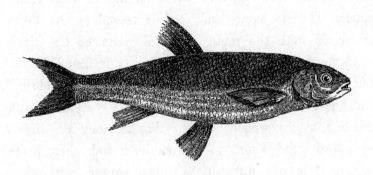

THE GRAINING.

Leuciscus Lancastriensis, Yarrell, Linn. Trans. vol. xvii. pt. 1, p. 5.
Cyprinus   ,,   *Graining,* Shaw, Gen. Zool. vol. v. p. 234.

Pennant, in his published account of a Tour in Scotland and Voyage to the Hebrides, pp. 11 and 12, has the following observation :—" In the Mersey, near Warrington, and in the river Alt, which runs by Sephton, Lancashire, into the Mersey near Formby, a fish called the Graining is taken, which in some respects resembles the Dace, yet it is a distinct and perhaps new species." A short description of this fish, occupying a few lines only, appears in the quarto edition, as well as in two octavo editions of the British Zoology ; and the Graining is also shortly characterized by Shaw in his General Zoology as already quoted.

One of the streams which produce the Graining rises in Knowsley Park ; and I have by the kindness of Lord Derby, the President of the Zoological Society, been most liberally supplied with specimens of this fish, and have thus been enabled to point out its specific distinctions.

2 a 2

Several streams in the township of Burton Wood and Sankey, which flow into the Mersey below Warrington, and others in or near the township of Knowsley, which also form the Alt, produce the Graining in considerable numbers. In its habits and food it resembles the Trout, frequenting both the rapid and still parts of the rivers, but is not known to exist in ponds. It is fished for with artificial flies, like the Dace or Trout; and Mr. Bainbridge, an enthusiastic fisherman, in his excellent Fly-fisher's Guide, published at Liverpool, says, "that as they rise freely, they afford good sport to the angler; and when in the humour, it is not difficult to fill a pannier with them. They sometimes, though not commonly, exceed half a pound in weight, and are much better eating than the Dace."

The Graining has not, that I am aware, been found in any other locality in this country; but on showing this fish to M. Agassiz, the Ichthyologist, of Neufchatel, he recognized it immediately as a species inhabiting some of the lakes of Switzerland, a detailed account of which will appear in his promised work on the Fishes of Central Europe.

The length of the head compared to the whole length of head, body, and tail, is as one to six; the depth of the body compared to the whole length, as one to five; the nose is more rounded than in the Dace, the upper line of the head being straighter; the eye rather larger; the inferior edge of the preoperculum less angular; the dorsal line less convex: the dorsal fin commencing exactly half-way between the point of the nose and the end of the fleshy portion of the tail; the dorsal fin in the Dace arises behind the middle. The first dorsal fin-ray in the Graining is short, the second ray the longest; the pectoral fins longer in proportion than in the Dace; the ventral fins placed, on a vertical line, but little in advance of the first ray of the dorsal fin; the anal commences, on a vertical line, under the termination of the dorsal fin-rays when that fin is depressed—the first ray is short, the second ray the longest, the last double; the fleshy pórtion of the tail is long and slender, the caudal rays are also long and deeply forked; all the fins a little longer than those of the Dace.

The scales are of moderate size, rather larger than those of the Dace, the diameter across the line of the tube greater, and the radiating lines less numerous, as the vignette representing a scale from the same part of the lateral line in two fishes of equal size will show: the number of scales of the series forming the lateral line forty-eight, those in an oblique line up to the base of the dorsal fin eight, and downwards to the origin of the ventral fins four; the lateral line descends from the upper edge of the operculum by a gentle curve to the middle of the body, and thence to the centre of the tail in a straight line. The fin-rays in number are—

D. 9 : P. 17 : V. 10 : A. 11 : C. 19.

The Graining has the top of the head, the back, and upper part of the sides of a pale drab colour, tinged with bluish red, separated from the lighter coloured inferior parts by a well-defined boundary line; the irides are yellowish white; cheeks and gill-covers shining silvery white, tinged with yellow; all the fins pale yellowish white.

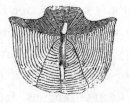 

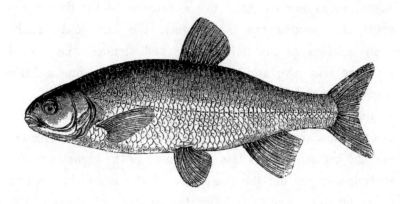

THE CHUB.

THE SKELLY. *Cumberland.*

Leuciscus cephalus,　*Chub,*　FLEM. Brit. An. p.187, sp. 64.
Cyprinus　　　,,　　LINNÆUS. BLOCH, pt. i. pl. 6.
　　,,　　,,　　*Chub,*　PENN. Brit. Zool. vol. iii. p. 485.
　　,,　　*Jeses,*　　,,　　DON. Brit. Fish. pl. 115.

THE CHUB is a well-known fish that is common in the Thames, and many other rivers of England : it is said to be plentiful in the Wye, and other rivers of Wales : it is the Skelly of Cumberland, so called on account of the large size of its scales ; it is also recorded as an inhabitant of the Annan, and other rivers in the south of Scotland.

In its nature the Chub is timid, frequenting deep holes in the more quiet parts of the sides of the stream, and sheltering itself generally under or near any bush or tree that will screen it from view. It feeds on worms, and on insects in their various stages ; and is mostly caught by anglers with a fly or other bait at the surface of the water, by a mode of fishing called dibbing. The Thames anglers for Chub

" cause themselves to be landed on an ait in the river, from
the banks of which, under the shelter of the willows, with a
long rod, a short line, and a lively cockchafer, they often
have good sport." The insect used as bait, whatever it
happens to be, whether large moth or cockchafer, hangs
pendent from the point of the rod, just touching the water;
and by repeatedly but gently tapping the butt-end of the
rod, the bait is moved in a manner exactly resembling the
struggles of a living insect that had by accident fallen into
the water.

Mr. Jesse says of the Chub, detained with other fish
where their actions could be noticed, that they were always
restless and shy, but could never resist a cockchafer when
thrown to them.

The Chub spawns about the end of April or the beginning of May, but does not acquire a large size; five pounds'
weight is the most that I can find recorded. It is considered
a coarse fish, and broiling with the scales on is one of the
best modes of preparing it for table.

The length of the head compared to the length of the
head and body to the end of the fleshy portion of the tail is
as one to four; the depth of the body rather greater than
the length of the head; the mouth large; the head wide
or chubby; the nostrils in a circular depression, one-third
nearer the eye than the end of the nose; the diameter of the
eye equal to one-fifth of the length of the whole head : the
scales on the body large; the number forming the lateral
line forty-four, with an oblique line of six scales above it to
the base of the dorsal fin, and of three scales below it to the
origin of the ventral fin; the lateral line descending by a
gentle curve till even with the end of the pectoral fin-rays,
then straight to its end.

The dorsal fin commences half-way between the point of

the nose and the end of the fleshy portion of the tail; the first ray short, the second the longest, both simple; the others branched, diminishing in length gradually to the last, which is the shortest; the seventh ray as long as the base of the dorsal fin: the pectoral fin rather small; the ventral fin arising on the same vertical plane as the dorsal; the anal fin large, commencing in a line with the end of the dorsal fin when its rays are pressed: the tail large and forked. The fin-rays in number are—

D. 10 : P. 16 : V. 9 : A. 11 : C. 19.

The colour of the top of the head blackish brown, with a streak of the same dark colour passing down behind the free edge of each operculum as far as the origin of the pectoral fin: the whole of the upper part of the back bluish black, the edge of each scale the darkest part; the sides bluish white, passing into silvery white on the belly: the dorsal and caudal fins dusky; the pectoral fins reddish brown; the ventral and anal fins reddish white; the irides golden yellow, the upper part dusky; cheeks and gill-covers rich golden yellow.

The fish described was a male of thirteen inches in length, and the milt appeared on pressure; the season backward. May 9th, 1835.

ABDOMINAL
MALACOPTERYGII.   CYPRINIDÆ.

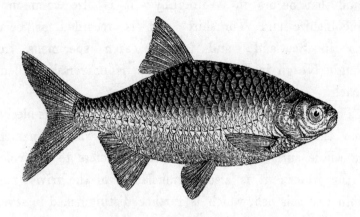

THE RED-EYE. RUDD.

ROUD. *Norfolk*. — FINSCALE. SHALLOW.

Leuciscus erythropthalmus,	Cuvier, Règne An. t. ii. p. 276.	
,, ,,	*Red-eye,* Flem. Brit. An. p. 188, sp. 66.	
Cyprinus ,,	,, Linnæus. Bloch, pt. i. pl. 1.	
,, ,,	*Rudd* and *Shallow,* Penn. Brit. Zool. vol. iii.	
	p. 479, pl. 83.	
,, ,,	*Red-eye,* Don. Brit. Fish. pl. 40.	
,, ,,	,, *Rudd, Roud,* and *Finscale,* Willughby,	
	249 & 252, Q. 3, f. 1.	

The species belonging to the second division of the genus *Leuciscus* of Klein have the dorsal fin placed so far behind the line of the ventrals as to bring it over the space between the ventral and anal fins.

To this second division belong four British species, the largest of which, the Rudd, or Red-eye, is a very common fish in Europe, as well as in various localities in this country. It is found in the Thames, and other waters near London; and I have seen some dozens together for sale in Hungerford fish-market. It is found in Oxfordshire and Warwickshire. It is the Shallow of the Cam and the Lode, two rivers in Cambridgeshire. It is abundant in the broads of Norfolk, where it is called Roud, a name that occurs in Willughby: it is also common in Lincolnshire and Yorkshire. It is recorded as occurring in Scotland; and I have seen specimens from Lough Neagh in Ireland, where it is universally called a Roach.

The name of Rudd attached to this species is derived from the prevailing golden coppery tint which ornaments the whole surface: the term Red-eye refers to the colour of the irides; it is also a translation of the trivial name applied to this fish, which is further distinguished in several countries of Europe by names that have reference to the red colour of its scales or its eyes.

The Rudd, in addition to its vivid colours, is also tenacious of life, and is on that account preferred by trollers as a bait for Pike. It breeds freely without requiring any care to be bestowed upon it, and is therefore useful as food for large Perch, Trout, or Pike. It is said to be a much better fish to eat than the Roach, but does not attain more than two pounds' weight. The food of the Rudd is worms, molluscous animals, and insects, with some vegetable matter: it spawns in April, or early in May, on or about aquatic plants, and the scales at this period are rough to the hand.

The length of the head compared to the length of the head and body is as two to seven: if measured to the forked centre of the caudal rays, as one to four: the depth of the body is to the same length as one to three: the head small; the nose rather blunt; the diameter of the eye one-fourth of the length of the whole head; nostrils in a circular depression half-way between the point of the

nose and the anterior edge of the orbit ; the nape and
back rise suddenly, the whole dorsal line very convex ;
the fleshy portion of the tail narrow ; abdominal line also
very convex, the depth of the body decreasing suddenly
from the commencement of the anal fin. The scales large ;
the number in the series forming the lateral line about
forty, in an oblique line ascending to the dorsal fin seven,
and descending to the line of the ventral fin four; the
scales having numerous concentric striæ, and two or three
radiating lines. The dorsal fin is placed very far back ;
the first ray arises half-way between the point of the nose
and the end of the short central caudal rays ; the base
of the whole dorsal fin over the space between the ventral and anal fins; the base of the dorsal fin equal to
the length of the sixth ray. From the point of the nose
to the commencement of the pectoral fin, from thence
to the origin of the ventral fin, and thence to the anal
aperture, are three very nearly equal distances; the anal
fin commences in a vertical line but little behind the
origin of the last ray of the dorsal fin, the base of the
fin equal to the length of the second or longest ray ;
the caudal rays rather long, the longest as long again
as the central short rays. The fin-rays in number are—

D. 10 : P. 15 : V. 9 : A. 13 : C. 19.

The Irish specimens of Rudd from Lough Neagh had
one ray more in the dorsal and anal fins.

The irides are orange red ; the cheeks and gill-covers
golden yellow ; upper part of the back brown, tinged
with green and blue ; the sides more pale ; the belly light
golden yellow ; the whole surface of the body tinged with
a brilliant reddish golden hue, varying when viewed in
different positions in reference to the light, which it is

difficult to name correctly ; the fins more or less bright cinnabar red, particularly in those specimens which I have seen from the Thames, Cambridgeshire, and Lough Neagh ; dorsal and caudal fins not so bright in colour as the fins of the under surface, but more inclining to reddish brown.

Walton says, " There is a kind of bastard small Roach, that breeds in ponds, with a very forked tail, and of a very small size ; which some say is bred by the Bream and right Roach ; and some ponds are stored with these beyond belief ; and knowing men that know their difference call them Ruds : they differ from the true Roach as much as a Herring from a Pilchard. And these bastard breed of Roach are now scattered in many rivers ; but I think not in the Thames." Under the account of the Bream, he adds— " Some say that Bream and Roaches will mix their eggs and melt together ; and so there is in many places a bastard breed of Breams, that never come to be either large or good, but very numerous."

It is probable that the fishes here alluded to were the true Rudd, and the second species of Bream, which have been already described; and an opinion apparently prevailed, notwithstanding the numbers in which they existed, that they were hybrids. The instances in which animals in a truly unlimited natural state make selections beyond their own species are probably very rare. Hybrids and permanent varieties are the consequence of restriction and domestication, and I confess my doubts of the existence of hybrid fishes.

THE AZURINE.

BLUE ROACH. *Lancashire.*

Leuciscus cœruleus, YARRELL, Linn. Trans. vol. xvii. pt. 1, p. 8.

AT the time I was favoured by Lord Derby with specimens of the Graining, which has been already noticed,
his Lordship also sent me examples of another fish, known
provincially by the name of the Blue Roach ; which is not
only new to our British catalogue, but which, like the
Graining, is not described, as far as I have been able to
ascertain, in any of the different works of European ichthyologists. M. Agassiz, however, assured me that this fish,
like the Graining, is an inhabitant of some of the Swiss
lakes, and will be described in his forthcoming work already
referred to.

The localities from which this species is obtained, within
the township of Knowsley, are but limited. It is hardy,
tenacious of life, and spawns in May. The flesh is said
to be firm, of good flavour, and to resemble that of the

Perch. The food, and the baits used for its capture, are the same as those taken by the Carp ; and the largest specimen known was not supposed to exceed one pound in weight.

The depth of this fish is to its length as two to seven ; and it is therefore in shape something similar to the Rudd, but is at once distinguished from that species by the slate blue colour of the back, the silvery whiteness of the abdomen, and also by its white fins, which in the Rudd are of a fine vermilion, or cinnabar red. The nose is blunt ; the mouth small ; the nostrils pierced on the upper surface of the nose, midway between the eye and the upper lip ; the eye moderate in size ; the whole head small, depressed ; the back arched, the abdomen also convex : the scales rather large, and differing in the number of concentric and radiating striæ from those of the Rudd ; the number of punctured scales forming the lateral line about forty-two, in an oblique line from thence to the base of the dorsal fin seven, and downwards to the origin of the ventral fin three : the lateral line, descending rapidly from the upper edge of the operculum, takes a curve parallel to the deep convex line of the abdomen. The fin-rays in number are—

D. 10 : P. 15 : V. 9 : A. 12 : C. 19.

The dorsal fin commences half-way between the eye and the end of the fleshy portion of the tail ; the first ray is short, the second ray the longest, the last ray double ; the base of the fin equal to the length of the seventh ray : the pectoral fin rather long, reaching nearly to the origin of the ventral fins, which arise in a vertical line considerably in advance of the dorsal fin, and thus bring that fin over the interval between the ventral and anal fins. From the vent the body diminishes rapidly, and the anal fin is situated

on the obliquity thus produced. The first ray of the anal
fin is short, the second the longest, and as long as the base
of the fin; the last ray double. The fleshy part of the
tail is narrow; the caudal rays forked, the central rays being
only half as long as those which are terminal.

The Azurine has the upper part of the head, back, and
sides of slate blue, passing into silvery white beneath, and
both shining with metallic lustre; the irides white, tinged
with pale straw yellow; all the fins white.

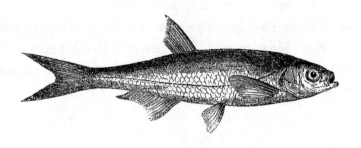

THE BLEAK.　BLICK, *Merrett.*

Leuciscus alburnus,　Cuvier, Règne An. t. ii. p. 276.
　　,,　　　　,,　　*Bleak,*　Flem. Brit. An. p. 188, sp. 67.
Cyprinus　　,,　　Linnæus.　Bloch, pt. i. pl. 8, f. 4.
　　,,　　　　,,　　*Bleak,*　Penn. Brit. Zool. vol. iii. p. 487, pl. 84.
　　,,　　　　,,　　　　,,　　Don. Brit. Fish. pl. 18.

The Bleak is a well-known small species inhabiting
many of the rivers of Europe, and is found in this country
in most, if not all, those which produce the Roach and the
Dace.　The Thames, the Lea, and the New River produce
the Bleak in considerable numbers.　They swim in large
shoals, spawning in May; and at that time the head and
gill-covers are rough to the touch.　Though not of sufficient
consequence to claim any attention as an article of food, or
at all superior as an eatable, the Bleak affords excellent
amusement to young fly-fishers, rising eagerly to almost any
small fly, and sporting incessantly on a fine day at the surface of the water.　Mr. Jesse, in his Gleanings, says : " But
of all the fish confined in a vivarium in Bushy Park, the
Bleak were the most amusing and playful.　Their activity

could not be exceeded; and it gave me much pleasure to
see them, on a still summer's evening, dart at every little
fly that settled on the water near them, appearing always
restless, and yet always happy."

This fish is frequently found to have the intestines partly
occupied by a species of tapeworm. I have taken them
out of much greater length than that of the whole fish
itself; and the name of Mad-Bleak has been bestowed upon
those fish that are seen occasionally swimming in an agitated
uneasy manner on the surface of the water;—an unnatural
action, not observed to occur in other species, and referred to
the annoyance or pain supposed to be produced by these
internal disturbers.

On the inner surface of the scales of Roach, Dace, Bleak,
Whitebait, and other fishes, is found a silvery pigment,
which gives the lustre these scales possess. Advantage has
been taken of the colouring matter thus afforded to imitate
artificially the Oriental pearl. When this practice was most
in fashion, the manufactured ornaments bore the name of
patent pearl, and the use was universal in the bead-trade
for necklaces, eardrops, &c. At present, it seems confined
to ornaments attached to combs, or small beads arranged with
flowers for head-dresses. So great was the demand formerly
at particular times, that the price of a quart measure of fish-
scales has varied from one guinea to five. The Thames
fishermen gave themselves no trouble beyond taking off the
side scales, throwing the fish into the river again; and it
was the custom for hawkers regularly before selling any
white-fish, as they were called, to supply the beadmakers
with the scales.

The method of obtaining and using the colouring matter
was, first carrying off the slime and dirt from the scales by
a run of water; then soaking them for a time, the pigment

VOL. I.   2 B was found at the bottom of the vessel. When thus produced, small glass tubes were dipped in, and the pigment injected into thin blown hollow glass beads of various forms and sizes. These were then spread on sieves, and dried in a current of air. If greater weight and firmness were required, a further injection of wax was necessary. Of this pigment, that obtained from the scales of Roach and Dace was the least valuable; that from the Bleak was in much greater request; but the Whitebait afforded the most delicate and beautiful silver, and obtained the highest price, partly from the prohibitory regulations affecting the capture of this little fish, the difficulty of transmission, and rapid decomposition.

This art of forming artificial pearls is said to have been first practised by the French. Dr. Lister, in his Journey to Paris, says, that when he was in that city, a manufacturer used in one winter thirty hampers of Bleak. Our term Bleak, or Blick, according to Merrett, which has reference to the whiteness of the fish, is derived from a Northern word, which signifies to bleach or whiten.

In a specimen seven inches long, the length of the head compared to the length of the head and body, without including the caudal rays, was as two to nine; the depth of the body compared in the same way was as one to four: but in a younger male specimen of five inches long, the depth of the body was only equal to the length of the head; and both measurements, therefore, were as two to nine. In the large specimen, the body was Dace-like in form and general appearance, but immediately distinguishable by the backward position of the dorsal fin, and the greater length of the base of the anal fin; the body elongated, the abdominal line rather more convex than the line of the back; the nose pointed; the under jaw the longest; the eye rather large:

the scales, beautifully striated, of moderate size, the number making up the series on the lateral line about forty-seven ; above it, to the base of the dorsal fin, six ; and below it, to the origin of the ventral fin, four : the dorsal fin commences half-way between the anterior edge of the eye and the end of the short central rays of the tail ; the first ray is but one-third of the length of the second ray, which is the longest in the fin ; the base of the fin as long as the sixth ray : the anal fin commences, in a vertical line, under the origin of the last ray of the dorsal fin, and occupies half the space between its commencement and the end of the fleshy portion of the tail ; the first ray not half as long as the second, which is the longest ; the base of the fin one-third longer than its longest ray : the caudal rays elongated, and deeply forked. The fin-rays in number are—

D. 10 : P. 17 : V. 9 : A. 18 : C. 19.

The colour of the back is a light greenish or ash brown tinged with blue ; the sides, belly, cheeks, and gill-covers shining silvery white ; the irides silvery, in large sized specimens tinged with yellow ; all the fins nearly white.

2 в 2

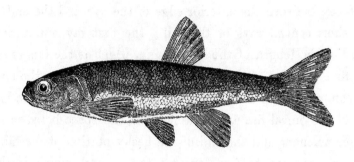

THE MINNOW, MINIM, OR PINK.

Leuciscus phoxinus, Cuvier, Règne An. t. ii. p. 276.
    ,,   ,,   *Minnow,* Flem. Brit. An. p. 188, sp. 68.
Cyprinus   ,,   Linnæus. Bloch, pt. i. pl. 8, fig. 5.
   ,,   ,,   *Minnow,* Penn. Brit. Zool. vol. iii. p. 489.
   ,,   ,,   ,,   Don. Brit. Fish. pl. 60.

THIS very prettily marked species, one of the smallest of
the British *Cyprinidæ,* rarely exceeding three inches in
length, inhabits many of the rivers, brooks, and canals in
England; it is also common in the Waters of Leith : it is
generally found in the same streams with Trout, preferring
gravelly bottoms, and swimming in shoals.   In its habits
this little fish is active and amusing ; many persons have
tried how long they could keep them in confinement, but
none have succeeded in preserving them beyond three years.
Its food consists of aquatic plants, worms, and small portions
of animal substance.   A writer in the fifth volume of Mr.
Loudon's Magazine of Natural History relates that, crossing a
brook, he " saw from the foot-bridge something at the bottom

of the water which had the appearance of a flower. Observing it attentively," he proceeds, " I found that it consisted of a circular assemblage of Minnows: their heads all met in a centre, and their tails diverging at equal distances, and being elevated above their heads, gave them the appearance of a flower half-blown. One was longer than the rest; and as often as a straggler came in sight, he quitted his place to pursue him; and having driven him away, he returned to it again; no other Minnow offering to take it in his absence. This I saw him do several times. The object that had attracted them all was a dead Minnow, which they seemed to be devouring."

The Minnow is very prolific, spawning in June: at that time the head is covered with small tubercles: the young are soon alive, and I have taken them three-quarters of an inch long by the first week of August.

A detailed account of many particulars interesting to the naturalist relating to the habits of this species when spawning will be found in the fifth volume of the Magazine of Natural History, page 681.

A gravelly spot was chosen; each female was attended by two males, one on each side; several ova, nevertheless, it appears, escape fœcundation, which the writer believes takes place at the moment of exclusion: the ova that received the influence of the male were obviously different, when examined under a microscope, from those that were unimpregnated: the membrane forming the investing capsule was more tough, and resisted injury longer, not breaking down so easily when touched: some attempts at artificial impregnation did not succeed. The spawning season with them is short, seldom exceeding two or three days, and the eggs become young fish in a very few days afterwards. The young Minnows are quite transparent, except the eyes, which are large:

in this state the larvæ of the angler's May-fly, and other ephemera, were their greatest enemies, and the diminutive fry seemed to be perfectly aware that they owed their safety to concealment; when exposed they immediately buried themselves again in the gravel.

The Minnow affords amusement to young anglers, biting readily at a small piece of a red-worm: it is principally used as a bait for Pike, Trout, and large Perch. The flesh is considered of good flavour, and when a sufficient quantity can be obtained, for which a small casting-net affords the best chance, they make an excellent fry.

The terms Minnow and Minim are said to be derived from the Latin *minimus*, in reference to the small size of the fish: they are called Pink on account, probably, of the bright red colour that pervades the belly and under parts in summer.

The length of the head compared to the length of the head and body, without including the caudal rays, is as one to four; the depth of the body not quite equal to the length of the head; the body elongated and slender; the dorsal and ventral outline but slightly convex; the surface smooth, covered with numerous minute scales; the lateral line straight from the tail as far as the plane of the origin of the ventral fin, then rising gradually to the upper edge of the operculum. The dorsal fin commences half-way between the anterior edge of the eye and the end of the fleshy portion of the tail: the rays of this fin and those of the anal fin partake of the character of the rays of the *Cyprinidæ* generally, in the first ray being short, the second the longest of the fin, both simple, all the others articulated and branched. The anal fin commences in a line under the origin of the last dorsal fin-ray; the tail rather large and forked, the outer rays being double the length of those in the centre. The fin-rays in number are—

D. 9 : P. 16 : V. 8 : A. 9 : C. 19.

The top of the head and back are a dusky olive, mottled, and lighter in colour on the sides; the belly white, and of a fine rosy or pink tint in summer, varying in intensity according to the vigour of the fish; the irides and gill-covers silvery: dorsal fin pale brown; pectoral, ventral, and anal fins lighter; the tail light brown, with a dark brown spot at the base of the caudal rays.

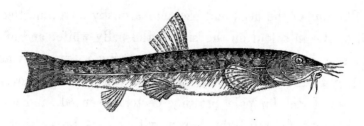

THE LOACH, LOCHE, OR BEARDIE.

Cobitis barbatula, Linnæus. Bloch, pt. i. pl. 31, fig. 3.
  ,,   ,,   *Loche,*   Penn. Brit. Zool. vol. iii. p. 379.
  ,,   ,,   *Bearded Loche,* Don. Brit. Fish. pl. 22.
  ,,   ,,   ,,   ,,   Flem. Brit. An. p. 189, sp. 69.

Generic Characters.—Head small; body elongated, covered with minute scales, invested with a thick mucous secretion; the ventral fins placed far back, in a line under a small dorsal fin; mouth small, without teeth; upper lip furnished with six barbules, four of them in front, and one at each corner; gill-openings small; branchiostegous rays 3.

The Loach is not uncommon in our rivers and brooks; but its habit of lurking under stones often prevents its being observed. Mr. Neill says it is also frequent in the Waters of Leith, and other rivers of Scotland in general; and Dr. Rutty includes it in his Essay towards a Natural History of the County of Dublin. The Loach delights in small, shallow, clear streams, and swims rapidly when disturbed by moving the stone under which it secretes itself. As they are difficult to catch from their slimy smoothness and activity, country boys are in the habit of surrounding any small pool, known to contain some, with a bank of clay, and then making sure of them by lading out the water.

Mr. Booth, in his Analytical Dictionary, considers that our term Loche is derived from the French *locher*, to be uneasy; alluding to the restless habits of the species of this genus, and their almost constantly moving from place to place. They are said to be particularly restless before and during stormy weather, and have been preserved in vessels, like the Leach, as living barometers,* from a notion that certain movements and alterations of position or situation indicated particular changes about to take place in the weather.

The species of this genus are remarkable in having six barbules about the mouth. Fishes thus provided are known to feed at or near the bottom of the water; and it has been stated in this work, at page 22, that those species which reside constantly so near the bottom as to acquire the name of ground-fish, have a low standard of respiration, and a high degree of muscular irritability. In the animals possessing this duration of the power of muscular contractility, as the Eels, flat-fish, and many others, there is reason to believe there exists also great susceptibility of any change that occurs in the electrical relations of the medium in which they reside: the restless movements of Eels and other ground-fish during thunder receive at least a probable explanation in the belief that no alteration in the weather takes place without some previous change in the electrical state of the atmosphere, which, by quality or quantity, may affect the water.

The Chinese, who breed and rear great quantities of

---

* The Lake Loche of the European Continent, *Cobitis fossilis* of authors, is in an old Continental Naturalist's Miscellany called *Thermometrum vivum*.

Gold Carp, find that thunder does them harm, and even sometimes kills them. Pennant says, Lobsters fear thunder, and are apt to cast their claws on a loud clap. These effects may be referred to spasmodic action of the muscles induced by electrical influence. If fishes of opposite habits, such as surface-swimmers and ground-fish, are put together into the same vessel of water, and a slight galvanic discharge passed through the fluid, the ground-fish with the lowest degree of respiration will be the most agitated.

Worms and aquatic insects are the food of the Loach. It spawns in March or early in April, and is very prolific, but seldom exceeds four inches in length. The flesh is accounted excellent; and in some parts of Europe these little fishes are in such high estimation for their exquisite delicacy and flavour, that they are often transported with considerable trouble from the rivers they naturally inhabit to waters contiguous to the estates of the wealthy. Linnæus, in his *Fauna Suecica*, says that Frederick the First, King of Sweden, had them brought from Germany, and naturalized in his own country.

Some peculiarities in the skeleton of the Loach will be pointed out after the description of its external appearance.

The length of the head compared with the length of the body alone is as one to four; the depth of the body is to the length of the head and body, without the caudal rays, as two to eleven; the nose is rounded, pointing downwards; the top of the head flat; the nostrils double, the most anterior tubular, the second pierced in a depression just before the eye; the lips large: the mouth small, placed underneath, the lower jaw the shortest; the form and situation of the mouth very similar to that of the Barbel, with four barbules or cirri over it on the upper lip in the front, and one at each lateral angle: the eye small; the body elongated, smooth, covered with a mucous secretion, rounded in form before the dorsal fin, compressed behind it : the dorsal fin commences half-way between the point of the nose and the end of the fleshy portion of the tail ; the ventral fins under the dorsal ; the anal fin commences half-way between the origin of the ventral fin and the end of the fleshy portion of the tail : the caudal rays slightly rounded. The fin-rays in number are—

D. 9 : P. 12 : V. 7 : A. 6 : C. 19 : vertebræ 36.

The head, body, and sides are clouded and spotted with brown on a yellowish white ground ; the belly and under surface white or yellowish white ; all the fins spotted with dark brown, the dorsal fin and the tail the most so ; the irides blue.

I am indebted to the kindness of Mr. George Daniell for the knowledge of two peculiarities in the structure of the bones of the Loach, which are represented in the vignette, and also for the use of a skeleton to draw and describe from.

Attached to each outer side of the first and second vertebræ is a hollow sphere of bone of equal size, between which, on the upper surface, the vertebræ are distinctly seen ; but the union of the two spheres underneath hides the vertebræ when looked towards from below. These circular bones, which are hollow, and the smooth insides of which can be seen through a horizontally elongated aperture that exists on the outer side of each,—these bones are analogous to the scapulæ, to their outer surfaces the bones of the proximal extremity of the pectoral fins are articulated, and the fin moved by powerful muscles, which assist in producing the rapid motion observable in this little fish. Another peculiarity existing in the upper surface of the head, is the want of union in the two parietal bones at the top; a deficiency which has been noticed by the late Rev. Lansdown Guilding to occur in the *Iguana tuberculata*, or common Guana,* and to which aperture that lamented naturalist applied the term *foramen Homianum*, in honour of Sir Everard Home, observing that the opening did not afford a passage to any nerve or blood-vessel. This peculiarity in the Loach, it will be observed, is another instance of a relation in structure between the fishes and reptiles, some of which have been already adverted to at page 40.

The vignette exhibits a magnified representation, four times larger than the natural size.

* Zoological Journal, vol. i. p. 130.

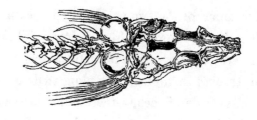

ABDOMINAL
MALACOPTERYGII.   *CYPRINIDÆ.*

THE SPINED LOCHE. GROUNDLING.

Botia tænia,   J. E. GRAY, Zool. Misc. p. 8.
Cobitis ,,   LINNÆUS. BLOCH, pt. i. pl. 31, fig. 2.
   ,,   ,,   BERKENHOUT'S Syn. 3rd edit. vol. i. p. 79.
   ,,   ,,   *Spinous Loche,* PENN. Brit. Zool. vol. iii. p. 381.
   ,,   ,,   *Groundling,* TURTON, Brit. Faun. p. 103, sp. 90.
   ,,   ,,   ,,   FLEM. Brit. An. p. 189, sp. 70.

Generic Characters.—Body ovate, lanceolate, compressed, with small scales; head and operculum naked, with a large spine just behind each nostril; mouth small; nose produced; dorsal fin moderate, medial, opposite the ventral fins; anal fin short.

I HAVE adopted the generic distinction proposed by Mr. J. E. Gray as it applies to one European species, and to the first eight out of the twelve species described by Dr. F. B. Hamilton, in his account of the Fishes of the Ganges, pages 350 to 359. The spine, which is forked and movable, situated behind the nostril and below each eye in the species of the genus *Botia*, is an organic difference formed by the suborbital bone, which distinguishes them from the unarmed species of the old genus *Cobitis*.

The Spined Loche is much more rare than that last described. Berkenhout, in his Synopsis of the Natural His-

tory of Great Britain and Ireland, says it is found in the
Trent, near Nottingham ; Dr. Turton, in his British Fauna, states that it inhabits the clear streams of Wiltshire ; and
the Rev. Leonard Jenyns has found it in the Lode, a small
river in Cambridgeshire, which runs into the Cam. Its
habits are but little known, or have not been distinguished
from those of the more common Loach. It is called Groundling from its habit of lurking under stones in search of
larvæ and insects. Bloch says it spawns in April or May,
and deposits its ova among stones on the bottom. It seldom
exceeds three inches in length. By the kindness of Mr.
Jenyns, I possess two specimens from the Lode, from one
of which the representation, of the natural size, at the head
of this article, was taken. The fin-rays in number are—

D. 8 : P. 9 : V. 7 : A. 6 : C. 15.

The form of the body is still more elongated, slender, and
compressed, than that of the Loach ; the nose more pointed ;
the mouth and the eyes smaller in proportion ; the pectoral
fin longer and narrower ; all the fins occupying the same
relative situation. The colours are similar, both of the
body and fins ; but a row of dark brown spots ranged along
the side are the most conspicuous.

THE PIKE.

PICKERELL. JACK. LUCE. — GEDD. *Scotland.*

Esox lucius, Linnæus. Bloch, pt. i. pl. 32.
   ,,   ,,   *Pike,*   Penn. Brit. Zool. vol. iii. p. 424, pl. 74.
   ,,   ,,   *Brochet,* Cuvier, Règne An. t. ii. p. 282.
   ,,   ,,   *Pike,*   Don. Brit. Fish. pl. 109.
   ,,   ,,   ,,   Flem. Brit. An. p. 184, sp. 55.

Generic Characters.—Head depressed, large, oblong, blunt; jaws, palatine bones, and vomer, furnished with teeth of various sizes; body elongated, rounded on the back; sides compressed, covered with scales; dorsal fin placed very far back, over the anal fin.

The Pike is a well-known inhabitant of the principal rivers and lakes of Europe; and although probably an introduced fish in this country, and for a long time rare, it is now exceedingly common in many of our rivers, and in almost all the lakes and large ornamental waters of England, Scotland, and Ireland.

* The family of the Pikes.

That Pike were rare formerly, may be inferred from the fact that, in the latter part of the thirteenth century, Edward the First, who condescended to regulate the prices of the different sorts of fish then brought to market, that his subjects might not be left to the mercy of the venders, fixed the value of Pike higher than that of fresh Salmon, and more than ten times greater than that of the best Turbot or Cod. In proof of the estimation in which Pike were held in the reign of Edward the Third, I may again refer to the lines of Chaucer, already quoted at page 336. Pikes are mentioned in an Act of the Sixth year of the reign of Richard the Second, 1382, which relates to the forestalling of fish. Pike were dressed in the year 1466, at the great feast given by George Nevil, Archbishop of York. Pike are mentioned in the famous " Boke of St. Albans," in the treatise on the art of fishing with an angle; the first edition of which is said to have been printed at St. Albans in 1481, and again at Westminster, by W. de Worde, in 1496.* Pike were so rare in the reign of Henry the Eighth, that a large one sold for double the price of a house-lamb in February, and a Pickerel, or small Pike, for more than a fat capon.

The Pike is strong, fierce, and active; swims rapidly, and occasionally darts along with the rapidity of lightning. The spawn is deposited among weeds in March or early in April ; and at this season the spawning fish will be found in narrow creeks or ditches that are connected with the larger waters they at other times inhabit.

The Rev. Revett Sheppard has noticed " an annual migration of Pikes which takes place in spring in the

---

* At the sale of the library of the late Duke of Roxburgh, an imperfect copy of this edition produced 147*l.*

Cam, " into which river," he says, " they come in great shoals, doubtless from the fens in the neighbourhood of Ely, where they are bred."

Bloch says the young reach the length of eight to ten inches the first year; twelve to fourteen the second; eighteen to twenty inches the third; and there are proofs on record, that from this last size, Pike, if well supplied with food, will grow at the rate of four pounds' weight a year, for six or seven successive years. Rapid growth requires to be sustained by a corresponding proportion of food, and the Pike has always been remarkable for extraordinary voracity. " Eight Pike, of about five pounds' weight each, consumed nearly eight hundred Gudgeons in three weeks; and the appetite of one of these Pike," says Mr. Jesse, " was almost insatiable. One morning I threw to him, one after another, five Roach, each about four inches in length : he swallowed four of them, and kept the fifth in his mouth for about a quarter of an hour, when it also disappeared." Digestion in the Pike goes on very rapidly, and they are therefore most expensive fish to maintain. In default of a sufficient quantity of other fishes to satisfy them, moor-hens, ducks, and indeed any animals of small size, whether alive or dead, are constantly consumed : their boldness and voracity are equally proverbial. Dr. Plot relates, that at Lord Gower's canal at Trentham, a Pike seized the head of a swan as she was feeding under water, and gorged so much of it as killed them both : the servants perceiving the swan with its head under water for a longer time than usual, took the boat, and found both swan and Pike dead. Gesner relates that a Pike in the Rhone seized on the lips of a mule that was brought to water, and that the beast drew the fish out before it could disengage itself. Walton was assured by his friend Mr. Segrave, who kept tame otters, that he had known a Pike, in extreme hunger, fight with one of his otters for a Carp that the otter had caught, and was then bringing out of the water ; and, with the old adage, adds, "it is a hard thing to persuade the belly, because it has no ears." A woman in Poland had her foot seized by a Pike as she was washing clothes in a pond ; and the same thing is said to have happened at Killingworth pond, near Coventry. The present head-keeper of Richmond Park was once washing his hand over the side of a boat in the great pond in that park, when a Pike made a dart at it, and he had but just time to withdraw it. Mr. Jesse adds, " that a gentleman now residing at Weybridge in Surrey, walking one day by the side of the river Wey, near that town, saw a large Pike in a shallow creek. He immediately pulled off his coat, tucked up his shirt-sleeves, and went into the water to intercept the return of the fish to the river, and to endeavour to throw it out upon the bank by getting his hands under it. During this attempt, the Pike, finding he could not make his escape, seized one of the arms of the gentleman, and lacerated it so much that the marks of the wound are still visible."

Pliny considered the Pike as the longest lived, and likely to attain the largest size, of any fresh-water fish. Pennant refers to one that was ninety years old ; but Gesner relates that, in the year 1497, a Pike was taken at Hailbrun in Suabia, with a brazen ring attached to it, on which were these words in Greek characters :—" I am the fish which was first of all put into this lake by the hands of the Governor of the Universe, Frederick the Second, the 5th of October 1230." This fish was therefore two hundred and sixty-seven years old, and was said to have weighed three hundred and fifty pounds. The skeleton, nineteen feet in length, was long preserved at Manheim as a great curiosity in natural history. The lakes of Scotland have produced Pike of fifty-five pounds' weight; and some of the Irish lakes are said to have afforded Pike of seventy pounds: but it is observed, says honest Isaac Walton, "that such old or very great Pikes have in them more of state than goodness; the smaller or middle-sized Pikes being, by the most and choicest palates, observed to be the best meat." The flesh of the Pike is of good quality; and those of the Medway, when feeding on the Smelt, acquire excellent condition, with peculiarly fine flavour. In Lapland, and some other Northern countries of Europe, large quantities of Pike are caught during the spawning season, being then most easily taken, and are dried for future use.

Among the various localities in England remarkable for the quality as well as the quantity of their Pike, Horsea Mere and Heigham Sounds, two large pieces of water in the county of Norfolk, a few miles north of Yarmouth, have been long celebrated. Camden, in his " Britannia," first printed in 1586, says, " Horsey Pike, none like ;" and Horsea Pike still preserve their former good character. I have been favoured, by a gentleman of acknowledged celebrity in field sports, with the returns of four days' Pike-fishing with trimmers — or liggers, as they are provincially called—in March 1834, in the waters just named; viz. on the 11th, at Heigham Sounds, sixty Pike, the weight altogether two hundred and eighty pounds; on the 13th, at Horsea Mere, eighty-nine Pike, three hundred and seventy-nine pounds; on the 18th, again at Horsea Mere, forty-nine Pike, two hundred and thirteen pounds; on the 19th, at Heigham Sounds, fifty-eight Pike, two hundred and sixty-three pounds : together, four days' sport, producing two hundred and fifty-six Pike, weighing altogether eleven hundred and thirty-five pounds. Pike have been killed

2 c 2

in Horsea Mere weighing from twenty-eight to thirty-four pounds each. These meres, or broads, as they are called in Norfolk, are of great extent: Horsea Mere and Heigham Sounds, with the waters connected, are calculated to include a surface of six hundred acres. As the mode of fishing for Pike with liggers on these extensive waters is considered to be peculiar, and affords great diversion, I may state that the ligger or trimmer is a long cylindrical float, made of wood or cork, or rushes tied together at each end: to the middle of this float a string is fixed, in length from eight to fifteen feet; this string is wound round the float except two or three feet, when the trimmer is to be put into the water, and slightly fixed by a notch in the wood or cork, or by putting it between the ends of the rushes. The bait is fixed on the hook, and the hook fastened to the end of the pendent string, and the whole then dropped into the water. By this arrangement, the bait floats at any required depth, which should have some reference to the temperature of the season; Pike swimming near the surface in fine warm weather, and deeper when it is colder, but generally keeping near its peculiar haunts. When the bait is seized by a Pike, the jerk looses the fastening, and the whole string unwinds; the wood, cork, or rushes, floating at the top, indicating what has occurred. Floats of wood or cork are generally painted in order to render them more distinctly visible on the water to the fishers who pursue their amusement and the liggers in boats. Floats of rushes are preferred to others, as least calculated to excite suspicion in the fish.

The body of the Pike is elongated, nearly uniform in depth from the head to the commencement of the dorsal fin, then becoming narrower; the surface covered with small scales, the lateral line indistinct: the length of the head compared to the whole length of head, body, and tail, as one to four : the dorsal fin, placed very far back, commences in a vertical line over the vent ; the first ray short ; the second and third increasing in length, but shorter than the fourth ; the length of the base of the fin about equal to the length of the longest of its rays : the dorsal and anal fins terminate on the same plane. From the point of the nose to the origin of the pectoral fin, from thence to the origin of the ventral fin, thence to the commencement of the anal fin, and from the vent to the end of the fleshy portion of the tail, are four nearly equal distances : the pectoral and ventral fins small ; the rays of the anal fin elongated, exceeding the length of the base of the fin ; the first three rays shorter than the fourth : caudal rays long and forked. The fin-rays in number are—

D. 19 : P. 14 : V. 10 : A. 17 : C. 19.

The head is elongated, depressed, wide ; gape extensive : the teeth on the vomer small ; those on the palatine bones larger and longer, particularly those on the line of the inner edges ; none on the superior maxillary bones : the lower jaw the longest, with numerous small teeth round the front, the sides with five or six, at a distance from each other, very long and sharp ; the nostrils in a groove at three-fourths of the distance between the point of the nose and the eyes ; the upper surface of the head exhibits various mucous orifices, placed in pairs ; the eyes near the frontal line, and half-way between the point of the nose and the end of the gill-cover ; cheeks and upper part of the operculum covered with scales ; preoperculum and operculum smooth and silvery, closing upon a corresponding smooth, circular, silvery disk. The colour of the head and upper part of the back dusky olive brown, becoming lighter and mottled with green and yellow on the sides, passing into silvery white on the

belly ; pectoral and ventral fins pale brown ; dorsal, anal,
and caudal fins darker brown, mottled with white, yellow,
and dark green ; irides yellow.

The Pike of the fisherman is the Lucie of heraldry, from
the Latin or old French name.

Three silver Pikes in a red field were the arms of the
ancient baronial families of Lucie of Cockermouth and Egremont.   The character of Justice Shallow, it is well known,
was drawn for Sir Thomas Lucy of Charlecote in Warwickshire ; but in the following line,

" They may give the dozen white Lucies in their coat," *

Shakspeare has somewhat amplified the charge ; for the arms
of Lucy, according to the heralds, were, gules crusilly or,
three lucies or pikes hauriant, argent ; numerous instances
of which bearing may be seen in the windows of the hall.

* Merry Wives of Windsor, Act i. scene 1.

ABDOMINAL
MALACOPTERYGII.   *ESOCIDÆ.*

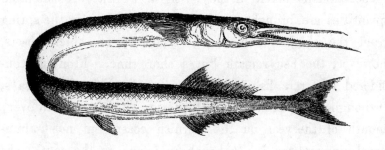

THE GARFISH. SEA-PIKE. MACKEREL-GUIDE.

GREENBONE. HORNFISH. LONG-NOSE. GOREBILL. SEA-NEEDLE.

Belone vulgaris,	Cuvier, Règne An. t. ii. p. 284.	
,, ,,	*Gar.* Flem. Brit. An. p. 184, sp. 56.	
Esox belone,	Linnæus. Bloch, pt. i. pl. 33.	
,, ,,	*Garpike,* Penn. Brit. Zool. vol. iii. p. 429, pl. 74.	
,, ,,	*Sea-Pike, Garfish,* Don. Brit. Fish. pl. 64.	

Generic Characters.—Head and body greatly elongated ; the latter covered with minute scales ; both jaws very much produced, straight, narrow, and pointed ; armed with numerous small teeth : the dorsal fin placed over the anal fin.

The Garfish, included by Linnæus in the genus *Esox*, and thus associated with the true Pike, was called Sea-Pike ; and, on account of its leaving the deep water in spring to spawn near the shore in the months of April or May, preceding the Mackerel in their annual visit to shallow water for the same purpose, it has received also the name of Mackerel-Guide. Other names, and they are not a few, have been suggested and bestowed upon it, either in reference to internal peculiarities or external form.

The Garfish occurs on the coasts of Norway and Sweden, and is mentioned by Pennant in his Arctic Zoology: it is occasionally taken in the Frith of Forth. Considerable quantities are brought to the London markets in the spring from the shores of Kent and Sussex; on which coasts, however, the fish remain but a short time. Montagu considered it a rare fish in Devonshire; but Mr. Couch says, " though considered a fish of passage, it is caught in every month of the year on the Cornish coast, but most abundantly in summer." It has been taken on the south, the east, and the northern shores of Ireland, from Cork to Londonderry. Considerable quantities are eaten in London in the spring; some from curiosity, but the larger portion from the moderate price at which they are sold: the flesh has some of the flavour of Mackerel, but is more dry: the bones are green. Great numbers are said to be caught off the coast of Holland; but they are only used as bait for other fish. I have obtained the young of the year seven inches long in December.

The Garfish, Mr. Couch says, " swims near the surface at all distances from land, and is seen not unfrequently to spring out of its element; its vivacity being such that it will for a long time play about a floating straw, and leap over it many times in succession. When it has taken the hook, it mounts to the surface, often before the fisherman has felt the bite; and there, with its slender body half out of water, it struggles, with the most violent contortions, to wrench the hook from its jaws. It emits a strong smell when newly taken."—The elongated, narrow, beak-like mandibles of this fish make a knowledge of its food a subject of some interest; but I have found only a thick mucus in the stomach, without any remains that I could name. In the works to which I have access, I can find no mention of the nature of its food.

The usual size of this fish is about twenty-four inches; the specimen described measured three inches less. The length from the point of the upper jaw to the end of the operculum, compared to the whole length, was as one to four; the depth of the body compared to the whole length, as one to sixteen: both jaws straight and very much elongated, the under one the most so; the teeth numerous, minute; the eye large, placed at the commencement of the last third portion of the head; the body uniform in depth to the anal fin, then tapering to the tail: dorsal and anal fin beginning and ending nearly on the same plane, the anterior rays of each of these fins longer than the other rays; pectoral fins small, immediately behind the free edge of the operculum; the ventral fins small, situated rather behind the middle of the whole length of the body; vent immediately in advance of the anal fin; the tail forked, the external long rays as long again as those of the centre. The number of fin-rays are—

D. 17 : P. 13 : V. 6 : A. 22 : C. 15.

The upper part of the head and back is of a dark greenish blue Mackerel-like tint, becoming lighter towards the sides, which, with the whole of the belly, are silvery white; irides pale yellow, pupil dark blue; cheeks and operculum brilliant silvery white; dorsal fin and tail greenish brown; pectoral, ventral, and anal fins white.

The great length of the upper jaw is produced by an elongation of the intermaxillary bones: great flexibility is obtained by ligamentous union; the gape is extensive, both jaws separating simultaneously; and this fish probably seizes its prey with quickness and certainty.

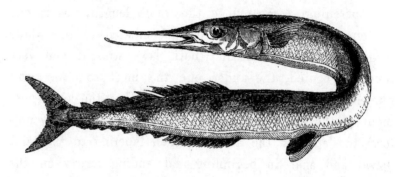

THE SAURY PIKE. SKIPPER.

GOWDNOOK.　*Scotland.*

Scomberesox saurus,　*Saury,*　　Flem. Brit. An. p. 184, sp. 57.
　　　,,　　　　,,　　Cuvier, Règne An. t. ii. p. 285.
Esox　　　　,,　　*Saury Pike,*　Penn. Brit. Zool. vol. iii. p. 430, pl. 75.
　　,,　　　　,,　　*Skipper Pike,*　Don. Brit. Fish. pl. 116.

Generic Characters.—The species of this genus, instituted by Lacépède, have the same structure of the jaws as those of *Belone,* last described ; and are similar also in the form of body and scales, with a keel-like edge to the belly ; but the posterior portions of the dorsal and anal fins are divided, forming finlets, as in the Mackerel.

The Saury Pike, or Skipper, was first described as a British species by Ray: those he saw were taken on the Cornish coast. It was also known to Rondeletius and Gesner; but has not been noticed either by Linnæus, Gmelin, or Bloch. The Rev. Mr. Low, in his Natural History of Orkney, says, that the year preceding that in which he wrote his *Fauna Orcadensis,* such a glut of these fish set into the head of Kerston Bay, that they could be taken

by pailfuls : numbers were caught, and heaps flung ashore.
According to Mr. Neill, the Saury is not at all an uncommon fish in the Frith of Forth, numbers running up with
the flood-tide in the autumn ; but they do not, like other
fishes, retire from the shallows at the ebbing of the tide,
but are then found by hundreds, having their long noses
stuck in the sludge, and are picked up by people from
Kincardine, Alloa, and other places. Mr. Pennant mentions that great numbers of Sauries were thrown ashore at
Leith, by a storm, in November 1768. The Saury has
been taken at Yarmouth on the east, and off Portland Island
on the south ; being, on some occasions, even plentiful in
Cornwall. Mr. Couch in his MS. says—

" The Skipper is more strictly than the Gar-Pike a migratory fish, never being seen in the Channel until the
month of June, and it commonly departs before the end
of autumn. It does not swim deep in the water ; and in
its harmless manners resembles the Flying Fish, as well as
in the persecution it experiences from the ravenous inhabitants of the ocean, and the method it adopts to escape from
their pursuit. It is gregarious, and is sometimes seen to
rise to the surface in large shoals, and flit over a considerable
space. But the most interesting spectacle, and that which
best displays their great agility, is when they are followed
by a company of Porpoises, or their still more active and
persevering enemies the Tunny and Bonito. Multitudes
then mount to the surface, and crowd on each other as they
press forward. When still more closely pursued, they
singly spring to the height of several feet, leap over each
other in singular confusion, and again sink beneath. Still
further urged, they mount again, and rush along the surface
by repeated starts for more than a hundred feet, without
once dipping beneath, or scarcely seeming to touch the

water. At last, the pursuer springs after them, usually across their course; and again they all disappear together. Amidst such multitudes—for more than twenty thousand have been judged to be out of the water together—some must fall a prey to the enemy; but as many hunt in company, it may be long before the pursuit is abandoned. From inspection, we should scarcely judge the fish to be capable of such considerable flights; for the fins, though numerous, are small, and the pectorals far from large—though the angle of their articulation is well fitted to raise the fish by the direction of their motions to the surface; the force of its spring must therefore be chiefly ascribed to the tail and finlets. It rarely takes a bait; and when this has happened, the boat has been under sail, the men fishing with a lask, or slice of Mackerel made to imitate a living bait.* The Skipper has not been commonly taken since the drift fishermen began the practice of sinking their nets a fathom or two beneath the surface—a circumstance which marks the depth to which they swim; but before this, it was usual to take them, sometimes to the amount of a few hundreds, at almost every shoot of the Pilchard nets."

The specimen from which the representation and description here given were taken, measured fourteen inches and three quarters; the head and jaws three inches and three quarters; of this, the narrow portion of the jaws, which curved slightly upwards towards the point, was about equal to the length of the other parts of the head; the lower jaw the longest: the body elongated, but considerably deeper for its length than that of the Garfish; the length of the jaws and head compared to the whole length of the fish, as one to four; the depth of the body two inches, or, as compared to the whole length, as two to seven. Pectoral fins

* See pages 128 and 129.

small ; a keel-like edge, commencing on each side in a line
with the lower edge of the gill-cover, passes the whole length
of the body ; the space between these lines not wider than
one quarter of an inch, except where they dilate a little to
include or pass outside of the ventral fins : the dorsal and
anal fins are placed far back, and commence on the same
plane ; the dorsal fin with five finlets behind it ; the anal
fin with seven finlets behind it : the tail deeply forked; the
two portions divided as far as the posterior edge of a scale-
like appendage, with which the fleshy portion terminates.

The numbers of the fin-rays are—

D. 9. V : P. 13 : V. 6 : A. 11. VII : C. 19.

The cheeks and gill-covers are brilliant silvery white ; the
irides golden yellow, the pupil rather elongated vertically ;
the upper part of the head, and the back, throughout its
whole length, are of a fine dark blue, lighter on the sides
below, and tinged with green ; lower part of the sides and
the belly silvery white ; all the fins dusky brown.

Mr. Couch is, as far as I am aware, the only naturalist
who has recorded a notice of a fish belonging to the genus
Hemiramphus of Cuvier on the British coast ; but the particular species it is not easy now to determine. The head
of a fish of that genus is made the subject of the following
vignette to show its peculiarity, and invite the future investigations of those admirers of the marine productions of
nature who reside on the coast.

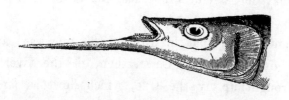

THE FLYING FISH.

Exocetus,	LINNÆUS.
,,	CUVIER, Règne An. t. ii. p. 286.
,, *volitans,*	*Winged Flying Fish,* PENN. Brit. Zool. vol. iii. p. 441, pl. 78.
,, ,,	*Common Flying Fish,* DON. Brit. Fish. pl. 31.
,, ,,	LINN.? JENYNS, Syst. Cat. p. 29, sp. 101.

Generic Characters.—Head and body covered with scales; pectoral fins very large, nearly as long as the body; dorsal fin placed over the anal; upper half of the tail the smallest; both jaws furnished with small teeth.

SEVERAL instances being on record of a species of Flying Fish having been either captured or seen at different parts of our coast, the subject requires to be noticed here; although the exact species, or even whether more than one species may not have occurred, has not as yet been positively decided.

Pennant states, that in June 1765 one was caught at a small distance below Carmarthen, in the river Towy, being brought up by the tide, which flows as far as the town.

Dr. Heysham, in his Catalogue of Cumberland Animals, prefixed to Hutchinson's History of that county, says at page 32—" Another Flying Fish was seen at Allonby, in September 1796, by Mr. Chancellor Carlyle, when he was bathing : it was near the shore, and upon the surface of the water, and came within a yard of him." According to Dr. Fleming, another occurred in July 1823, ten miles from Bridgewater, in the Bristol Channel, a notice of which was communicated to the Linnean Society by the Rev. S. L. Jacob.

The following letter appears in the fortieth number of the Royal Institution Journal, addressed to the editor.

" Sir,

" In going down Channel on the 23rd of August last, with light winds from the E.N.E. inclinable to calm, when off Portland, we were surprised by the appearance of a rather large shoal of what is commonly called the Flying Fish. They were evidently closely pursued by some one of their numerous enemies, from the frequent and long flights which they took ; but it was impossible to discover what that enemy was, though passing close to the vessel. The fact may possibly interest some of your numerous scientific readers. J. C. W."

" *Sunderland, Dec. 2nd,* 1825."

From the MS. of Mr. Couch another instance may be quoted of a Flying Fish " which threw itself on shore on the sandy margin of Helford River, near Falmouth, at full two miles from the open sea, where it was found while yet living. I was informed by Mr. John Fox, of Plymouth, in whose collection this specimen was in 1828, that it measured sixteen inches in extreme length, and that the pectoral fin was eight and a half inches long : a size which caused me to suppose it might be the *E. evolans*, of which I possess a specimen twenty and a half inches long. There can be little doubt that this fish had been chased out of its usual haunt by some one of those voracious inhabitants of the deep by which they are continually persecuted."

In illustration of its habits, Pennant states that the Flying Fish " in its own element is perpetually harassed by the Dorados, and other fishes of prey. If it endeavours to avoid them by having recourse to the air, it either meets its fate from the gulls or the albatross, or is forced down again into the mouth of the inhabitants of the water, who below keep pace with its aerial excursion. Neither is it unfrequent that whole shoals of them fall on board of those ships that navigate the seas of warm climates."

The most recent observations on the habits and powers of the Flying Fish are those by Mr. George Bennett, the author of Wanderings in New South Wales, and other countries, who appears to have devoted particular attention to the subject. " I have never," observes this gentleman, " been able to see any percussion of the pectoral fins during flight ; and the greatest length of time that I have seen these *volatile* fish on the *fin* has been thirty seconds by the watch, and their longest flight mentioned by Captain Hall has been two hundred yards, but he thinks that subsequent observation has extended the space. The most usual height of flight, as seen above the surface of the water, is from two to three feet ; but I have known them come on board at a height of fourteen feet and upwards ; and they have been well ascertained to come into the channels of a line-of-battle ship, which is considered as high as twenty feet and upwards."

" But it must not be supposed they have the power of elevating themselves in the air, after having left their native element; for, on watching them, I have often seen them fall much below the elevation at which they first rose from the water, but never in any one instance could I observe them raise themselves from the height at which they first sprang; for I regard the elevation they take to depend on the power of the first spring or leap they make on leaving their native element."

The writer of the supplementary part to the class Fishes in Mr. Griffith's edition of Cuvier's Animal Kingdom agrees with Mr. George Bennett. He states that the Flying Fishes " rise into the air by thousands at once, and in all possible directions. Their flight, as it is called, carries them fifteen or eighteen feet out of the water: but it is an error to call them Flying Fishes; they do not in reality fly— they only leap into the air, where they have not the power of sustaining themselves at will. They never come forth from the water except after a rapid course of swimming. When put alive into a vessel of sea water, in which there was not sufficient space to acquire momentum, they were only able to rise out of it a few inches. The lines which they traverse when they enjoy full liberty of motion are very low curves, and always in the direction of their previous progress in the water."

The recent observations of both these writers confirm the view taken by Cuvier of the powers of Flying Fishes, as described in the *Règne Animal** of that author; who, using the words *flight* and *wings* figuratively only, says, their flight is never very long, and their wings only serve them as parachutes.

" The Flying Fishes themselves feed on mollusca and any

* " Leur vol n'est jamais bien long; s'élevant pour fuir les poissons voraces, ils retombent bientôt, parce que leurs ailes ne leur servent que de parachutes." —*Règne Animal*, tom. ii. p. 287.

small fish. Their flesh has an agreeable flavour, and is often eaten by mariners on long voyages."

For the reasons before stated, that some doubts exist as to the exact species which have been taken on our coast, no description is attempted, and the attention of Ichthyologists is invited to the subject. The figure is taken from Bloch's representation of *Exocetus volitans.*

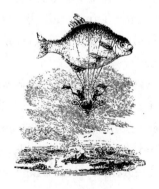

ABDOMINAL
MALACOPTERYGII.   *SILURIDÆ.*

THE SLY SILURUS.

SHEAT FISH.

Silurus glanis, LINNÆUS.
    ,,   ,,   *Sheat Fish,* BLOCH, pt. i. pl. 34.
    ,,   ,,   *Sly Silurus,* STEWART's Nat. Hist. vol. i. p. 354.
    ,,   ,,   JENYNS, Syst. Cat. p. 27.

Generic Characters.—Head large, naked, broad, depressed ; mouth furnished
with barbules or cirri ; lips thick, crenated ; eyes small ; body elongated, compressed, without scales, covered with a copious mucous secretion.

DR. FLEMING, in his History of British Animals, notices
a remark of Sibbald, leading to the conclusion that the
Silurus glanis may have occurred in his day in some of
the Scottish rivers. At the end of his list of river fishes,
he adds *Silurus sive Glanis.* (*Scotia Illustrata,* p. 25.)
A figure and a short description of this fish, derived
from Bloch, is therefore introduced here to enable observers
to identify a specimen should it again occur. It has, however, been suspected that the supposed *Silurus* might have
been the Burbot.

Bloch says, a single fin on the back, and six barbules at the mouth, of which those on the upper lip are the longest, form the distinctive characters of this fish.

The head is shaped like a shovel, flattened, and of a dark green; the mouth very large and wide; the jaws circular, the lower one the longest, both furnished with numerous small incurved teeth; the pharyngeal bones are also furnished with similar teeth. On each side of the upper lip a long barbule; the nostrils are round; placed between these long barbules, and behind them, are the eyes, small, the pupils black, the irides white.

The back is round, of a dark green: the sides, above the lateral line, of the same colour; paler green below it; and the whole body covered with dark spots not very determinate in shape. The body is thick and long; the belly short, expanded, and of a pale yellow colour; the whole of the body covered with slime: the pectoral fins are dark blue at the base and the extremities, the middle portions yellow; the first ray strong, bony, and serrated on the inner surface: the dorsal and ventral fins are yellow at the base, bluish towards the ends, and both placed much nearer the head than the tail; the anal fin is long, the tail rather rounded, both greyish yellow, with violet-coloured edges. The fin-rays in number are—

D. 5 : P. 18 : V. 13 : A. 24 : C. 17.

This fish is found not only in almost all the fresh waters of Europe, but even in those of Africa and Asia. Pliny states that it exists in the Nile. Bloch mentions that examples weighing from seventy to eighty pounds' weight have been taken in canals near Berlin. It has been found of very large size in the Wolga and the Danube, and is included by Nillson and others among the fishes of Norway and Sweden.

It is found also, though rarely, in the sea, and has been taken in the Baltic.

Dr. Smith includes this species of Silurus in his Natural History of the Fishes of Massachusets ;* and mentions that Dr. Flint of Boston, an accurate naturalist, had seen specimens measuring eight feet in length in the Ohio and Mississippi rivers.

The Silurus is represented as sluggish in its habits, and a slow swimmer, taking its prey by lying in wait for it, in a manner somewhat similar to the Angler, *Lophius*, already described ; hiding itself in holes or soft mud, and apparently depending upon the accidental approach of fishes or other animals, of which its long and numerous barbules may be at the same time the source of attraction to the victims, and the means of warning to the devourer. From its own formidable size, it can have but few enemies in the fresh water ; and from them its dark colour, in addition to its habit of secreting itself either in holes or soft mud, would be a sufficient security. In spring, the male and female may be seen together, about the middle of the day, near the banks or edges of the water, but soon return to their usual retreats. The ova when deposited are green ; and the young are excluded between the sixteenth and nineteenth days.

The flesh of the Silurus is white, fat, and agreeable to many persons as food, particularly the part of the fish near the tail ; but on account of its being luscious, soft, and difficult to digest, it is not recommended to those who have weak stomachs. In the Northern countries of Europe, the flesh is preserved by drying, and the fat is used as lard.

The two very elongated barbules of the upper lip are supported by extensions of the intermaxillary bones, which increases the sphere of action and consequent utility of these

* Page 189.

organs of touch, by extending their influence beyond the range of the shorter cirri of the lower lip. Fishes furnished with these oral appendages are known by their habits to be ground-feeders ; and it may be mentioned, as affording additional proof of certain powers or qualities supposed to be possessed by such fishes, that Bloch says of this Silurus, it seldom leaves its hole except during storms. Another writer observes, that it lives long after being taken out of the water, and comes up to the surface on the approach of stormy weather.

Searching Beckwith's enlarged edition of Blount's Tenures, I found those that here follow ; which, as they refer to fishes or fishing, may be considered entitled to a place in this work.

" In the simplicity of older times, when gold and silver were scarce, the household of the king was supported by provisions furnished from his demesnes. By degrees the servants here employed obtained a fixed tenure of the estates, rendering certain services, and supplying certain provisions. Many lands were from time to time granted on condition of yielding such supplies ; but these reservations were small, and many of them only to be rendered when the king travelled into the country where the land lay. In some, special care was taken that he should not make this service burthensome by coming too often.

"*Aylesbury.*—William, son of William of Alesbury, holds three yard-lands of our lord the king in Alesbury, in the county of Bucks, by the serjeanty of paying three Eels to our lord the king, when he should come to Alesbury in winter.

" *Conway Castle*—Is now held of the crown by Owen Holland, Esq. at the annual rent of six shillings and eight pence, and a dish of fish to Lord Hertford as often as he passes through the town.

"*Degemue and Eglosderi, county of Cornwall.*—William Trevelle holds one Cornish acre of land in Degemue and Eglosderi, by the serjeanty of finding one boat and nets for fishing in Hellestone Lake, whensoever our lord the king should come to Hellestone, and so long as he should stay there.

"*Gloucester.*— Pennant states that it has been an old custom for the city of Gloucester annually to present his Majesty with a Lamprey pie, covered with a large raised crust.

"*Rodeley, county of Gloucester.*—Certain tenants of the manor of Rodeley pay to this day, to the lord thereof, a rent called Pridgavel, in duty and acknowledgment to him for their liberty and privilege of fishing for Lampreys in the river Severn. Pridgavel: Prid, for brevity, being the latter syllable of Lamprid, as this fish was anciently called; and gavel, a rent or tribute.

"*Stafford.*—Ralph de Waymer held of the king in fee and inheritance the stew or fish-pond without the eastern gate of the town of Stafford, in this manner, that when the king should please to fish, he was to have the Pikes and Breams; and the said Ralph and his heirs were to have all the other fishes with the Eels coming to the hooks, rendering therefore to the king half a mark at the feast of St. Michael.

"*Yarmouth.*—The town of Yarmouth in Norfolk is bound to send to the sheriffs of Norwich a hundred Herrings, which are to be baked in twenty-four pies or pasties, and thence delivered to the lord of the manor of East Carlton, who is to convey them to the king. They are still sent to the clerk of the kitchen's office at St. James's. In 1778, the sheriffs of Norwich attended with them in person, and claimed the following allowance in return, viz.—' Six white

loaves, six dishes of meat, (out of the king's kitchen) ; one
flaggon of wine ; one flaggon of beer ; one truss of hay ; one
bushell of oats ; one pricket of wax ; six tallow candles.'
But no precedent appearing of these things having been delivered, they were refused.—*Records of the Board of Green
Cloth.*"

The vignette below, with which. this volume concludes,
represents the Fish-market at Newcastle-upon-Tyne.

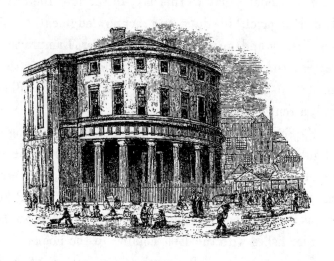

END OF THE FIRST VOLUME.

LONDON:
PRINTED BY SAMUEL BENTLEY,
Dorset Street, Fleet Street.

SUPPLEMENT

TO THE

HISTORY

OF

BRITISH FISHES.

BY

WILLIAM YARRELL, F.L.S. V.P.Z.S.

ILLUSTRATED WITH WOODCUTS.

IN TWO PARTS.

LONDON:

JOHN VAN VOORST, 1, PATERNOSTER ROW.

M.DCCC.XXXIX.

LONDON:
PRINTED BY SAMUEL BENTLEY,
Bangor House, Shoe Lane.

PREFACE TO THE SUPPLEMENT.

On publishing this Supplement to the History of British Fishes, I have only respectfully and very sincerely to return my best thanks to those friends and naturalists, who have, either by their private communications or public announcements, supplied the novelties contained herein.

These additions to the British Catalogue of Fishes are so many gratifying testimonials of the increasing number of observers, whose attention is directed towards the inhabitants of our seas; and I feel a sincere pleasure in the prospect of the many new subjects, and more correct illustrations, which our Ichthyology is likely to derive from the great interest now taken in this branch of Natural History.

To render the pictorial part of this Supplement as useful as its size and character would admit, I have introduced, as vignettes, representations of the bones of the cranium of several well-known fishes, derived from the works of Cuvier, Rosenthall, and others : and should this part of the plan be approved as a worthy mode of occupying a portion of that space usually devoted to lighter subjects, it may, on some future occasion be so enlarged upon as to include an illustration of one cranium in almost all the principal genera. In the present instance, however, not to interfere with the ornamental appearance of these crania, as vignettes, by a repetition of letters or numbers in reference to each particular bone, I have confined the markings to the Perch only, as here introduced, premising, that a little useful perseverance will lead to a knowledge of the analogous bones in other crania.

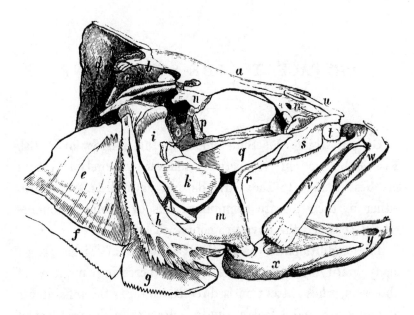

a. Principal frontal bone.
b. Parietal.
c. Inter occipital.
d. Inter parietal.
e. Operculum.
f. Suboperculum.
g. Interoperculum.
h. Preoperculum.
i. Temporal.
k. Tympanal.
l. Sympletic.
m. Jugal.

n. Posterior frontal.
n*. Anterior frontal.
o. Great ala.
p. Sphenoid.
q. Internal pterygoid.
r. Transverse.
s. Palatal bone.
t. Vomer.
u. Nasal.
v. Superior maxillary.
w. Inter maxillary.
x. Articular portion, and

y. Dental portion of the lower jaw, or inferior maxillary bone.

This Supplement is divided into two parts that each separate part may be bound up, if required, with the particular volume to which it more exclusively belongs. All the wood engravings in the Supplement have been executed by **Mr. Vasey.**

Ryder Street, March, 1839.

SUPPLEMENT

TO THE FIRST VOLUME OF THE

HISTORY OF BRITISH FISHES.

ACANTHOPTERYGII.   *PERCIDÆ.*

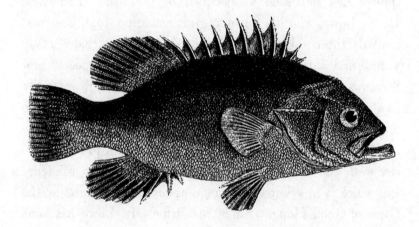

COUCH'S POLYPRION.

Polyprion cernium.

Polyprion cernium,	Cuv. et Val. Hist. des Poiss. t. iii. p. 21, pl. 42.
,, ,,	M. A. Val. Mem. du Mus. t. xi. p. 265, pl. 17.
Amphiprion Americanus,	Schneider, Syst. Ichth. p. 205.
,, *Australe,*	,, ,, pl. 47.
Scorpæna Massiliensis,	Risso, Ichth. p. 184.
	Stone Basse, Couch, Linn. Trans. vol. xiv. p. 81.
Serranus Couchii,	*Couch's Serranus,* Brit. Fish. vol. i. p. 12.

Generic Characters.—A single elongated dorsal fin, the rays of the anterior portion rather short and spinous, those of the secondary portion longer and flexible : branchiostegous rays 7 ; small incurved teeth on the bones of both jaws, on the palatine bones, and on the vomer, with some elongated teeth among the smaller ones ; cheeks, operculum, the whole of the body, the base of the flexible portion of the dorsal and anal fin, and the base of the tail covered with small rigid scales, serrated at the free margin ; suborbital bone, pre-

B

operculum and operculum, below the line of the pectoral fin, denticulated ; operculum, above the line of the pectoral fin traversed by a single strong, horizontal bony ridge, ending in a point directed backwards ; over the eye, over the operculum, and over the origin of the pectoral fin, a semicircular row of short spines ; the first ray of the ventral fin, and the first three rays of the anal fin, furnished also with small short spines.

In the first edition of the History of British Fishes, I ventured to consider the Stone Basse of Mr. Couch, of which that gentleman had favoured me with a drawing, as an undescribed species of the genus *Serranus* of Cuvier. At that time I had not seen a specimen of the fish. The Rev. R. T. Lowe, who has devoted great attention to fishes, particularly those taken at Madeira, where he has resided many years, first intimated to me that this, my supposed new Serranus,—which I had called Couch's Serranus, in reference to a naturalist and a friend, from whom I had received so much valuable assistance,—was in fact the *Polyprion cernium* of Cuv. and Val. *Hist des Poiss.* t. iii. p. 21, a species well known to him, being a common fish at Madeira, and which is now known to range as far to the south as the Cape of Good Hope. Since that time Mr. Lowe has sent me from Madeira a fine and perfect specimen of this fish, which I have shown to several good observers on our southern coast, where Mr. Couch's Stone Basse occurs, who have no doubt that this fish is the same as the Stone Basse of Mr. Couch, and it therefore now appears in its place among the British Fishes under its most recent systematic appellation. I am still, however, anxious to identify this species with the name of Mr. Couch, who first made it known as a British fish, and have therefore now called it Couch's Polyprion.

This species was the subject of a particular memoir by M. A. Valenciennes, published in the *Mem. du Mus.* t. xi. as already quoted, and is remarkable in having escaped the observation and record of all the early Schthyological writers, although the fish is common in the Mediterranean, attains a large size,—sometimes weighing one hundred pounds,—and measuring five or six feet in length. Mr. Baker of Bridge-water tells me, that this fish, of three feet in length, is not uncommon in the Bristol Channel. Mr. Couch, in reference to its habits, says, " this species approaches the Cornish coast under peculiar circumstances. When a piece of tim-ber, covered with barnacles, is brought by the currents from the more southern regions, which these fishes inhabit, consi-derable numbers of them sometimes accompany it. In the alacrity of their exertions, they pass over the wreck in pur-suit of each other, and sometimes, for a short space, are left dry on the top, until a succeeding wave bears them off again. From the circumstance of their being usually found near floating wood covered with barnacles, it might be supposed that this shell-fish forms their food ; but this does not appear to be the case, since, in many that were opened, nothing was found but small fishes. Perhaps these young fishes follow the floating wood for the sake of the insects that accompany it, and thus draw the Stone Basse after them."

The Rev. Robert Holdsworth of Brixham, who has fur-nished me with many interesting notes on British fishes, sends me word that on the Devonshire coast this fish is also called Stone Basse and Wreck-fish, thus illustrating the habits of the species as noticed by Mr. Couch, by a refer-ence to the floating timbers to which the barnacles adhere, and float along with them. Two paragraphs from Mr. Holdsworth's letter on this fish, are as follows :—" October 7, 1824. The crew of the Providence smack found a large log of mahogany in Start Bay, covered with long barnacles, and surrounded by a shoal of these fish. They jigged,—that is, caught with a pole, having a barbed hook at the end, four or five. I had two cooked, which I purchased of the crew of the Providence, and found them excellent." Captain Ni-cholls, in a voyage from St. John's, Newfoundland, to the coast of Portugal, " having his ship's bottom very foul, and

B 2

covered with barnacles, was becalmed for many days within a hundred leagues of Oporto, and was for a fortnight surrounded with these fish, which followed the ship, and were caught by the crew. He fed his men upon them for twelve or fourteen days, and considered them excellent food."

As before noticed, according to M. Valenciennes, Savigny, and Risso, this Polyprion,—the only species of the genus,—is common in the Mediterranean, where it lives throughout the year over rocky bottoms in deep water. The flesh is white, tender, and of good flavour. M. Valenciennes says it feeds on mollusca and small fishes ; he found sardines in the stomach.

The Rev. R. T. Lowe says this Polyprion is one of the most common fish in the market at Madeira ; where, when small, it is called Chernotte, and when large, Cherne, (pronounced Shareny by the Portuguese,) and Jew-fish by the English. It is there, also, deservedly held in esteem for the table.

Specimens taken at the Cape of Good Hope were sent by M. Delaland to Baron Cuvier at Paris, who could perceive no difference between them and specimens from the Mediterranean or the Channel.

There is good reason to believe, on the authority of Dr. Latham, as recorded by Schneider, that this fish also inhabits the shores of America.

In the fish here described, the length from the point of the upper jaw to the posterior end of the horizontal bony ridge on the operculum, is to the whole length of the fish, exclusive of the caudal rays, as one to three ; the depth of the fish in the vertical line of the origin of the ventral and pectoral fins, is to the whole length, from the point of the lower jaw, when the mouth is open, to the end of the caudal rays, also as one to three ; the thickness of the fish equal to half its height ; the lower jaw is the longest ; the nostrils double, the openings circular ; the eyes dark brown ; the peculiarities of the head, teeth, and gill-covers, are detailed in the generic characters ; the ventral and pectoral fins have their origin in a vertical line under the fourth spinous ray of the dorsal fin : the upper half of this fish is of a dark purplish brown, the under part almost silvery white ; the membranes connecting the various fin-rays dark brown ; the extreme margin of the tail is nearly white.   Young specimens are described and figured as marbled over with two shades of brown ; the lateral line rises high over the base of the pectoral fin, afterwards following a course nearly parallel with the outline of the back.   The figure here given was taken from the specimen of this fish sent me by Mr. Lowe, which measured sixteen inches in length.   The fin-ray formula is as follows :—

D. 11 + 12 : P. 16 : V. 1 + 5 : A. 3 + 9 : C. 17 : Vert. 26.

A representation of the bones forming the cranium of this Polyprion is here added as a vignette.

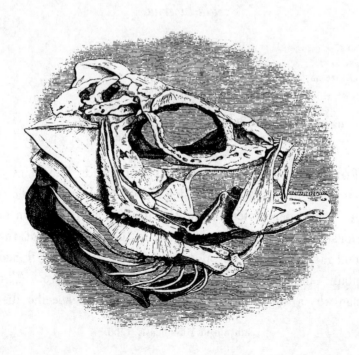

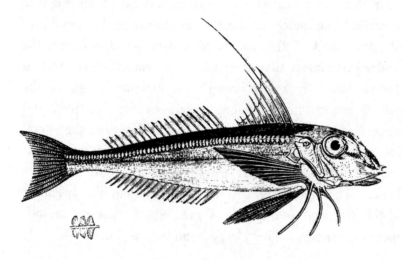

THE SHINING GURNARD,

OR LONG-FINNED CAPTAIN.

Trigla lucerna.

Cuculus	Rondelet, Latin edition, p. 287.
Rouget,	,, French edition, p. 227.
Trigla lucerna,	*Brigotte,* Brunnich, p. 76.*
,, ,,	*Orghe,* Risso, Ichth. p. 209.
,, *milvus,*	,, ,, Hist. p. 395.
,, *lucerna,*	*L'Orgue,* Cuv. et Val. Hist. des. Poiss. t. iv. p. 72.
,, ,,	*Long-finned Captain,* Mag. Zool. and Bot. vol. i, p. 526.

The Gurnard figured above has been made known as a
species new to the British Catalogue by Dr. Parnell, who obtained several specimens from the fishermen of Brixham in
Devonshire, by whom, in reference to the elongation of the
second ray of the first dorsal fin, it is called the Long-finned
Captain, and by whom also it is not considered rare. The
reason why a species so strongly marked as to specific dis-

* Ichthyologia Massiliensis, 1768.

tinction should have remained till lately unnoticed on our shores, will probably be found in the circumstance that this Gurnard does not generally exceed nine inches in length, which not being considered by the fishermen a marketable size, the fish is not often brought on shore ; yet its flesh is esteemed as sweet and delicate.

The capture of several examples of this fish at Brixham, and the announcement of the circumstance in the first volume of the Magazine of Zoology and Botany, page 526, with a description and figure, has not, that I am aware, elicited any notice of its occurrence on other parts of our coast, yet it may be presumed to be plentiful as a species ; Dr. Parnell saw seven taken at once in a trawl net, and it is decidedly common in most parts of the Mediterranean. Brunnich, who described it in 1768, as quoted under the representation of the fish, found it at Marseilles. Savigny, according to M. Cuvier and Valenciennes, found it at Naples. Dr. Leach sent specimens to Paris from Malta. M. Risso includes it in both his volumes among the fishes taken in the environs of Nice, and mentions it even as one known to Aldrovandus, quoting lib. ii. cap. 58, page 279. But little appears to be known of the particular habits or food of this species ; but it is supposed to spawn about June, from the large size of the roe in a female fish taken in that month. Dr. Parnell's specimens were obtained in the month of September.

I have followed M. Cuvier and Valenciennes in including references to the work of Rondelet, but with some doubt whether the fish there represented and described is not rather a different species of Gurnard. Our fish was probably called *lucerna,* from the brilliant and shining longitudinal silvery band which pervades the whole length of each side. I am indebted to Dr. Parnell for the specimen from which the following description was taken.

The whole length nine inches and one quarter. From

the point of the nose to the end of the occipital spine, is to
the whole length of the fish as one to four; the depth of the
head is to the whole length of the fish as one to six and a
half; the depth of the body is to the whole length as one to
six; the nose is rather short and blunt; at the superior anterior edge of each orbit is a single short bony spine directed
upwards; at the inferior anterior edge of each orbit there is a
groove directed downwards and forwards to the base of the
external nasal bone, in which groove, about half way between
the eye and the nose, the nostril is pierced; the exterior surface of the head granulated and hard; the posterior margin
on each side furnished with two spines directed backwards,
one from the edge of the operculum, the other from the occipital bone above it; the region of the scapula, behind the
operculum, is furnished with another spine, also directed
backwards. The fin-ray formula is as follows:—

D. 9 — 18   P. 10 — 3 : V. 6 : A. 17 : C. 14.

The first dorsal fin commences in a line over the base of
the pectoral fin, the second ray is more than as long again as
the first ray, and the third ray is also a little longer than the
first ray; afterwards the rays decrease in length gradually, the
last ray being the shortest; the second dorsal fin commences
in a vertical line over the anal aperture; the rays of this fin
are nearly uniform in length throughout, the fin ending on
the same plane with the anal fin, the rays of which commencing immediately behind the anal aperture, are also
nearly uniform in length throughout; the tail in shape is
lunate; the dorsal ridge contains from twenty-four to twenty-
six plates, each ending in a single point; the lateral row of
scales, peculiar to the Gurnards, are in this species formed
like wings, and are represented of an enlarged comparative
size below the tail of the figure of the fish. The head and
upper part of the body are of a fine vermilion colour; the

irides silvery ; along the side of the body a broad and shining silvery band ; the belly below reddish white ; the pectoral fins of a deep blue ; all the other fins rosy red.

The characters of this Gurnard are so well marked that it is not likely to be confounded with any other species.

The vignette below represents the cranium of the Sapphirine Gurnard.

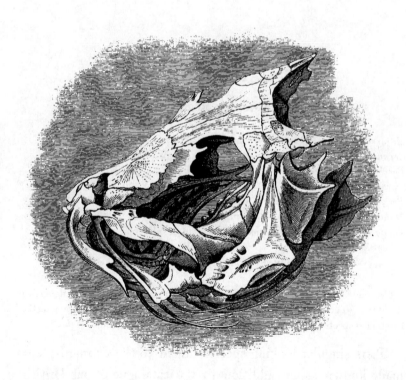

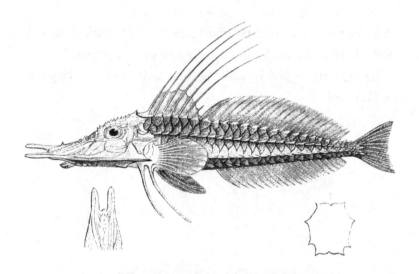

THE MAILED GURNARD.

Peristedion Malarmat.

Malarmat,	BELON, p. 209.
Cornutus, sive Lyra altera,	RONDELET, Lat. Edit. p. 299.
Forchato, *Malarmat,*	,, Fr. ,, p. 237.
Lyra altera,	WILLOUGHBY, p. 283, tab. S. 3.
Trigla cataphracta, Malarmat,	BRUNNICH, p. 72.
Malarmat,	DUHAMEL, t. iii. Sect. 5, p. 113, pl. 9. f. 2.
Trigla cataphracta, Le Malarmat,	BLOCH, pt. x. pl. 349.
Peristedion malarmat,	CUV. et VAL. Hist. Pois. t. iv. p. 101.
,, ,, *Mailed Gurnard,*	Mag. Nat. Hist. vol. i. N. S. p. 17.

Generic Characters.—Body covered with bony plates, forming a defensive armature. The nasal bone divided into two points. The mouth has no teeth. In other respects the characters are similar to those of the genus *Trigla*.

THIS singular-looking species, allied to the Gurnards, was made known as an addition to the catalogue of our British Fishes by Dr. Edward Moore of Plymouth, in the Magazine of Natural History for 1837, conducted by Mr. Charlesworth, as quoted among the references placed below the figure: it was caught on the fishing ground between Plymouth and the Eddystone in the autumn of 1836. It will be observed by the synonymes quoted, which are arranged chronologically, that this fish has been known from the time of Belon, who published in 1553, and has given a figure from an engraving on wood, which is easily recognised. This fish is also figured and described in the work of Rondelet, who from a resemblance which it bears to *Trigla lyra*, the systematic name of our English Piper Gurnard, British Fishes, vol. i. p. 44, called this fish *Lyra altera*, and also *Forchato*, from its elongated and bifurcated nasal bones. Brunnich, after Rondelet, called it *cataphracta*, in reference to the armour-like scales with which the body is defended. The term *Malarmat* applied to a fish so well armed, at least defensively, could only have been bestowed in joke by way of antiphrase.

M. Risso, who has briefly described some of its habits, says, it frequents deep water over rocky ground, approaching the shallows only at the period of spawning. It swims with rapidity, occasionally breaking off portions of the extended nasal bones against the rocks among which it harbours. It is said to be solitary in its habits, and feeds upon such animals as the medusæ, the beroe, and the thinner skinned crustacea. This fish inhabits all the western parts of the Mediterranean, and is rather common on most of the shores, where it attains the length of two feet. The British specimen recorded by Dr. E. Moore was about eleven inches long. It is said to be a rare species in the Adriatic, but has been taken at Venice. Duhamel, in his *Traité des Peches*, says, that this fish, though so rare on the coasts of the Channel as to be almost unknown, is common on the coasts of Spain and Provence, where it is caught in deep water. It is fished for all the year; but as an article of food it is in the greatest estimation in Lent. As there is but little

to eat upon this fish when it is small, those of the largest
size are the most in request. Duhamel gives the following
instructions for preparing this fish for the table: if it is intended for stewing, it is necessary to soak it in warm water
in order to get off the skin and scales, which is most easily
effected by commencing the removal at the tail; if it is
preferred to broil it, it is then only necessary to open the
body of the fish, and put inside fresh butter, fine herbs, and
seasoning to increase the flavour of the meat, which is white
and delicate. When it is sufficiently cooked the scales come
off easily.

Dr. Moore very obligingly sent his British specimen of
this fish up to London that I might see it, and I found that
it exactly resembled an example from the Mediterranean in
my own collection, with which I compared it.

The bones of the nose are very much elongated, forming
a projecting and forked snout of two broad and flattened
processes, which are each an inch in length, and parallel to
each other, half an inch apart at the base, on the upper surface of which there are one large and two smaller mam-
millary protuberances. From the end of the elongated nasal
bone to the posterior end of the ridge on the cheek at the
base of the pectoral fin, the length is three inches and a half
in a fish of eleven inches, or rather less. The nasal, orbital,
and occipital ridges, are armed with numerous sharp tooth-
like processes. The orbit of the eye is oval, its greatest
length horizontal, the irides silvery; the jaws are semicircular in shape; the form of the opening of the mouth, which
is without teeth, is also semicircular; the length of the head,
from the point of the nasal bone to the end of the suborbital
ridge, is to the whole length of head, body, and tail together,
as one to three.

The body is octagonal, covered with bony scales, or plates,
laid over each other like a coat of mail; from the centre of

the scales, forming in continuous lines the eight angles of the body, projects a sharp-pointed process directed backwards ; the scales vary in number on the different angles from twenty-three to thirty.

The fin-ray formula, according to Cuvier, is as follows :—

D. 7. 19 : P. 12. 2 : V. 1 $+$ 5 : A. 18 : C 11 : Vert. 43.

The first dorsal fin has seven rays, but the point of distinction between the first and second dorsal fins is liable to some misconception, as it is only indicated by a decrease in the extent or elevation of the connecting membrane. Five or six of the rays of the first dorsal fin end in elongated flexible filaments, as shown in the figure. It is supposed that the males only in this species have these filaments elongated, the rays in the females remaining short, and this may account for some differences that appear in the representations given by some of the authors herein referred to. The second dorsal fin usually contains eighteen or nineteen short rays. The pectoral fin is stated by Cuvier to contain twelve rays, but his figure in illustration exhibits but ten rays, and I find there are ten rays in the pectoral fin in the Mediterranean specimen before referred to ; Dr. Moore's fish is described as possessing but eight rays ; they appear therefore liable to variation ; the free rays common to the Gurnards are in this species limited to two ; between the ventral fins is an elongated and flattened sternum ; the body ends at the tail in three short projecting spines on each side of the base of the caudal rays ; the form of the tail is lunate. Dr. Moore says of his fish that " its colour, when fresh, was of a uniform scarlet, like the Red Gurnard, gradually softening to pale flesh colour towards the abdomen ; the anal and dorsal fins were crimson ; but the others pale and greyish.

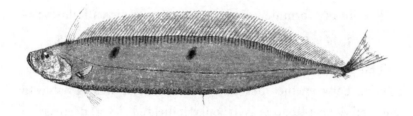

THE VAAGMAER,

OR DEALFISH.

Trachypterus vogmarus.

Trachypterus, Bogmarus, Cuv. et Val. Hist. Nat. des Poiss. t. x. p. 346.

THE publication of the History of British Fishes has brought me into communication with Professor John Reinhardt, Curator of the Royal Museum, and also of the University Museum at Copenhagen. This gentleman, desirous of supplying the deficiency, both as to figure and description, which existed at the time of publishing the account of the Vaagmaer, or Dealfish, British Fishes, vol. i. p. 191, has very obligingly forwarded to me a copy of his memoir, printed in the Transactions of the Royal Society of Copenhagen, containing a detailed account and a figure of this fish, from a specimen obtained in Iceland. By the kindness of Dr. Cantor, the friend and countryman of M. Reinhardt, I am enabled to present a free translation of so much of this Danish paper as refers to the description of this very rare fish, with a reduced figure from the plate which accompanied the memoir.

The specimen of the Vaagmaer, from which the drawing and description were taken, was during the summer of 1828 thrown up alive on the beach near Thorshavn in Iceland, and was procured by Mr. Möller for the Royal Museum of Natural History. Fortunately, a ship at the time was ready to sail for Copenhagen, by which the fish, preserved in spirits, was forwarded. It arrived in about ten days, and in such beautiful condition that the brilliant red colour of the fins had not faded, nor had the membrane connecting the fin-rays been torn ; only the anterior dorsal and the ventral fins were injured, so as to leave but short roots ; the continuation of which is therefore indicated by fine lines.

A previous account of this, as well as of another less perfect specimen, found thrown on shore near Frederikshavn in Jutland, was laid before the Royal Society of Copenhagen in the winter of 1829. As I have not been able to procure a better specimen, and a useful delineation of this fish is wanted, while we, through the figures given by M. Valenciennes, are enabled to compare several species from the Mediterranean, I have thought it right to supply this deficiency by having an engraving made under my own superintendence of the Icelandic Vaagmaer, to the description of which the following paper is devoted.

The result of the account of the two specimens above mentioned, as communicated in 1829 to the Royal Society, was, that the Northern Vaagmaer, contrary to the opinion of its former describers, is indeed provided with ventral fins, by which its generic relation to those of the Mediterranean has been decided, as well as its systematic rank : while a comparison with one of the Mediterranean species preserved in the Museum, established its specific difference.

M. Valenciennes, in his excellent account of the genus *Trachypterus* in his tenth volume, has added a few remarks to the previous history. Although the specimen he examined

was dried and partly defective, the relative dimensions and
the number of the dorsal rays nevertheless agree. Some difference between the short description of M. Valenciennes
and that which follows, will be pointed out hereafter.

The body of the Vaagmaer is compressed, or sword-blade
like throughout, more than half of its whole length, or, in
the present specimen, from the occiput to within eleven
inches of the caudal extremity of the dorsal column ; the
height is nearly the same at both extremities, and only one
seventh part less than the height at the central part of the
body, where it is greatest. In this particular it differs from
the two species from the Mediterranean, with more than
one hundred and sixty dorsal rays, according to their dimensions given by M. Valenciennes,—namely, those of *Trachypterus falx*, and *Tr. iris*, a difference distinctly shown,
particularly in the latter species. In those two species the
greatest height is at, or near, the occiput, from whence it
more or less rapidly decreases towards the caudal fin. Of
the *Tr. leiopterus* I am uncertain, as the author has given
no dimensions of the height, although he elsewhere states
that this species has a caudal fin much thinner than that of
the *Vagmarus*.

The colour of the head and body is silvery, varied only
by the blackish grey of the head, and by two obliquely oval
spots of the same colour on each side. The long dorsal fin,
and the almost vertical triangular caudal fin, are of a light
red. The silvery colour arises from a thin layer on the
epidermis, of the same nature as that of the ventral membrane observed in several other fishes. I have not been able
to observe any traces of scales. The skin underneath the
silvery cover is divided or furrowed by diagonal lines, forming small flat elevations, some of which are round, and others
angular. Towards the abdominal margin, particularly on
each side of the sharp edge, these elevations appear as papil-

lary warts of remarkable firmness, but by no means osseous, which, decreasing in size behind the anus, are lost entirely towards the tail.

In the number of its lateral dark spots, the Vaagmaer resembles the *Tr. leiopterus*, which, according to M. Valenciennes, has only two ; but, in reference to the position of these spots, there exists a difference between these two species. In the Vaagmaer they are placed farther backwards, the situation of the most anterior spot being at the commencement of the second fourth part of the whole length of the fish, the posterior being situated about half way, or near the middle. Both spots are nearer each other in the *Tr. leiopterus* than in the present species. The total length of the specimen represented, measured from the point of the nose to the end of the dorsal column, is forty-three inches six lines ; with the upper jaw protruded the whole length is forty-four inches seven lines. The greatest height of the body in the present specimen, twenty inches from the angle of the mouth, or four inches in advance of the anus, is contained five times and a half in the length, while the height at the nuchal region, about six inches from the end of the nose, is contained nearly seven times in the total length. The height at a distance of thirty-six inches is but a little more than one eleventh of the total length, and at the distance of forty inches is little more than one thirtieth.

The greatest diameter is near the part where the gill-cover is attached to the head, and is contained four times in the height of that region, or five times in the greatest height, the diameter of which is scarcely one-tenth. The diameter decreases towards the narrow part of the tail. The greatest diameter of the body is in the region of the lateral line, and decreases towards the dorsal and ventral profile, particularly towards the former, where it becomes sharp like the edge of a knife, by which the spinal processes and the intervening

c

bones of the dorsal rays become apparent on the surface of the thin external covering.

The head from the end of the nose to the posterior margin of the gill-cover is contained seven times and a quarter in the total length; the length of the head is therefore nearly equal to the height of the fish at the nuchal region.   The outline of the lower jaw forms an ascending arch, which at the angle of the mouth meets the straight and slightly declining profile of the forehead, by which the lower jaw, when the mouth is closed, becomes much elevated, and the opening of the mouth turned upwards.   When the lower jaw sinks into a horizontal position, the upper jaw is much projected, and becomes some-what longer than the lower.

The formation of the jaws, the form and position of the gill-covers, and the radiating grooves on the latter, on the jaws and frontal bones, agree with the description of those parts in the *Tr. Falx*, as given by M. Valenciennes, to which I beg to refer as far as regards the Vaagmaer.

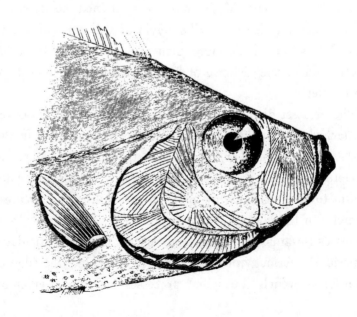

The dentition in this species appears to exhibit some deviations from that of *Tr. Iris* and *Tr. Spinola,* in which the teeth of the upper and lower jaw are nearly vertical, and are seen, although the mouth is more than half closed. In the description of *Tr. Falx* no mention is made of the position of the teeth. In the Vaagmaer the maxillary teeth are thin, conical, and pointed, nearly recumbent, with the apex turned towards the pharynx. On the intermaxillary bones only four teeth appear, two on each bone, somewhat within the margin : the inner teeth do not exceed two lines in length. In the lower jaw the teeth are placed nearer the outer margin, and towards the front, four on one side, three on the other, with some variation in size. A single-pointed tooth, three lines in length, is placed vertically on the central line of the vomer, but no other sharp teeth appear either behind this tooth, or on the palatine bones, which, according to M. Valenciennes, is the case in *Tr. Falx.* The superior pharyngeal bones are studded with pointed curved teeth, one line in length ; the inferior pharyngeal bones are wanting altogether.

The large eyes, lodged in a circular orbit, are situated near the frontal profile. The longitudinal diameter of the orbit is, compared to the length of the head, as one to three and a half; the iris is silvery white, its breadth somewhat greater than the diameter of the pupil.

The nostrils are very small, opening into narrow cavities, situated above the anterior and superior part of the orbital margin ; the larger nostril, a small rima, is situated close upon the margin ; the smaller one is oval, and is placed a little higher up.

The anterior extremity of the tongue is somewhat broad, with a rounded margin, concave above, flat and keeled underneath ; the tongue is entirely free, and may easily be placed in a horizontal position, as if intended to throw small bodies towards the pharynx.

c 2

The lateral line, commencing from the nuchal region, descends nearly vertically opposite the middle of the orbit, from whence it proceeds obliquely downwards, until behind the pectoral fin, it reaches a distance from the ventral profile somewhat shorter than the distance of the dorsal profile. It now continues straight towards the extremity of the tail, approaching the lower caudal margin. This line is covered by a series of small oblong osseous shields, from the middle of which rises a small spine directed forwards. The shields and their spines increase in size towards the thin part of the tail, from whence they again decrease, although the last shield is much larger than those of the central part.

The short pectoral fins are situated nearer the ventral margin than to the lateral line, and nearly opposite the apex of the gill-cover. The number of the rays is in the right pectoral fin eleven, in the left only ten.

Of the ventral fins, there remain only some short roots of the rays, situated close to the ventral margin, in a direction nearly parallel with, but a little further back, than the pectoral fins. The number of the rays is six.

Of the rays of the anterior dorsal fin only five roots are left, the first of which is somewhat thicker than the rest, and situated five inches eight lines from the edge of the closed jaws. The interval between this fin and the commencement of the posterior dorsal fin, is twice the distance between two rays. The posterior, or long dorsal fin, has one hundred and seventy-two rays, of which the first ray is situated six inches and one line from the point of the jaw; the last ray half an inch from the last vertebra. The anterior part is very low, increasing in height by degrees until it reaches the commencement of the last fourth part of the total length, where the height of the present specimen amounts to three inches eleven lines, or about one half of the greatest height of the body; from thence it decreases rapidly, so that the last ray is only a little longer than the first. The rays are slender, flexible spines, without the slightest trace of transverse marks; their articulating surface dilates into a saddle-shaped shield, with a short curved point in the centre, by which a number of small sharp bodies appear along the root of the fin. The rays themselves, however, are quite smooth to the touch, and, under a lens, are, as M. Valenciennes in his own specimen found them, a little sharp.

The more or less vertically raised caudal fin contains eight rays; the length of the upper and under ray is to the length of the two central rays as four to three. The latter named rays are sharp to the touch, and viewed through a lens are observed to be studded over with a number of small spines.

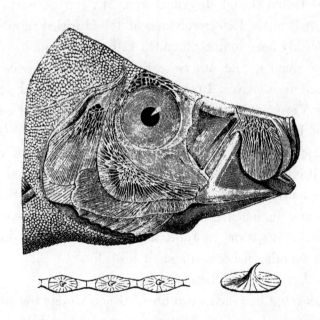

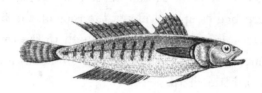

THE SLENDER GOBY.

Gobius gracilis.

Gobius gracilis, *Slender Goby,* Jenyns, Man. Brit. Vert. p. 387.
　　,,　　　,,　　　　,,　　　,, Parnell, Wern. Mem. vol. vii. p. 245.

THis Goby, though described from Mr. Jenyns' work, was not figured in the former volumes of the British Fishes. It has probably been long confounded with *Gobius minutus,* but is more slender, and otherwise distinguished. It was first described by the Rev. Leonard Jenyns in his Manual of the British Vertebrate Animals, from specimens obtained on the coast of Essex. Dr. Parnell says, " This well-marked Goby is occasionally found in the Firth of Forth, but is not common ; it inhabits the same situations as the *minutus,* and they are frequently taken together. I have found it in the Solway Firth, and in much greater plenty on the southern coast of England. It spawns in June, and is of little value except as food for other fishes and aquatic birds."

Mr. Jenyns' description is as follows :—

" Length, three inches two lines. Form closely resembling the *minutus,* but more elongated and slender throughout ; greatest depth barely one-seventh of the whole length : snout rather longer : opercle approaching more to triangular, the lower angle being more cut away, and the ascending margin more oblique; a larger space between it and the pectorals: the two dorsals further asunder: rays of the second dorsal longer; these rays also gradually *increasing* in length, instead of *decreasing*, the posterior ones being the longest in the fin, and rather more than equalling the whole depth: rays of the anal in like manner longer than in *G. minutus*.

The fin-rays in number are—

D. 6. 12 : P. 21 : V. 12 : A. 12. : C. 13, and two short rays.

In all other respects similar. The colours also resembling those of *minutus*, with the exception of the anal and ventral fins, which are dusky, approaching to black in some places, instead of plain white, as in the *minutus*."

The vignette below represents the cranium of *Gobius niger*.

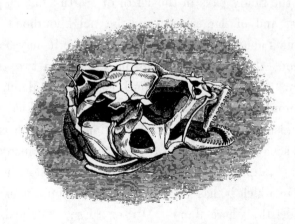

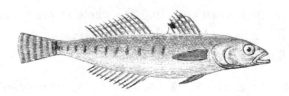

THE ONE-SPOTTED GOBY.

Gobius unipunctatus.

Gobius unipunctatus, *One-spotted Goby*, Parnell, Memoirs of the Wernerian
Nat. Hist. Soc. vol. vii. p. 243.

This Goby, says Dr. Parnell, " does not appear to have
been noticed by previous authors. I have observed it in
most of the sandy bays in the Firth of Forth ; but in greater
numbers, and of larger size, in the neighbourhood of the
salmon-nets above South Queensferry, where it may be found
throughout the summer months in water from two to three
feet deep. I found it on the south coast of England, equally common with the *Gobius minutus*, or Freckled Goby.
I have also found it in many situations where the *minutus*
was not seen ; and the *minutus* has been taken in many places
where the *unipunctatus* did not exist. The most northern
locality in which it has yet been observed appears to be the
Moray Firth, where James Wilson, Esq. obtained a fine
specimen of three and a half inches in length."

" This fish, although closely allied to the other species of
the same genus, is undoubtedly quite distinct from them ; the
black spot on the first dorsal fin being far more constant and

conspicuous than any character which distinguishes the rest
of the British Gobies. The only species it can well be mistaken for is the *G. minutus*, but differs from it in having a
black spot between the fifth and sixth ray of the first dorsal
fin; the second dorsal with eleven rays, and the tail fin even
at the extremity. Whereas the *G. minutus* has no black
spot between the fifth and sixth ray of the first dorsal fin;
the rays of the second dorsal ten in number, and the tail fin
rounded at the end."

A specimen, two inches and a half in length, is thus described by Dr. Parnell. " Body rather elongated, rounded
in front, compressed at the tail; flattened on the nape; head
long in proportion to its depth, one fourth of the length, including half the caudal rays; operculum and preoperculum
rounded. Colour of the head, back, and sides, pale brownish
yellow; throat and belly white; dorsal and caudal fins
freckled and barred with pale brown; first dorsal fin with a
black spot between the two last rays, which assumes a beautiful appearance when newly taken from the water; lateral
line crossed by six or seven dark spots, the one at the base
of the tail being most conspicuous. First dorsal fin with fine
flexible spiny rays, of which the second and third are rather
the longest, commencing behind the base of the pectorals,
and ending in a line over the end of the pectoral rays; second dorsal fin remote from the first, commencing in a line
over the vent, and ending over the last ray of the anal; the
anterior rays longer than the terminal ones; all flexible and
branched, except the first, which is simple; anal fin similar
to the second dorsal, leaving a wide space between its termination and the base of the caudal rays; ventral fins united so
as to form but one fin; the middle rays the longest, extending to the vent; each ray is branched except the first and
last, which are very short and simple; between each is
stretched a membrane, forming the base of the ventral disk.

Pectorals, when turned forward, reaching to the middle of the orbit; the middle rays the longest; tail even at the end. Eyes rather large, placed high on the head, approximating; cheeks tumid; under jaw the longest; teeth small and sharp, placed in two rows in each jaw, none on the tongue, palatine bones, or vomer; a small tubercle in front of the anal fin. Number of fin-rays :—

D. 6. 11 : P. 16 : V. 10 : A. 11 : C. 15.

The vignette below is a representation of the barnacle.

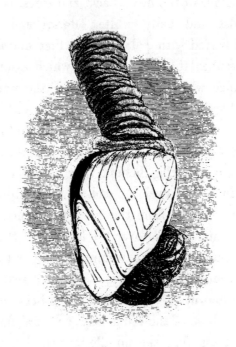

ACANTHOPTERYGII. *GOBIOIDÆ.*

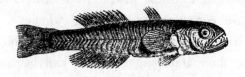

THE WHITE GOBY.
Gobius albus.

Gobius albus, *The White Goby,* PARNELL, Transactions of the Royal Society of Edinburgh, vol. xiv.

THIS species of Goby, Dr. Parnell observes, " holds such a conspicuous place in the genus, that it cannot well be mistaken for any other. I first noticed it in the Solway Firth, in June 1836, where I obtained in one day, after the recess of the tide, fifty specimens. They are evidently the fry of a large species. When first taken from the water they are soft and transparent ; the eyes are large and prominent ; the scales which cover their body are large, thin, and very deciduous. The length is about two inches ; the head is large ; the gape is wide ; the teeth are long and sharp, placed in a single row in each jaw. The first dorsal fin commences over the anterior third of the pectorals ; the second dorsal fin commences over the vent, and ends opposite to the base of the last anal rays. The cheeks are tumid ; the border of the operculum rounded ; the body is transparent, and marked by a number of fine depressed lines, placed in an oblique direction ; the lateral line is straight throughout its length. The number of the fin-rays are—

D. 5. 13 : P. 16 : V. 13 : A. 13 : C. 12.

The last ray of the anal and second dorsal fin is longer than the first, and reaches, when folded down, to the base of the tail rays. These fishes are supposed (erroneously) by the fishermen to be the young of the Sting-fish, *Trachinus vipera,* and are consequently destroyed whenever they come within their reach. On transferring them to a bottle of alcohol they lose their transparent aspect, and become hard and opaque. In the month of July, when I had occasion to revisit the Solway Frith, I endeavoured to obtain additional specimens, presuming that by this time they would have somewhat increased in size ; but not a single specimen could be found, nor has the parent fish ever come within the observation of the fishermen.

" The first dorsal fin of this fish, as possessing but five rays, is sufficient to distinguish it from every other British species of the same genus."

The teeth in this species are also more formidable in proportion to the size of the fish than those of any other British Goby.

ACANTHOPTERYGII.   *LABRIDÆ.*

JAGO'S GOLDSINNY.

Crenilabrus rupestris.

	Jago's Goldsinny,	RAY, Syn. Pisc. p. 163, tab. 1, f. 3.
Sciæna rupestris,		MUS. Adol. Fr. pl. 31, f. 65.
Labrus ,,		LINN. Syst. Nat. p. 478, sp. 27.
,, ,,		MULLER, Prod. Zool. Dan. p. 45, sp. 382.
Perca ,,		MULLER, Zool. Dan. tab. 107.
Lutjanus ,,		BLOCH, pt. vii. tab. 250, f. 1.
Labrus ,,		NILS. Prod. Icht. Scand. p. 76, sp. 5.
Perca ,,		RETZ. Faun. Suec. p. 337, sp. 73.
Crenilabrus ,,	Jago's Goldsinny,	SELBY, Mag. Zool. and Bot. vol. i. p. 167.
,, ,,	,, ,,	THOMPSON, Mag. Zool. and Bot. vol. ii. p. 445.
,, ,,	,, ,,	THOMPSON, Zool. Proc. 1837, p. 57.
Labrus ,,		FRIES and EKSTRÖM, Scandinavian Fishes, pt. ii. pl. 3, fig. 1.

In the month of February 1836, Dr. George Johnston obtained three specimens of the *Lutjanus rupestris* of Bloch, two of which were picked up in Berwick Bay, and the third near Barncleugh ; these specimens were thrown on shore after a violent storm, and having been sent by Dr. Johnston to his friend Mr. Selby, became the subject of a notice in the first volume of the Magazine of Zoology and Botany, as quoted under the figure of the fish here given.

This fish Mr. Selby most correctly referred to the Gold-sinny of Jago, in the Synopsis of our countryman and naturalist John Ray, who appears to have been the first to make it known ; but this fish being also a northern species, was afterwards figured and described in the various works here quoted among the synonymes. Since the occurrence of the specimens on our eastern coast, Mr. Thompson of Belfast has obtained two others at Bangor, County of Down, where they were caught, with one or two other species of Wrasse, by angling boys. I have received from T. S. Rudd, Esq. two beautifully coloured examples of this fish, which were taken on the Yorkshire coast, from the finest of which the figure here engraved was drawn ; one specimen has also been taken on the coast of North Wales by my friend Mr. Thomas Eyton. Among some *Labri* supplied me by Mr. Couch from Cornwall, before the occurrence of the specimens in Berwick Bay, was a small fish of this species, but being by accident somewhat discoloured and distorted, and this species differing in colour when young, I did not recognize it as the *Lutjanus rupestris* of Bloch, but figured it as a vignette to the Scale-rayed of the British Fishes, vol. i. p. 300. Since that time Mr. Couch has very kindly supplied me with more small specimens, which will enable me to describe this fish as it appears at different stages of its growth, premising, however, that I have seen no examples of more than seven inches in length.

This species is taken occasionally in the Baltic ; in Sweden, Denmark, and Norway, where it is sometimes caught by angling from rocks, as in this country. Another coloured figure of this fish has recently appeared in the new work of

MM. Fries and Ekström, on the Fishes of Scandinavia, now in course of publication, in parts, at Stockholm.

The length of the specimen here described was six inches and a half. The length of the head one inch and three quarters; the diameter of the eye three eighths and a half, or one fourth of the length of the head; the irides silvery; the teeth, long, strong, curved, and pointed, particularly in the anterior part of the upper jaw; both preoperculum and operculum covered with scales; the preoperculum distinctly crenated throughout the greater part of its ascending edge; the dorsal and pectoral fin commence on the same vertical line; the membrane connecting the first four spinous dorsal rays black; the spinous rays shortest at the commencement of the fin, becoming gradually, but slightly, more elongated towards their union with the soft rays, and in length about equal to one fourth of the depth of the body of the fish; the soft rays more lengthened; from the base of the last of which to the end of the caudal rays, is about the same length as that of the head, and about one fourth of the whole length of the fish. Half way between the base of the last soft dorsal ray and the extreme end of the caudal rays, there is on the upper edge of the body and tail a conspicuous roundish black spot, equally visible on either side; the caudal fin-rays scaled from their base on a line with this black spot half way along, the ends of the caudal rays slightly rounded; the anal fin with three spiny rays, and ending with elongated soft rays, the base of the last of which is a little in advance of the base of the last soft dorsal ray in a vertical line; the ventral fin begins a little behind and below the base of the pectoral fin; the pectoral is in length, compared to the length of the fish, as one to seven. The prevailing colour in the largest specimen is orange, the free edge of each scale being of a light golden yellow; the colour is darkest over the three or four lines of scales along the highest part of the back, and lightest on the lower part of the sides and belly ; the body is also indistinctly marked with five transverse bands, the first of which descends from below the more anterior spinous rays of the dorsal fin, and the fifth from below the elongated soft rays of the dorsal fin ; but I have never seen these bands near so strongly marked as they are made to appear in Bloch's coloured figure, the ground colour of the body of which resembles that of one of my specimens. Young examples of this species are of a uniform yellowish flesh colour ; the fins still lighter ; but the black spot at the commencement of the dorsal fin, and on the upper part of the base of the tail, are very conspicuous from the uniform paleness of the body and fins generally, and, but for these two constant spots, are not unlike the *Labrus pusillus* of Mr. Jenyns, as figured in this Supplement. These spots appear to be good distinctions ; very young specimens of *Crenilabrus cornubicus,* which in the British Fishes should have been called the Corkwing, are constantly marked with the spot on the middle of the side of the tail, in specimens measuring only one inch and a half in length. The fin-ray formula in Jago's Goldsinny is—

D. 17 + 9 : P. 14 : V. 1 + 5 : A. 3 + 7 : C. 13.

The number of scales along the lateral line is thirty-two, and four or five more extend along the basal half of the rays of the caudal fin ; there are four rows of scales between the lateral line and the dorsal ridge, and eleven rows of scales between the lateral line and the anal aperture.

M. Nilsson says, this species is liable to variations in colour, and some of the species taken in Northern localities are tinged with green.

ACANTHOPTERYGII. *LABRIDÆ.*

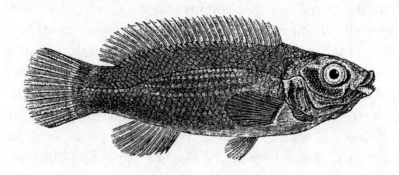

THE CORKLING.

Crenilabrus pusillus.

Turdus minor,   Corkling,   Ray, Syn. Pisc. p. 165.
Labrus pusillus,   ,,   Jenyns, Brit. Vert. p. 392, sp. 70.
Crenilabrus multidentatus, Ball's Wrasse, Thompson, Proc. Zool. Soc. 1837,
                   p. 56.

THIS species, of which no examples more than four inches in length have been as yet recorded, was obtained by Professor Henslow at Weymouth, and four or five specimens are now preserved in the Museum of the Cambridge Philosophical Society. I possess one which was sent me by Mr. Couch from Cornwall ; and Mr. Thompson of Belfast has recorded the occurrence of three others, which were taken at Youghal in Ireland, by Mr. Ball, in the summer of 1835.

These last specimens were characterised by Mr. Thompson in the Proceedings of the Zoological Society for 1837, page 56—not without some hesitation—as a new species, under the name of *Crenilabrus multidentatus ;* but subsequent comparative examinations of the specimens of the two countries, appear to show that they are identical, and they are here therefore brought together.

D

Mr. Jenyns' description of a specimen, four inches in length, is as follows :—" Distinguished by its small size. Back but little elevated, sloping very gradually towards the snout; ventral line more convex than the dorsal; sides compressed : depth contained about three times and three quarters in the entire length; thickness half the depth, or barely so much; head one-fourth of the entire length : snout rather sharp; jaws equal : teeth of moderate size, conical, regular, about sixteen or eighteen in each jaw : eyes rather high in the cheeks, situate half-way between the upper angle of the preopercle and the margin of the first upper lip; the space between about equal to their diameter, marked with a depression; a row of elevated pores above each orbit : preopercle with the ascending margin very oblique: the basal angle, which falls a little anterior to a vertical line from the posterior part of the orbit, very obtuse, and remarkably characterised by a few minute denticulations, which further on become obsolete, and in some specimens are scarcely anywhere obvious : lateral line a little below one-fourth of the depth; nearly straight till opposite the end of the dorsal, then bending rather suddenly downwards, and again passing off straight to the caudal; number of scales on the lateral line about forty-five : dorsal commencing at one-third of the length, excluding caudal; spinous portion nearly three-fourths of the whole fin, the spines very slightly increasing in length from the first to the last, which last is not quite one-third of the depth of the body; soft portion a little higher than the spinous, of a somewhat rounded form, the middle rays equalling nearly half the depth : anal commencing a little anterior to the soft portion of the dorsal, and terminating a little before it; the first three rays spinous, the third being the longest, but the second the stoutest spine; soft rays resembling those of the dorsal : caudal nearly even, with rows of scales between the rays for nearly half their length : pectorals rounded, about two-thirds the length of the head, immediately beneath the commencement of the dorsal ; all the rays soft and articulated, and, except the first, branched : ventrals a little shorter ; the first ray spinous, shorter than the second and third, which are longest ; all the soft rays branched ; the last ray united to the abdomen by a membrane for half its length.

B. 5 : D. 20 + 10 or 11 : P. 14 : V. 1 + 5 : A. 3 + 9. : C. 13.

Colours of specimens in spirits yellowish brown, with irregular transverse bands ; dorsal irregularly spotted with fuscous ; anal light brown ; the other fins pale."

" It is apparently," says Mr. Jenyns, " quite distinct from any of those described by other authors. Though belonging to the present section (*Labrus*), which it is convenient to retain, it would seem to form the transition to the *Crenilabri*, to which its near affinity is indicated by the rudimentary denticulations on the margin of the preopercle."

The vignette below represents the bones of the head in the genus Labrus.

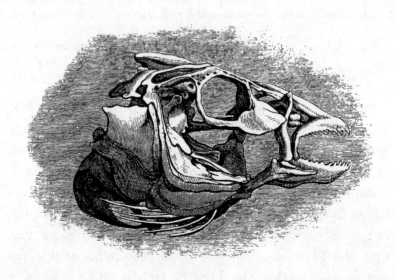

D 2

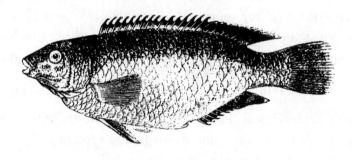

THE SMALL-MOUTHED WRASSE,

OR ROCK COOK.

Crenilabrus exoletus.

Labrus exoletus,	Linn. Syst. Nat. p. 479, sp. 33.
„ „	„ Faun. Suec. p. 117, sp. 331.
„ „	Muller, Prod. Zool. Dan. p. 46, sp. 386.
„ „	Fab. Faun. Gicœnl. p. 166, sp. 120.
„ „	Retz, Faun. Suec. p. 335, sp. 67.
„ „	Nils. Prod. Icht. Scand. p. 77, sp. 7.
„ „	Fries et Ekst. Scand. Fish. pt. ii. pl. 3, fig. 2.
Crenilabrus microstoma,	*Small-mouthed Wrasse,* Thompson, Zool. Proc. 1837. p. 55.
„ „	„ „ „ Mag. Zool. & Bot. vol. ii. p. 446, pl. 14.
„ „	*Rock Cook,* Couch, Cornish Fauna, p. 39.

Soon after the publication of the British Fishes, Mr. Couch very kindly supplied me with two examples of this Small-mouthed Wrasse, a species which I had not till then seen, and which on the Cornish coast is called the Rock Cook, where it is not so common as the Corkwing (*Crenilabrus Cornubicus*), nor does it take a bait like that fish, but is generally caught in the pots set for crabs. Since that time Mr. Thompson has recorded the occurrence of this species in two northern localities in Ireland, at Cairnlough in the county of Antrim, and at Lough Foyle in the county of Londonderry. At the former place the fish was found by Dr. Drummond, and at both places by Captain Portlock.

Although this fish was most appropriately called *microstoma*, for it may be immediately distinguished when among other *Crenilabri* by this very obvious peculiarity, it proves to be a species long known to more northern naturalists. Mr. Thompson has given a coloured representation of this fish in the second volume of the Magazine of Zoology and Botany, as previously quoted, and the recent publication at Stockholm of another coloured figure in the second part of the Fishes of Scandinavia, by MM. Fries and Ekström, leave no doubt of the two fishes being the same, and enable us to identify our species as the *Labrus exoletus* of Linnæus. It is a fish of small size, seldom exceeding four inches or four inches and a half in length, and is taken occasionally on the coasts of Sweden, Denmark, and Norway, and, according to Fabricius, as far north as Greenland, where, however, it is said to be rare.

The specimen from which the figure and description were taken, measured four inches in length, and one inch and one quarter in depth; the length of the head compared to that of the whole fish, is as one to four, or rather less. This species exhibits a slight elevation over the eye in the line of the frontal profile; the figure here given marks the true position and relative length of the various fins. The teeth are flat, even, and incisor-like, with the corners slightly rounded; some light-coloured lines extend from the mouth to the orbit, and over part of the cheek; the irides are silvery; the colour of the head and body is dark brown on the upper part. passing into pale wood-brown underneath, and on the sides and belly; the colour of the dorsal, caudal, and anal fins dark brown; the pectoral and ventral fins lighter; and my specimens having been many months preserved in spirits have lost some of the lighter tints which the coloured figure of Mr. Thompson's fish, and that also of MM. Fries and Ekström exhibit. The formula of the fin-rays is—

D. 19 + 6 : P. 13 : V. 1 + 5 : A. 6 + 7 : C. 12, and 2 shorter rays.

The number of Scales forming the lateral line are thirty-two, with four rows above to the dorsal edge, and eleven below to the anal aperture.

The vignette represents a mode of fishing practised in South America.

ABDOMINAL
MALACOPTERYGII.   CYPRINIDÆ.

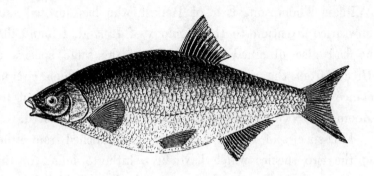

THE POMERANIAN BREAM.

Abramis Buggenhagii.

Cyprinus Buggenhagii,	*Carpe de Buggenhagen,*	BLOCH, vol. iii. pl. 95.
Abramis,	„ *Large Scaled Bream,*	THOMPSON, Zool. Proc. 1837, p. 56.

I AM indebted to Mr. William Brandon of Chancery Lane for a fine specimen of this fish which was sent me in the year 1836 from Dagenham in Essex. Mr. Brandon who is the renter of the waters at Dagenham Breach, so well known to the London anglers, and who has frequently favoured me with examples of other species from this locality, having taken this Bream in his net with other fish, very kindly sent it to me with a note stating that it differed from the Bream he had usually caught in that water; and finding when he reached home and made closer examination, that it did not accord with the characters of either of the Bream figured and described in the British Fishes, he begged my acceptance of it, hoped it might prove of some interest, and requested to know what it was. I understand from Mr. Brandon that he has since at different times taken from twenty to thirty of the same sort.

The characters of this species are so decided, that I had no difficulty in identifying it as the *Cyprinus Buggenhagii* of Bloch; and on the next visit to London of my friend William Thompson, Esq. of Belfast, who has devoted such unwearied attention to the Zoology of Ireland, I found that he had also obtained an example of the same species of Bream from the river Lagan, near Belfast, which circumstance was made public in the printed Proceedings of the Zoological Society for 1837, page 56, as already quoted.

This species of Bream is at once distinguished from either of the two species which have been hitherto found in this country, by the greater thickness of its body, which is equal to half its depth; while in either of our other Bream the thickness of the body is only equal to one third of its depth; the scales of this species are also larger in proportion, although the figure here given, not having been drawn on a comparative scale with them, does not exhibit this peculiarity. The anal fin is shorter and has a smaller number of rays than that of *Abramis blicca*, which in its turn has its anal fin smaller, and with fewer rays than that of *Abramis vulgaris*, which is the Bream most generally known in this country.

This new species was first described by Bloch from specimens found in Swedish Pomerania, in the river Péne, and in the lakes communicating with it. The specimens were sent to Bloch by M. Buggenhagen, and hence the trivial name which has been devoted to it for specific distinction. I have also called it the Pomeranian Bream, considering it no objection to attach to this fish the name of the country in which it was first discovered, although it may happen to have been afterwards found elsewhere. The fish attains to the length of twelve or fourteen inches in that country according to Bloch; the flesh is white, but not much in request on account of the number of small bones which are found in it. It is taken in the same manner and by the same means as the common

Bream; and Bloch reports that the fishermen are greatly pleased when they take this fish in their nets: they have learned by experience that when this Bream appears they shall have a successful fishery: they believe that the other Bream follow this fish, and the name they have accordingly bestowed upon it in that country signifies guide or conductor. Except in Bloch I do not find this species included in either of the works I possess, or have yet gained access to, which treat of the fresh-water fishes of the different countries of the continent of Europe.

The specimen of this fish from Dagenham, from which the following description was taken, measured fifteen inches in length, of which the head was three inches, or, compared to the whole length of the fish, as one to five; the depth of the body a little in advance of the line of the first ray of the dorsal fin, where the body is deepest, five inches, or one third of the whole length; the thickness of the body two inches and a half, equal to half the depth, or one sixth of the whole length; the head is rather small and pointed, the mouth is also small; the diameter of the eye about one fifth of the length of the head, the iris silvery and about the same breadth as the pupil; the operculum rather large and angular; the pectoral fin rather small; half the ventral fin, in advance of a vertical line falling from the origin of the first dorsal fin ray; the dorsal fin commences exactly half way between the point of the nose and the end of the caudal fin; but the base of the dorsal fin in this fish is longer than the base of the same fin in either of our other species of Bream; the anal fin is shorter than that of the shortest of the other Bream, and has three rays less; it is also less falcate in form, or more equal in the length of its rays; the tail in shape at its posterior edge rather lunate, the outer rays elongated; the formula of the fin rays is

D. 12: P. 17: V. 9: A. 19: C. 19: Vertebræ 41.

The number of punctured scales forming the lateral line fifty-two ; from the anterior edge of the dorsal fin to the lateral line, following the oblique direction of the scales, there are eleven scales ; from the lateral line downwards to the base of the pectoral fin, four scales, not including in either enumeration the punctured scale of the lateral line itself.

Upper part of head and back dark blackish blue, becoming lighter on the upper part of the sides, and passing into silvery white on the lower part of the sides and belly ; pectoral fin, dorsal fin and tail, bluish brown, tinged with pale red ; ventral and anal fins with less brown and more pale red.

The vignette represents the bones of the head in the common Bream.

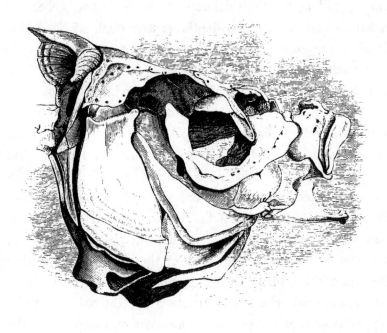

*ABDOMINAL
MALACOPTERYGII.*   *ESOCIDÆ.*

EUROPEAN HEMIRAMPHUS.

Hemiramphus Europæus.

Hemiramphus Europæus, *European Hemiramphus*, Mag. Nat. Hist. 1837, p.505.

IN a valuable communication on the Fishes of Cornwall, made to the Linnean Society some years ago by Jonathan Couch, Esq. of Polperro, which was published in the fourteenth volume of the Transactions of that Society, the author thus expresses himself in reference to a small fish which appeared to be a species of the genus *Hemiramphus :*—" I have met with a species which I have never seen described, unless it be the *Esox Brasiliensis Linn. Syst. Nat.* (Hemiramphus Brasiliensis Cuv.) It was taken by me in the harbour at Polperro, in July 1818, as it was swimming with agility near the surface of the water. It was about an inch in length, the head somewhat flattened at the top, the upper jaw short and pointed, the inferior jaw much protruded, being at least as long as from the extremity of the upper jaw to the back part of the gill-covers. The mouth opened obliquely downwards ; but that part of the under jaw which protruded beyond the extremity of the upper, passed straight forward in a right line with the top of the head. The body was compressed, lengthened, and resembled that of the Garpike, *Esox belone.* It had one dorsal and one anal fin, placed far behind and opposite to each other. The tail was straight ; the colour of the back was a bluish green, with a few spots ; the belly silvery."

In August 1837, Dr. Clarke of Ipswich favoured me with a letter, of which the following is an extract :—" My brother, Mr. Edward Clarke of Ipswich, who is particularly interested in the study of British fishes, was examining the sea-shore in the vicinity of Felixtow, a village in Suffolk, between Harwich and Orford, a few days ago, August 7th 1837, when he observed a shoal consisting of myriads of small fish, which, upon a nearer examination, he supposed to be the young of the Garfish. As he had previously not found any so small, he secured a few specimens ; and, upon bringing them home and examining them, they were found not to be the young of the Garfish, but those of a species of *Hemiramphus*. From their being so very young, it probably may be difficult to determine whether they belong to a described species ; but from the circumstance of their having been seen in great abundance in a small pool left by the retiring tide, it is, I think, pretty evident that the ova must have been deposited and vivified in the neighbourhood of our shores. I send you the fish, thinking that an examination of the specimens themselves will be far more satisfactory than any figures or description of my own. One specimen was taken about double the size of those now sent to you."

The representation of this fish is half as large again as the natural size. It can scarcely be doubted from the quantity of fry seen, as well as from their very small size, that the spawn from which they were produced must have been deposited on our shores by the parent fish ; and yet, as far as we are aware, these parent fish have hitherto escaped capture. This might not appear very extraordinary ; but from the circumstance that the size attained by the fry in the months of July and August, as well as the general similarity in the form and appearance of the *Hemiramphus* to our well-known Garfish and Saury-pike, would lead to the belief that the *Hemiramphus* visited our shores about the same time of the year as these fishes. The Garfish appears on the coast in April, and spawns in May; the Saury-pike makes its first appearance in June. For these fish, but particularly for the former, nets are worked on various parts of the coast, and considerable quantities are taken; but no adult specimens of *Hemiramphus*, unless we are to suppose they have remained hitherto unrecognised by the fishermen. It is also not a little singular, that up to the present time, with the exception of the small specimens already referred to, as taken at two places very distant from each other, no example of any species of Hemiramphus has been found, either in the Mediterranean, the Channel, or in the North seas. I have lately had an opportunity of conversing with two eminent foreign naturalists, to whom I showed the specimens, who agreed with me that no adult species of *Hemiramphus* had been recorded as found in the seas of Europe.

One question may be hazarded,—Is this fish, with its unequally developed jaws, the very young state of our common Garfish (*Belone vulgaris*)? Except in the peculiarity of the mouth, it is certainly very like it; but our young Garfish of the year taken in December, when they are about seven inches long, specimens of which I possess, have the upper jaw of the same comparative length as the lower one. Another season or two will probably decide the question, and it will be as interesting, in an ichthyological point of view to be able to determine this to be the young state of *Belone* as that there exists a true *Hemiramphus* in the seas of Europe.

The two examples obligingly sent me by Dr. Clarke, are too young and too minute to make any attempt to define specific characters desirable, beyond such as the remarks of Mr. Couch, and the representation here given will supply; and I only propose, for distinction's sake, that it should be called *Hemiramphus Europæus*.

ABDOMINAL
  MALACOPTERYGII.   *ESOCIDÆ.*

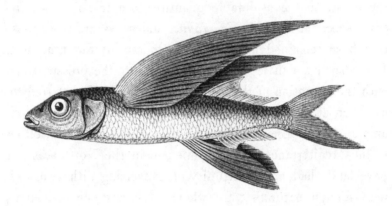

THE GREATER FLYING FISH.

Exocætus exiliens.

Hirundo,	BELON, p. 195.
Mugil alatus.	RONDELET, Lat. E. p. 267.
Muge volant,	" Fr. E. p. 211.
,, ,,	WILL, tab. P. f. 4.
Muge volant,	DUHAMEL, Pl. 2, Sec. 8, pl. 6, f. 3.
Hirondelle de mer,	,, Pl. 2, Sec. 3, pl. 22, f. 2.
Exocætus exiliens,	*Le Muge volant,* BLOCH, pt. 12, pl. 397.

IN a Cornish Fauna, by Jonathan Couch, Esq. which has recently been published for the Royal Institution of Cornwall, Mr. Couch has included a species of Flying Fish which threw itself on to the Quay at Plymouth, and the specimen is still preserved. From an inspection of this example Mr. Couch was enabled to determine that it was the Greater Flying Fish, *Exocætus exiliens,* or *Le Muge volant* of Bloch, the well-known species of the Mediterranean ; and Mr. Couch adds, that he has reason to believe, from the dimensions as given to him by the possessor, that the individual Flying Fish which was found at Helford, where it was discovered on the sand, having just then expired, was of the same species. This specimen, which is in the possession of Mr. John Fox of Plymouth, measures sixteen inches in length.

The elongated ventral fins, placed very far backwards, readily serve to distinguish this fish, which has long been well known in the Mediterranean, and was, I believe, first figured by Belon in the year 1553, by Rondelet in his Latin edition in 1554, and in the French edition printed at Lyons in 1558. For the general habits of the Flying Fish, the reader may consult the first volume of the History of British Fishes, page 398. Bloch says that the Greater Flying Fish attains the length of eighteen inches; and the specimen from which the representation in the work of Duhamel was taken, measured sixteen inches. Bloch says this fish is found in the Red Sea as well as in the Mediterranean. Our countryman Willoughby saw it in Calabria. Rondelet states that it is found in quantity at the mouth of the Rhone, and Duhamel mentions that, besides being plentiful in the Mediterranean, it had also been taken in the ocean. The flesh of this fish is rich, and is said to be more delicate than that of the herring.

The head is wide and flat on the top, but somewhat angular underneath; the mouth is small, the lower jaw rather longer than the upper; both jaws are furnished with pointed teeth, those in the lower jaw being the smaller of the two; the eyes are large, the irides silvery, the pupil dark blue; the nostrils large, and placed rather nearer to the eye than to the point of the nose; the operculum has the appearance of polished steel; the body of the fish is covered with large scales, which adhere but slightly; the upper part of the body is a fine blue colour, the lower part silvery white; the lateral line is placed very low down and runs throughout its whole length, but little above, and parallel to, the ventral profile; the pectoral

fins are very large and of a fine transparent blue colour; the
ventral fins long, and almost rounded at the end; the dorsal
and anal fins are falcate, beginning and ending nearly on the
same plane; the tail consists of two unequally sized lobes, of
which the lower lobe is the larger.   The fin ray formula, according to Bloch, is

B. 10: D. 11: P. 18: V. 6: A. 12: C. 22.

According to M. Risso, the female is heavy with roe in
the spring, and is remarkable for the variations that occur in
the number of the rays of her fins.

END OF THE SUPPLEMENT TO THE FIRST VOLUME.

London: Printed by Samuel Bentley, Bangor House, Shoe Lane.

SUPPLEMENT

TO THE SECOND VOLUME OF THE

HISTORY OF BRITISH FISHES.

ABDOMINAL
  MALACOPTERYGII.   *SALMONIDÆ.*

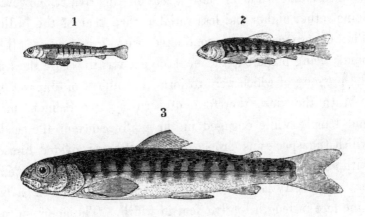

THE SALMON.

Salmo salar,   Auctorum,   British Fishes, vol. ii. p. **1.**

Since the publication of that part of the History of British Fishes which contains an account of the Salmon, Mr. John Shaw of Drumlanrig, Dumfriesshire has printed in the Edinburgh New Philosophical Journal for July 1836 and January 1838, detailed particulars of various interesting and valuable experiments, made by himself, on the developement and growth of the fry of the Salmon, from their exclusion from the *ova* to the age of seven months.

Three ponds, varying in size, one eighteen feet by twenty-two, the second eighteen feet by twenty-five, and the third thirty feet by fifty, were prepared at a convenient distance from a Salmon river, (the Nith,) the ponds two feet deep, thickly embedded with gravel, and supplied from a small stream of spring water, in which the larvæ of insects were abundant. The distance from the river to the ponds is stated as rather less than fifty yards, a proximity, it is observed, " sufficient to place the young fish confined in them on a similar footing with those in the river, so far as situation is concerned. The average temperature of the water is also nearly the same in both ; that of the rivulet, however, being rather higher and less variable than that of the Nith." The experiments were conducted with great care. The ponds being prepared, the next object was to secure the fish, the progeny of which were to form the subject of observation. " With the view, therefore, of securing two Salmon, male and female, while engaged in the performance of the act by which the species is propagated, Mr. Shaw provided himself with an iron hoop five feet in diameter, on which he fixed a net of a pretty large mesh, so constructed as to form a bag nine feet in length by five feet in width. The hoop and net were then attached to the end of a pole nine feet long, thus forming a landing net on a large scale. The weight of the net with its iron hoop being upwards of seven pounds, it instantly sunk to the bottom when thrown into the water."

" Being thus prepared with the means of carrying his experiment into execution, Mr. Shaw proceeded to the river Nith on the 27th January 1837, and readily discovered a pair of adult Salmon depositing their spawn. Before proceeding to take the fish, he formed a small trench in the shingle by the edge of the stream, through which he directed a small current of water from the river two inches deep. At the end of this trench was placed an earthenware basin of considerable size, for the purpose of ultimately receiving the *ova.* The fish were then, at one instant, both enclosed in the hoop, and allowed to find their way into the bag of the net by the aid of the stream. Having drawn them ashore, the female, while still alive, was placed in the trench, and a quantity of the *ova* pressed from her body. The male was then placed in the same situation, and a quantity of the milt being pressed from his body, passed down the stream, and thoroughly impregnated the *ova.* The spawn was then transferred to the basin, and deposited in the stream of the feeder to the first pond. The temperature of the stream was 40 deg., and that of the river from which the Salmon were taken 36 deg. The skins of the parent Salmon were preserved and exhibited, that no doubt as to the species might be entertained. The weight of the male when taken was sixteen pounds, and that of the female eight pounds."

Without following Mr. Shaw through the details on this, as on three or four other occasions, it may be sufficient to state, that the young fish ruptures the external capsule of the *ovum,* or may be said to be hatched in about

114 days when the temperature of the water is 36°
101   „   „   „   „   43°
90   „   „   „   „   45°

When first emerging from the membrane within which the young fish has been enclosed, the remains of the yolk or vitelline portion of the *ovum* is still attached by its own capsule to the abdomen of the fish as represented in the figure No. 1, which is taken from a specimen given me some years ago by Sir William Jardine. The remains of the yolk supplies nourishment to the young fish till it is able to take food by the mouth. Mr. Shaw has ascertained that the yolk is absorbed in twenty-seven days. At the end of two months the young fish is one inch and one quarter long, and the figure No. 2 is from Mr. Shaw's representation. At the end of four months the young fish measures two inches and a half

B 2

in length, and at the end of six months it had attained the length of three inches and three quarters.

From these experiments Mr. Shaw infers, that the growth of the young of the Salmon has been much overrated; that as the young Salmon in its progress assumes at a certain age the markings and colour of the Parr; that the Parr, as a distinct species, does not exist; and finally, that the young of the Salmon do not go down to the sea till they are more than twelve months old at the least, that is sometime during their second year, if not still later than that.

That the young of the Salmon, from their particular appearance at a certain age, have been constantly called Parrs, I readily admit; but so have also the young of two other migratory species, *S. trutta* and *S. eriox*; I think, therefore, that this is not conclusive evidence of the non-existence of a distinct small fish, to which the name of Parr ought to be exclusively applied; it rather shows the want of power among general observers to distinguish between the young of closely allied species, three or four of which are indiscriminately called Parrs.

That the rate of growth in the young of the Salmon has been exaggerated may be very true; but the rate of the growth of the fry in Mr. Shaw's ponds cannot be expected to equal that which would have taken place in the open river. Circumscribed in space over which to roam, and limited in food, as to variety at least, if not in quantity, in small ponds, the growth would be retarded in proportion; and this circumstance seems proved by Mr. Shaw's own remark, in which he states that the fish in the third pond (the largest pond of the three) " were considerably larger than those in the first pond, the difference in length at the age of six months amounting to an inch, or more than one fifth.

That the young fish do not go down to the sea till their second year, I am willing to believe on Mr. Shaw's authority, because he has devoted great attention to the subject, and has for years had opportunities for observation which give great weight to his opinion. I have thus purposely adverted to the experiments of Mr. Shaw on account of their great interest, merit, and value ; and because I am now enabled, through the kindness of Thomas Lister Parker, Esq., to offer a continuation of remarks on the growth of the Salmon in fresh water, which illustrate and confirm some of the views of Mr. Shaw ; and in order to prevent any misconception of the terms employed, I shall speak of the young Salmon of the first year as a Pink ; in its second year, till it goes to sea, as a Smolt ; in the autumn of the second year as Salmon Peal, or Grilse, and afterwards as adult Salmon.

In the autumn of the year 1835, Thomas Upton, Esq. of Ingmire Hall, situated between Sedbergh and Kendal, began to enlarge a lake on his property, and in the spring of 1836, some Pinks from the Lune, a Salmon river which runs through a valley not far from the lake, were put into it. This lake, called Lillymere, has no communication with the sea, nor any outlet by which fish from other waters can get in, or by which those put in can get out. The Pinks when put into Lillymere did not certainly weigh more than two or three ounces each. Sixteen months afterwards,—that is, in the month of August 1837, Thomas L. Parker, Esq. then visiting his friend, fished Lillymere, desirous of ascertaining the growth of the Pinks, and with a red palmer fly caught two Salmon Peal in excellent condition, silvery bright in colour, measuring fourteen inches in length, and weighing fourteen ounces. One was cooked and eaten, the flesh pink in colour, but not so red as those of the river ; well flavoured, and like that of a Peal. The other was sent to me in spirit of wine, and a drawing of it immediately taken. In the month of July 1838, eleven months after, another small Salmon was caught, equal to the first in condition and colour, about two inches longer and three ounces heavier. No doubt was entertained that these were two of the Pinks transferred to the lake in the spring of 1836, the first of which had been retained sixteen months, and the other twenty-seven months, in this fresh-water lake.

Desirous of ascertaining the appearance of the young Salmon at periods intermediate between the states as Pinks and Salmon Peal, other experiments were tried. Pinks in the river Hodder in the month of April are rather more than three inches long, and are considered to be the fry of that year: at this time, Smolts of six inches and a half are also taken. The smolts are considered as the fry of the previous year, and are distinguished by the blue colour on the upper half of their body, the silvery tint of the lower half, and the darker hue of the fins generally as compared with those of the Pink. In this state as to colour, the Smolts are said to have assumed their migratory dress and go down to the sea in May. In June the young Pink in the Hodder measures about four inches; in July it measures five inches, and no Smolts are then found in the river. To be further convinced of this change, and the length of time required to produce it, a Pink put into a well at Whitewell* in the forest of Bowland in November 1837, was taken out in the state of a Smolt of six inches and a quarter in July 1838. In another instance more Pinks by Mr. Upton's directions were put into Lillymere in September 1837, and Mr. Parker caught five or six in the state of Smolts of seven and a half inches in August 1838. In referring to the particular size of the Pinks in the river Hodder at stated periods, it may be necessary to remark that the Pinks of different rivers, and even in the same river, will be found to vary in size, depending on the time at which the spawn was deposited, the temperature of the season, and other causes.

* For a view of Whitewell, see British Fishes, vol. ii. p. 88.

I may here observe that I am indebted to the kindness and liberality of Thomas Lister Parker, Esq. for a variety of specimens, as well as for the requisite information concerning them. Of the various fishes, when received, accurate drawings were immediately made, and coloured representations of six examples at different ages, in illustration of this subject, are in preparation, and may be had distinct from this supplement.

A knowledge of the growth of young Salmon in a freshwater lake, as here described, and the experiment has succeeded elsewhere,* may be useful to those gentlemen who possess lakes near Salmon rivers from which they can supply them with Pinks: whether the Salmon thus prevented going to salt water will still retain sufficient constitutional power to mature their roe, and by depositing it in the usual manner, as far as circumstances permit, produce their species, would be a subject worthy of further investigation. That the rate of growth in young Salmon has some reference to the size of the place to which they are restricted, as hinted when describing Mr. Shaw's experiments, receives further confirmation in these river, lake, and well specimens. The Smolt taken from the well in July 1838, where it had been confined for eight months, was rather smaller in size at that time than the Smolts in the Hodder in the preceding April, though both were Pinks of the same year, namely 1837. The Smolt taken from the lake in August 1838, which then measured seven inches and a half, had also grown more rapidly than that in the well, but had not acquired the size it would have gained had it been allowed to go to sea. Further, it may be observed, that the Salmon Peal from the lake in August 1837, then eighteen months old, though perfect in colour, is small for its age; while that of July 1838, or twenty-nine months old, is comparatively still more defi-

* See British Fishes, vol. ii. p. 21.

cient in growth, supposing both fish to have resulted from
Pinks of the year 1836, and put into the lake at the same
time ; of which there was no doubt, since the lake, the formation of which, though commenced in the autumn of 1835,
was not finished till February 1836, soon after which the
first Pinks were put in.

In March 1839, Mr. Upton put six dozen Charr from
Windermere into his lake.

The vignette represents the bones of the head in the
Salmon.

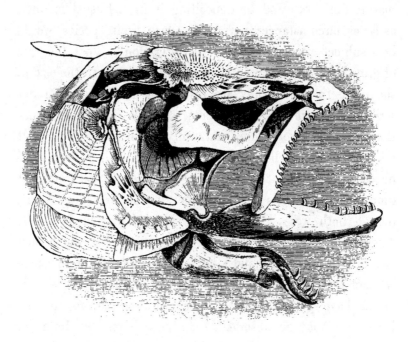

*ABDOMINAL
MALACOPTERYGII.*   *SALMONIDÆ.*

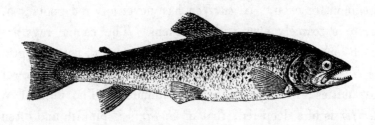

THE LOCHLEVEN TROUT.

Salmo Levenensis, WALKER.
    „ *cæcifer,*   PARNELL.

I AM indebted to Dr. Parnell for the loan of a beautiful specimen of this Trout from which the figure was taken, and the following account of it by Dr. Parnell is from the seventh volume of the Memoirs of the Wernerian Natural History Society of Edinburgh.

" This fish is considered by most writers on British Ichthyology to be identical with *Salmo fario,* the common Trout, differing from it only in the colour of the flesh, and in having no red spots on the sides. It is true that food and season may have a great share in diminishing or increasing the external markings and colour of the flesh ;* but they can have no effect in shortening or lengthening the rays of the fins, or in adding numbers to the cæcal appendages."

" The differences that exist between *S. cæcifer* and *S.*

---

* James Stuart Monteath, Esq. of Closeburn, caught a number of small river Trout, and transferred them to a lake (Loch Ettrick) where they grew rapidly ; their flesh, which previously exhibited a white chalky appearance, became in a short time of a deep red, while their external appearance remained the same from the time they were first put in.

fario are very striking. The pectorals in *S. cæcifer* when expanded are pointed, in *S. fario* they are rounded. The caudal fin in *S. cæcifer* is lunated at the end; in *S. fario* it is sinuous or even. *S. cæcifer* has never any red spots; *S. fario* is scarcely ever without them. The caudal rays are much longer in *cæcifer*, than in *fario*, in fish of equal length. In *S. cæcifer* the tail fin is pointed at the upper and lower extremities; in *S. fario* they are rounded. The flesh of *S. cæcifer* is of a deep red, that of *S. fario* is pinkish and often white. The cæcal appendages in *S. cæcifer* are from sixty to eighty in number; in *S. fario* I have never found them to exceed forty-six."

" Lochleven (of which the barren isle and now dismantled castle are famous in history as the prison-place of the beautiful Queen Mary) has long been celebrated for its breed of Trout. These, however, have fallen off of late considerably in their general flavour and condition, owing, it is said, to the partial drainage of the Loch having destroyed their best feeding ground, by exposing the beds of fresh-water shells, the animals of which form the greater portion of their food.* They spawn in January, February, and March."

" The fish described does not appear to be peculiar to this Loch, as I have seen specimens that were taken in some of the lakes in the county of Sutherland with several other Trout, which were too hastily considered as mere varieties of *S. fario.* It is more than probable that the Scottish lakes produce several species of Trout known at present by the name of *S. fario,* and which remain to be further investigated."

Dr. Richardson, who has had opportunities of examining very fine specimens of this celebrated Trout, considers it distinct from *S. fario,* and has pointed out some of the differences between them : the scales are thick, and when dry

* There are two or three varieties of *S. fario* in Lochleven with white and pinkish flesh, which are much inferior in flavour to *S. cæcifer.*—Encyc. Brit.

exhibit a small ridge in the centre of each, not perceived in other Trout : in its large and strong fins, and in its habit, as stated by Dr. Parnell, of spawning in spring, it differs from *S. fario*, which spawns in autumn, and resembles some of the large species of Trout of the great northern lakes. Three individuals of the Lochleven Trout dissected by Dr. Richardson had each seventy-three pyloric cæca, and in one of them fifty-nine vertebræ were counted. The largest of the specimens measured twenty inches and a quarter, including the caudal fin, and two inches less to the end of the scales.

Dr. Parnell's description, taken from a specimen measuring one foot in length, is as follows :—" Head rather more than one-fifth of the whole length ; caudal fin included ; depth between the dorsal and ventral fins less than the length of the head. Gill cover produced behind ; basal margin of the operculum oblique ; preoperculum rounded ; end of the maxillary extending back as far as the posterior margin of the orbit. Colour of the back deep olive green ; sides lighter ; belly inclining to yellow ; pectorals orange, tipped with grey ; dorsal and caudal fins dusky ; ventral and anal fins lighter ; gill cover with nine round dark spots ; body above the lateral line with seventy spots ; below it ten ; dorsal fin thickly marked with spots of a similar kind ; anterior extremities of the anal and dorsal fins without the oblique dark bands which are so conspicuous and constant in many individuals of *S. fario*. First dorsal fin placed half-way between the point of the upper jaw and a little beyond the fleshy portion of the caudal extremity of the body ; all the rays branched except the two first ; the third ray the longest, equalling the length of the long caudal ray ; the seventh as long as the base of the fin ; the last considerably more than half the length of the third, equalling the length of the middle caudal ray ; fin even at the end (in many specimens it is concave, with the last ray longer than the preceding one) ; caudal fin crescent-shaped, the middle ray rather more than half the length of the longest ray; third ray of the anal fin the longest, equalling the length of the fifth dorsal ray; the last ray as long as the base of the fin, ventral fin equalling the length of the fifth ray of the anal; the third ray the longest; third ray of the pectorals equalling the length of the long caudal ray; the last ray half the length of the fin. Teeth stout and sharp, curved slightly inwards; thirty-two in the upper jaw, eighteen on the lower; twelve on each palatine bone; thirteen on the vomer; and eight on the tongue. Scales small and adherent; twenty-four in an oblique row between the middle dorsal ray and the lateral line; flesh deep red; cæca eighty. The number of fin rays.

D. 12: P. 12: V. 9: A. 10: C. 19.

The vignette represents the castle and the island in Loch-leven.

ABDOMINAL
MALACOPTERYGII.   *SALMONIDÆ.*

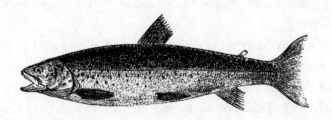

THE GREAT LAKE TROUT,

OR GREAT GREY TROUT.

Salmo ferox, JARDINE and SELBY, and British Fishes, vol. ii. p. 60.

SINCE the publication of the History of British Fishes, in which the existence of the Great Lake Trout in Lough Neagh, was recorded as ascertained by Mr. Thompson of Belfast, that gentleman, following up his zoological researches, has learned that this fish exists in Lough Corrib, in the county of Galway, and also in Lough Erne, in the county of Fermanagh, thus proving it, to use Mr. Thompson's words, to be an inhabitant of the three largest lakes in Ireland, and it will probably yet be found in most of the lakes of any considerable extent in that country. Mr. Thompson has very kindly supplied me with a young fish of this species from which our representation was taken, and which, differing from specimens of large size in having the spots more numerous, may be an acceptable addition. As mentioned in the former volume, this Lake Trout, when small, is in Ireland called a *Dolachan ;* when large a *Buddagh,* and they are usually caught on night lines baited with a perch or a pollan. The mode of taking this fish in the large Lochs of Scotland is given in the second volume of the British Fishes, page 61.

I have reason to believe that this same species of Great Grey Trout is an inhabitant of some of the large lakes of Scandinavia.

Sir Thomas Maryon Wilson, Bart. visited Sweden last summer, ascending the Gota river in his yacht, the Syren, and passing through the celebrated sluices of Tröllhattan, cruised and fished in Lake Wenern, visiting his friend Mr. Lloyd, who resides near the southern extremity of this noble lake.

Sir Thomas M. Wilson brought back with him five or six skins of the Great Trout of the lake, which were caught by spinning with a bleak, and must, from their large size, have afforded some excellent diversion. The largest of these specimens measured forty-two inches in length, and weighed about thirty-four pounds: the next largest weighed thirty-two pounds: the third twenty-seven pounds, besides others of smaller size. These large Trout, and larger than these are seldom seen, are observed to be males; the females, according to Mr. Lloyd, who has lived for some years on the borders of the lake, rarely exceed twenty or twenty-two pounds. The number of fin rays in these specimens averaged

D. 13: P. 14: V. 9: A. 11: C. 19.

Among other fish taken by Sir Thomas Wilson, was a large specimen of the Ide, *Leuciscus idus* of authors. This fish, which resembles our English Chub, was caught in the Gotha Elf, a short distance above the falls of Tröllhattan, whilst trolling for pike on a windy day: its weight was between four and five pounds. The skins of these various specimens were effectually preserved and mounted after they were brought to England.

Sir Thomas M. Wilson did me the favour to show me his numerous sketches of scenery, taken during this trip, which include views of the Gota river, the cities and country on its banks, the celebrated falls of Tröllhattan and parts of Lake Wenern at different points of view ; very kindly allowing me the use of a coloured drawing from which the vignette below, on a reduced scale, was taken. This view represents Mr. Lloyd's cottage on the eastern bank of the Gota ; the yacht of Sir Thomas Wilson lying at anchor immediately opposite ; with the remarkable and finely wooded hills of Hunneberg and Halleberg, so much celebrated for the peculiarity of their geological structure, bounding the distance.

ABDOMINAL
   MALACOPTERYGII.   *SALMONIDÆ.*

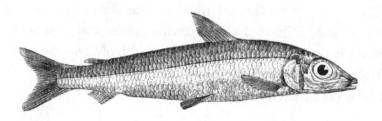

THE HEBRIDAL SMELT.

Osmerus Hebridicus,   *Hebridal Smelt*,   Yarrell, Supplement to Brit. Fishes.

I am indebted to Mr. William Euing of Glasgow for the opportunity of making known a new species of Smelt which that gentleman did me the kindness to send to me in the month of November 1837. This fish is at once clearly distinguishable from our long-known and highly-esteemed favourite, the common Smelt, and is the more interesting from the circumstance of its being—at least, as far as I am aware—entirely new to Ichthyology. Mr. Euing passed part of the summer of 1837 near Rothsay in the Isle of Bute; and the Smelt in question was brought to him by a fisherman, who stated that he caught it on a hand line in the bay of Rothsay, about two hundred yards from the shore, in twelve fathom water; that it was, though well known, but rarely seen. This specimen measured six inches and a half; but another example of the same sort, measuring eight inches in length, that was taken near the same place in June 1836, was full of roe, and when first caught the cucumber-like smell, so peculiar to the Smelt, was in this species also very apparent.

Unable to find any notice of a second species of Smelt in Europe in any Ichthyological work with which I am acquainted, I have little doubt that this fish has not been previously described; and in reference to the locality in which alone it has been as yet taken, I have ventured to name it the Smelt of the Hebrides, *Osmerus Hebridicus.*

The specimen sent me by Mr. Euing, measuring six inches and a half in length, is one inch and one eighth deep at the commencement of the dorsal fin, at which part the body is deepest; the thickness of the body compared to the depth is as one to two, or exactly half: the length of the head is one inch and three eighths, and is, in reference to the whole length of the head and body, without the tail, as one to four. The jaws are nearly equal in length, without teeth upon either; but there are four long teeth upon the tongue; the eye is very large, the diameter almost equal to one third of the whole length of the head, and placed at a distance of little more than its own diameter from the point of the nose: the upper surface of the head is flattened, descending by a rapid slope to the nose; the line of the lower jaw straight; the posterior edge of the operculum rounded; the back of the fish, or its dorsal outline, slightly arched; the abdominal line nearly straight; the sides compressed. The dorsal fin commences half way between the point of the nose and the anterior edge of the adipose or rayless dorsal fin, the longest ray nearly twice the length of the base of the fin; the last dorsal fin ray but three, the same length as the base of the whole fin. The adipose fin is placed very near the tail; the tail itself deeply forked. The pectoral fin reaches to the plane of the commencement of the dorsal, and its length, if turned forwards, would reach to the centre of the eye. The ventral fin is in a vertical line under the last ray of the dorsal fin; there is a slender axillary scale; but the ends of the ventral fin rays being injured, the length of the fin cannot be mentioned.

VOL. II.   C

The anal fin has its last ray underneath the posterior edge of the adipose fin ; but the rays of the anal fin are also broken. The formula of the fin rays is as follows :—

D. 11 : P. 14 : V. 12 : A. 12 : C. 19.

The scales are large and deciduous, the lateral line prominent and nearly straight.   Below the lateral line for the whole length of the body two rows of the scales are silvery white, forming a conspicuous elongated band, like that to be observed in the Atherine,* the rest of the body and fins dull amber colour, the gill covers silvery and iridescent.

The figure of our well-known common Smelt is inserted as a vignette to exhibit the comparative characters of the two species.

* British Fishes, vol. i. p. 214.

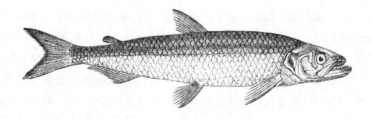

*ABDOMINAL
MALACOPTERYGII.*

SALMONIDÆ.

THE POWAN.

Coregonus La Cepedei,		The Powan, PARNELL, Annals of Nat. Hist. vol. i. p. 161.
,,	*clupeoides,*	*The Herring-like Coregonus,* LACEPEDE, Hist. Nat. du Poiss. 8vo edit. tom. x. p. 386.

DR. PARNELL, whose Ichthyological investigations in Scotland have not been confined to the "Fishes of the Forth," only, has described in the first volume of the Annals of Natural History a species of *Coregonus*, to which he has attached the name of *Lacepedei*, this species having been first noticed, or perhaps distinguished, by this celebrated French naturalist. This fish is found in Loch Lomond, one of the largest and most picturesque lakes in the west of Scotland. It is not unlikely that some of the species of *Coregoni* found in the northern lakes of England, Scotland, and Ireland, may exist in the lakes of Scandinavia, M. Nilsson, Professor of Natural History at Lund, describing in his *Prodromus Ichthyologiæ Scandinavicæ* no less than eight species as belonging to that country ; but from a certain general agreement in

c 2

the characters of the *Coregoni*, it is difficult to refer our species with certainty in the absence of foreign specimens with which to make actual comparison.

It appears, on reference to his Natural History of Fishes, that Lacépède became aware of the existence of this *Coregonus* in Loch Lomond by the communication of M. Noel, who visited Scotland in August 1802. Although some little differences appear in the descriptions of this fish, as given by Lacépède and Dr. Parnell, there is little doubt that both authors had the same species under consideration. This fish bears, as observed by Dr. Parnell, considerable resemblance in appearance and also in the number of its fin-rays to the *Salmo Wartmanni* of Bloch, part 3, tab. 105, a species of *Coregonus*, named after a learned physician, who first described it. It is found in some of the lakes of Switzerland, and also in lake Constance; but Lacépède, to whom the *Wartmanni* was known, considered the Loch Lomond *Coregonus* distinct. It is thus described by Dr. Parnell, from a specimen fourteen inches in length.

" Head long and narrow, of an oval form, about one-fifth the length of the whole fish, caudal fin included; depth of the body between the dorsal and ventral fins less than the length of the head. Colour of the back and sides dusky blue, with the margin of each scale well defined by a number of minute dark specks; belly dirty white; the lower portion of the dorsal, pectoral, ventral, and anal fins dark bluish grey; irides silvery, pupils blue. First ray of the dorsal fin commencing half-way between the point of the snout and the base of the short lateral caudal rays; the first ray simple, the rest branched; the second and third the longest, equalling the length of the pectorals; the seventh ray as long as the base of the fin; the last ray one-third the length of the fourth; adipose fin large and thin, situate midway between the base of the fourth dorsal fin-ray and the tip of the long upper ray of the caudal fin ; anal fin commencing half way between the origin of the ventral fin and the base of the middle caudal ray ; the first ray simple, the rest branched ; the second rather the longest ; the third as long as the base of the fin ; the last ray half the length of the fifth ; ventral fins commencing under the middle of the dorsal ; the third ray the longest, equalling the length of the same ray of the dorsal ; pectorals long and pointed, one-sixth the length of the whole fish, caudal fin included ; the first ray simple ; the second and third the longest, the last short, not one-fourth the length of the first ; tail deeply forked, with the long rays of the upper portion curving slightly downwards, giving the fin a peculiar form. Gill cover produced behind ; the basal line of union between the operculum and suboperculum oblique ; the free margin of the latter slightly rounded ; pre-operculum angular ; snout prominent, somewhat of a conical form, extending beyond the upper lip ; jaws of unequal length, the lower one the shortest. The maxillary bone broad, the free extremity extending back to beneath the anterior margin of the orbit. Teeth in the upper jaw long and slender, about six in number ; those on the tongue shorter and more numerous. Eyes large, extending below the middle of the cheeks ; lateral line commencing at the upper part of the operculum, and running down the middle of the sides to the base of the middle caudal ray. Scales large and deciduous, eighty-four forming the lateral line, eight between the dorsal fin and lateral line, and the same number between the lateral line and the base of the ventrals." The numbers of the fin-rays, including the two short rays at the commencement of the dorsal and anal fins, are

D. 14 : P. 16 : V. 12 : A. 13 : C. 20 : cæca 120.

" This fish grows occasionally to the length of sixteen inches. In the stomach of one of the specimens examined were found several species of *Entomostraca,* larva of insects, a few *Coleoptera,* a number of small tough red worms, little more than half an inch in length, and about the thickness of a coarse thread, besides a quantity of gravel, which the fish had probably accumulated when in search of the larva."

" These fish are found in Loch Lomond in great numbers, where they are called *Powans* or *Freshwater Herrings.* They are caught from the month of March until September with large drag-nets, and occasional instances have occurred in which a few have been taken with a small artificial fly : a minnow or bait they have never been known to touch. Early in the morning and late in the evening large shoals of them are observed approaching the shores in search of food, and rippling the surface of the water with their fins as they proceed. In this respect they resemble in their habits the Vendace of Lochmaben and the saltwater herring. They are never seen under any circumstances in the middle of the day. From the estimation these fish are held in by the neighbouring inhabitants, they are seldom sent far before they meet with a ready sale, and are entirely unknown in the markets of Glasgow. In the months of August and September they are in best condition for the table, when they are considered well flavoured, wholesome and delicate food. They shed their spawn in October to December, and remain out of condition until March."

Although agreeing in the number of fin-rays with the Pollan of Ireland, this Loch Lomond fish is at once distinguished from it by the peculiar form of its mouth, a representation of which, in two points of view, inserted as a vignette, and contrasted with the same parts in the Pollan, both of the natural size, will, better than description, convey the appearance in proof of distinction. The Loch Lomond fish being remarkable for the depth of the upper lip, and the large size of the lateral free portions of the superior-maxillary bones.

Dr. Parnell has described a second species of *Coregonus* found in Loch Lomond, which differs from the first in having a smaller head, yet agreeing exactly in the number of all the fin-rays ; but as I learn by communication with Dr. Parnell that since the publication of his paper he has obtained many specimens from Loch Lomond, the characters of which are intermediate in reference to the two fishes described, and appear to connect them, I have not figured it as a distinct species.

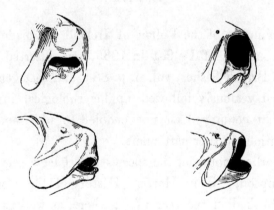

ABDOMINAL
MALACOPTERYGII.   *SALMONIDÆ.*

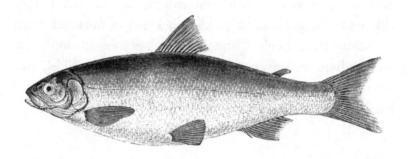

THE POLLAN.

Coregonus Pollan,   *The Pollan,* Thompson, Proceedings Zool. Soc. for 1835,
                         p. 77 ; and Magazine of Zool. and Bot.
                         vol. i. p. 247.

A short notice of the Pollan of Ireland, as made known
by Mr. Thompson of Belfast in 1835, was inserted in the
History of British Fishes, vol. ii. p. 88 ; and that gentleman
having most zealously followed up his zoological investigations in that country, I am now enabled to add from his researches various further particulars.

" The earliest notice of the species that I have seen," says
Mr. Thompson, " is in Harris's History of the County of
Down, published in the year 1744, where, as well as in the
statistical surveys of the counties of Armagh and Antrim, it
has subsequently been introduced as one of the fishes of
Lough Neagh, under the name of Pollan : but, as may be
expected in works of this nature, little more than its mere
existence is mentioned."

" The habits of this fish do not, with the exception of its
having been in some instances taken with the artificial fly,

differ in any marked respect from those of the Vendace of Scotland or the Gwyniad of Wales, and are in accordance with such species of continental Europe as are confined to inland waters, and of whose history we have been so fully informed by Bloch. The Pollan approaches the shore in large shoals, not only during spring and summer, but when the autumn is far advanced. The usual time of fishing for it is in the afternoon, the boats returning the same evening. On the days of the 23rd, 24th and 25th of September 1834, which I spent in visiting the fishing stations at Lough Neagh, it was along with the common and great lake trout, *Salmo fario* and *Salmo ferox*, caught plentifully in sweep-nets, cast at a very short distance from the shore. About a fortnight before this time, or in the first week in September, the greatest take of the Pollan ever recollected occurred at the bar-mouth, where the river Six-mile-water enters the lake. At either three or four draughts of the net, one hundred and forty hundreds,—one hundred and twenty-three fish to the hundred,*—or 17,220 fish were taken ; at one draught more were captured than the boat could with safety hold, and they had consequently to be emptied on the neighbouring pier. They altogether filled five one-horse carts, and were sold on the spot at the rate of 3s. 4d. a hundred, producing 23l. 6s. 8d. From 3s. 4d. to 4s. a hundred has been the ordinary price at the lake side, or directly from the fishermen ; some years ago it was so low as 1s. 8d. the hundred, but at that time the regular system of carriage to a distance, as now adopted, did not exist. At the former rates they are purchased by carriers, who convey them for sale to the more populous parts of the neighbouring country, and to the towns within a limited distance of the lake. They are brought in quantities to Belfast ; and when the supply is good, the cry of ' fresh pollan' prevails even to a greater ex-

* The English long hundred is six score, or one hundred and twenty.

tent than that of 'fresh herring,' though both fishes are in
season at the same period of the year. In the month of
June 1834, fifty hundreds,—six thousand one hundred and
fifty individuals—of pollan and one hundred and twenty-five
pounds weight of trout were taken at one draught of a net,
at another part of the lake near Ram's Island, which was the
most succeesful capture made there for twenty-four years. In
1834 this fish was more abundant than ever before known.
Like the Gwyniad and Vendace, the Pollan dies very soon
after being taken from the water, and likewise keeps for a
very short time. It is not in general estimation for the
table, but is, I think, a very good and well-flavoured fish."

"Though permanently resident, the pollan is very far from
being generally diffused throughout Lough Neagh. It rarely
occurs between the river Mayola and Toone ; while from the
Six-mile-water to Shane's Castle is so favorite a resort, that a
few houses that formerly stood near the latter locality, were
dignified with the name of Pollan's Town."

"In the months of November and December this fish deposits its spawn where the lake presents a hard or rocky bottom. On the 4th of December 1835, a quantity of the
largest Pollans I have seen were brought to Belfast market.
Several were thirteen inches in length, and all on dissection
proved to be females just ready to deposit their roe. On
the 11th of the same month several male specimens of full
size that I procured, and which contained milt most promi-
nently developed, measured but eleven inches and a half,—
thus showing that in maturity the female fish exceeds the
male in length in the proportion of thirteen to eleven and a
half. Its average weight when in season is about six ounces.
One specimen, mentioned to me as the largest taken within
the last ten years, weighed two pounds and a half. The
only food that I have, without resorting to the microscope,
detected in the stomach of the Pollan was a full grown speci-

men of the bivalve shell *Pisidium pulchellum.* A pebble of equal size was also found with it." In the stomach of a specimen given me by Mr. Thompson I found a species of *Gammarus.* Mr. Thompson, in some more recent examinations, has found mature individuals of *Gammarus aquaticus,* and the larvæ of various aquatic insects; some shells of the genus *Pisidium,* one of the fry of the three-spined stickleback, and a few fragments of stone. Others were found to contain minute *Entomostraca,* two *Pisidia,* and a *Limneus pereger,* this last was three lines in length.

Besides inhabiting Lough Neagh, the Pollan has also been found in Lough Derg, an expansion of the Shannon; and Lord Cole, who has most condescendingly interested himself in the History of British Fishes, had the kindness to send me a jar full of Pollan from Lough Erne in the county of Fermanagh, from one of which specimens our figure was taken. The Pollan of Lough Erne are rather deeper for their length than those of Lough Neagh. His lordship has also sent me two species of Charr from Ireland; some from Lough Eask being identical with the Charr of the Cumberland Lakes, while those from Lough Melvyn are short and deep fish with large fins exactly resembling the Charr found in two or three lakes in Wales, the particulars of both of which are described in the second volume of the British Fishes.

To return to the Pollan of Ireland, Mr. Thompson's description is as follows: " The relative length of the head to that of the body is about as one to three and a half; the depth of the body equal to the length of the head; the jaws equal in length, both occasionally furnished with a few delicate teeth; the tongue with many teeth; the lateral line sloping downwards for a short way from the operculum, and thence passing straight to the tail. Nine rows of scales from the dorsal fin to the lateral line, and the same number thence to the ventral fin, the row of scales on the back and that of the lateral line not included. The third ray of the pectoral fin the longest. The fin-ray formula is as follows—

B. 9 : D. 14 : P. 16 : V. 12 : A. 13 : C. 59 : vertebræ 59.

Of these, the first two rays of the dorsal fin, and the first two rays also of the anal fin are short.

" The colour to the lateral line dark blue, thence to the belly silvery; dorsal, anal, and caudal fins, towards the extremity, tinged with black ; pectoral and ventral fins of crystalline transparency, excepting at their extremities, which are faintly dotted with black. Irides silvery, pupil black."

In a number of these Pollan from Lough Erne as well as Lough Neagh, the base of the last ray of the dorsal fin is exactly half way between the point of the nose and the extreme end of the longest upper caudal ray. Nine rows of scales from the base of the first ray of the dorsal fin to the lateral line, and the same number from the lateral line to the origin of the ventral fin, with eighty-eight scales forming the lateral line. The fin-rays in number on several specimens exactly as stated by Mr. Thompson.

The vignette represents the bones of the cranium in the genus *Coregonus*.

SUBBRACHIAL
MALACOPTERYGII.   *GADIDÆ.*

THE FOUR-BEARDED ROCKLING.

Motella cimbria, The Four-bearded Rockling, PARNELL, Wern. Mem. vol. vii.
p. 449. pl. 44.
Gadus cimbrius,   LINNÆUS, Syst. Nat. p. 440, sp. 16.
      ,,   ,,   RETZ, Faun. Suec. p. 323.
Enchelyopus cimbricus, SCHNEIDER, Syst. Ichth. p. 50, sp. 1, tab. 9.
Motella cimbrica,   NILSSON, Prod. Ichth. Scand. p. 48, sp. 2.

THIS species of *Motella,* first described by Linnæus, is
included by Dr. Parnell in his description of the Fishes of
the Forth, a specimen, fourteen inches in length, having
been brought to him by a Newhaven fisherman, who had
caught it a little to the east of Inchkeith on a Haddock line
baited with muscles. It is a species perfectly distinct from
the Three or the Five-bearded Rocklings, so much more
common on various parts of the coast, and may at once be
distinguished from either by the greater length of the filament, which is placed in advance of the almost obsolete
first dorsal fin. This filament in a fish of nine inches
long, measures one inch and seven-eighths; and in another
fish of ten inches and a half in length, measures two inches
and a quarter, as I find from portions of two specimens

sent me by Mr. Euing of Glasgow, to whom I am indebted
for the opportunity of making known the new species of
Smelt. These two specimens of the Four-bearded Rockling were taken near Rothsay, and in reference to them
Mr. Euing's letters contain the following remarks : —" I
have never met with the Three or the Five-bearded Rockling, but small specimens of that with four cirri are frequently brought in on the long lines from deep water. It
is, indeed, by no means a very rare fish with us, and I have
seen it at almost every visit to the coast since 1827, the year
in which I first observed it."

This fish is rare in the Baltic, but is not uncommon on
the southern coast of Sweden ; it is found also among the
islands of the Catigat ; on the west coast of Norway, and in
the Atlantic.

Dr. Parnell says, on dissecting the specimen, I found
the stomach filled with shrimps and small crabs. The cæcal
appendages were few in number ; the roe was large ; the ova
small and numerous, and apparently in a fit state to be deposited. It is probable that the habits of this fish are similar
to those of the other species, but from its rarity it is difficult to determine."

Description by Dr. Parnell, from a specimen fourteen
inches in length : " Form closely resembling that of the
Five-bearded Rockling, but the length of the head somewhat
greater compared to that of the body. The body elongated,
rounded in front, compressed behind, tapering from the vent
to the caudal extremity ; greatest depth less than the length
of the head. Head one-sixth of the entire length, caudal
fin included, slightly depressed ; snout blunt, projecting considerably beyond the under jaw ; eye large, of an oval form,
placed high up, and about its own length from the point of
the nose ; operculum rounded, oblique ; gill-opening large ;

gape wide; maxillary extending in a line with the posterior margin of the orbit; teeth sharp and fine, forming two rows in the under jaw, and five rows in the upper; a few are also placed in a cluster on the anterior part of the vomer; barbules four, one a little in front of each nostril, one at the extremity of the upper lip, and one on the chin; tongue fleshy, smooth, and without teeth. Fins:—the first dorsal fin obsolete, scarcely discernible, commencing over the operculum, and terminating a little in front of the second dorsal, composed of a number of short, fine, capillary rays, of which the first is by far the largest; second dorsal taking its origin in a line over the ends of the pectorals, and terminating a little in advance of the caudal; anal fin commencing in a line under the twelfth ray of the second dorsal, and ending under the last ray but three of the same fin, in form similar to the second dorsal, but the rays scarcely more than one half the length; the first ray simple, the rest branched; caudal rounded at the extremity, the length of the middle rays equalling the space between the first and the twelfth rays of the anal, the lateral rays simple; ventral fins jugular, the second rays the longest, about two-thirds the length of the pectorals; the pectoral fins rounded at the extremities, equalling the length of the caudal; the first rays stout and simple, the rest branched. The fin-rays in number are,—

1st D. 50 : 2nd D. 50 : P. 16 : V. 5 : A. 43 : C 20. Vert. 52.

Scales small, smooth, and adherent, covering the head, body, and membranes of the dorsal, caudal, and anal fins; lateral line formed by a number of oval depressions, placed at intervals from each other, commencing over the operculum, taking a bend under the ninth, tenth, and eleventh rays of the second dorsal fin, from thence running straight to the middle ray of the caudal. Colours:—Back and sides of a greyish brown; belly dirty white; second dorsal fin lighter in colour at the edge; pectorals, caudal, and lower part of the dorsal, dark brown, approaching to black; anal and ventrals dusky."

The vignette represents the cranium of the Common Codfish.

SUBBRACHIAL
MALACOPTERYGII.   PLEURONECTIDÆ.

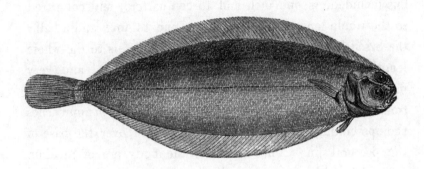

THE LONG FLOUNDER.

Platessa elongata, The Long Flounder, YARRELL, Suppl. to Brit. Fishes.

I AM indebted to Mr. Baker, of Bridgewater, for several interesting communications on Birds and Fishes, one of the most valuable of which is the opportunity afforded me of making known what appeared to that gentleman to be a species of Flounder undescribed as a British Fish, and which, after having made the usual search, I have reason to believe is not only undescribed as a British Fish, but is altogether new to Ichthyology. I have only as yet seen the single specimen sent me for my use on this occasion by Mr. Baker, from which a drawing has been made of the natural size, and the reduced representation here given engraved on wood; but I understand from Mr. Baker's son that his father had obtained a second example of the same fish. The specimen now before me was obtained at Stoford, in Bridgewater Bay, in the month of December. Little is of course known of the habits of so recent and so rare an acquisition.

VOL. II.   D

The whole length of this specimen is seven inches and three-quarters; the length of the head one inch and one quarter, and compared to the whole length of the fish, as one to six; the greatest breadth of the body, dorsal and anal fins included, is one inch and three-quarters, and compared to the whole length of the fish, as one to four and a half; the breadth, including the dorsal and anal fin, is to the whole length as three to eight. The body very thin, and very much elongated in form; the lateral line passing straight from the tail along the middle of the fish till it approaches the operculum, then rises in a slight curve over the base of the pectoral fin. The scales on the body are of medium size, oval, with numerous radiating striæ on the free portion. The fins deep, and the tail long.

The outline of the whole head is rather circular, the mouth oblique from below upwards, and below the line of the longitudinal axis of the body; the jaws nearly equal in length, each furnished with a single row of small and regular teeth; the eyes rather large, the upper eye, or that on the left side, being a little in advance of the lower, or that on the right side; the inter-orbital bony ridge prominent; the boundary lines of the preoperculum and operculum forming two concentric portions of circles. The pectoral fin, arising immediately behind the edge of the operculum, is about half as long as the head; the ventral fin, in a line under the edge of the operculum, is about half as long as the pectoral fin. The dorsal fin, commencing with short rays in a line over the eye, is at its greatest elevation about the middle of the fish, and from thence diminishes gradually to the end, which is on the fleshy portion of the tail, and short of the origin of the caudal rays; the anal fin begins close to the ventral fin, immediately behind the post anal spine; the first and last rays short, those in the middle of the fin the longest, and the fin ends on the same plane as the dorsal. The tail is elongated; its length equal to that of the head, and in form but slightly rounded at the end ; the sides parallel.

The fin-rays in number are,—

D. 110 : P. 11 : V. 6 : A. 96 : C. 24.

The colour of this specimen on the upper surface is a uniform pale brown, the membranes of the different fins being rather lighter in colour than the body of the fish ; the under surface of the body very pale wood-brown ; the irides yellow.

This specimen has been preserved dry.

The vignette below represents the cranium of the Common Flounder.

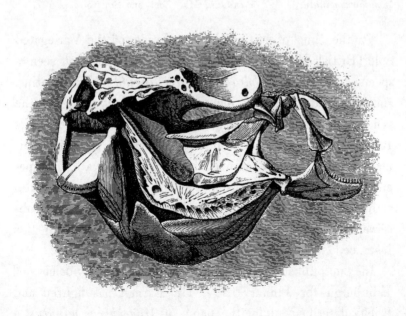

D 2

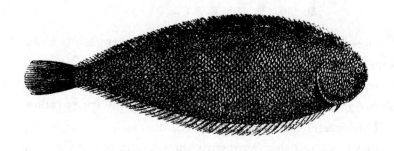

THE SOLENETTE,

OR LITTLE SOLE.

Monochirus linguatulus,	Cuvier, Reg. An. t. ii. p. 343.
Solea parva sive lingula,	Rondeletius, p. 324.
La petite Sole,	,, French Edit. Lyons, p. 260.
Solea parva sive lingula Rondeletii,	Willoughby, p. 102, F. 8, fig. 1.
Pleuronectes lingula,	Linn. Syst. Nat. p. 457, sp. 10.
Monochirus minutus,	Parnell, Mag. Zool. and Bot. vol. i. p. 527.

At the time of writing the description of the Variegated Sole (British Fishes, vol. ii. page 262), I had not seen a specimen of the true *Solea parva sive lingula* of Rondeletius, and now find that I have included two distinct species in the synonymes employed to designate the Variegated Sole. The Rev. L. Jenyns, in his Manual of British Vertebrate Animals, appears to have suspected that there was a fourth species of Sole on our coast, since, at the conclusion of the description of his third species, he has observed, " further observation is necessary in order to decide whether, in this instance, I have confounded two nearly allied species."

In the published proceedings of the Royal Society of Edinburgh for January 1837, Dr. Parnell has figured and briefly described, under the name of *Monochirus minutus,* a

small species of Sole obtained by him at Brixham on the
Devonshire coast, which appears to be the true *Solea parva
sive lingula* of Rondeletius. This small fish is at once distinguished from the Variegated Sole of Donovan, and other
English authors, by the tapering of the body towards the
tail, and more particularly by the dorsal and caudal fins being
united to the base of the tail, which is not the case in the
Variegated Sole. This union of the two fins with the tail
is shown in the figure given by Rondeletius, and again by
Willoughby, as referred to.

Dr. Parnell has obtained several examples of this interesting little species, which is not unfrequently taken in the
trawl-nets by the fishermen of Brixham, but on account of
its diminutive size it is seldom brought on shore. It has
evidently been confounded with the Variegated Sole ; but,
independently of other distinctions, the Variegated Sole has
the tail separated from the dorsal and caudal fins by a considerable interval.

The Variegated Sole of Donovan and of Montagu's MS.
the Red-backed Flounder of Pennant's Zoology, and the
Variegated Sole of Dr. Fleming, are so many specimens of
the truly Variegated Sole, and are each of them quite distinct from the true *lingula*. Duhamel appears to have distinguished and figured both species. Mr. Thompson has
obtained both species on the coast of the North of Ireland,
and by his kindness I have now his specimens before me
for comparative examination. Dr. Parnell has given me two
examples of his *Monochirus minutus*, which, as before observed, I believe to be the true *Solea parva sive lingula* of
Rondeletius; and I have also two specimens of the true
Variegated Sole ; one of these, from which the figure in the
British Fishes was drawn, has the dark clouded variation in
colour extending, as in Donovan's figure, over the back as
well as the fins : in a specimen belonging to Mr. Thompson,

in one of my own, and in Montagu's specimen, as described in his MS. the dark variations in colour are confined to patches on the fins, as in Pennant's figure; but without reference to colour, this species is immediately known by the space which occurs between the two elongated fins and the tail, which Montagu says was equal to half an inch in his specimen, which measured nine inches.

Both these species belong to the genus *Monochirus* of Cuvier, distinguished from those of the genus *Solea* by the very small size of the upper pectoral fin, and the very rudimentary state of the pectoral fin on the under side, and is, indeed, sometimes entirely wanting. Of our two British species of *Monochirus*, the *M. linguatulus* of Cuvier has the smaller upper pectoral fin of the two, as observed by Mr. Thompson, who has, in a recent number of the Annals of Natural History, published some interesting details on the two British species of the genus *Monochirus*.

From the numbers of these fishes which are taken in the trawl-nets off Brixham throughout the whole year, says Dr. Parnell, and from their never appearing to attain a large size, there can be little doubt but that they are arrived at their full growth. The fishermen, who appear perfectly familiar with their appearance, call them Red Soles; and scarcely a trawl-boat leaves Brixham Harbour that does not capture a dozen or more of these fish daily; but, from their diminutive size, they are either thrown overboard, or left to decay at the bottom of the vessels.

Description : — " Length five inches; the width at the upper third nearly two inches : the colour of the back light reddish brown, the under surface pale white; every sixth or seventh ray of the dorsal and anal fin black. In shape this fish is similar to the Common Sole, but is of a more wedge-shaped form, becoming narrow at the caudal extremity. The head is small, one-sixth of the whole length; the mouth is crooked; each jaw is furnished with a number of minute teeth, placed close together, and extending but half way round the mouth; the eyes are small; the upper, or left eye, a little in advance. The dorsal fin commences immediately over the upper lip, and runs down the back, to be connected with the caudal rays; the anal fin begins under the posterior margin of the operculum, and continues to the tail. The number of the fin-rays are,—

D. 73 : P. 4 : V. 4 : A. 54 : C 14.

The scales are small, with from twelve to fifteen denticles at their free extremity, rendering the whole surface of the fish rough to the touch when the finger is passed along from the tail to the head. The pectoral fin, on the eye-side, is small, with the lower half black, while the fin on the opposite side is very minute, and of a pale white; the lateral line is straight throughout; the tail is rounded at the end, and mottled with brown."

The vignette represents the fishing-house at Virginia Water.

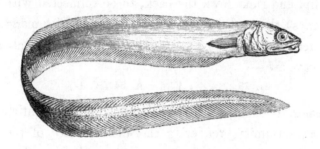

DRUMMOND'S ECHIODON.

Echiodon Drummondii, THOMPSON, Proceedings Zool. Soc. 1837, page 55.
      ,,   ,,   ,,   Transactions ,,   ,, vol. ii. part iii. p.
                                 207, plate 38.

Generic Characters.—Head oval : jaws furnished with large cylindrical teeth
in front, other smaller teeth on the palatal bones and on the vomer. Gill
apertures large ; branchiostegous membrane with seven rays. Body smooth,
without scales, elongated, compressed. Dorsal and anal fins nearly as long as
the body ; all the rays soft ; no ventral fins ; anal aperture near the head.

A DEAD specimen of the fish figured above was found
by Dr. J. L. Drummond on the beach at Carnlough, near
Glenarm in the county of Antrim, in the month of June
1836, and from its appearance when found it was conjectured
that it had been cast ashore by the tide of the preceding
night, when a strong easterly wind prevailed. The specimen
was given by Dr. Drummond to his friend Mr. W. Thompson of Belfast, and being new in form, was made by the latter
gentleman the subject of a communication to the Zoological
Society, which appeared in the Proceedings and Transactions
of that Society as here quoted.

This specimen, Mr. Thompson observes, "being, so far as known to me, unique, I have been unwilling to injure its appearance by dissection. In external characters it is excluded from the *ophidia* proper in consequence of not having the barbules ; and though agreeing with the *Fierasfers* in the negative character of wanting these appendages, yet, by having the dorsal fin strongly developed and elevated, it ranges not with them."

" Its want of the very obvious character of the *Ophidia*, renders all comparison with them unnecessary ; but of two species belonging to the *Fierasfers*, and which approach the present specimen most nearly, I may state that it possesses many of the characters of the *Ophidium fierasfer* of Risso, but differs from that species in the teeth, (both jaws are described as armed with three rows of sharp and hooked teeth,) number of fin-rays, and some minor characters ; besides, there is nothing said of the remarkable teeth terminating both jaws, as exhibited in my specimen. In the Règne Animal we again find an *Ophidium dentatum* described as having in each jaw " *deux dents en crochets*," but no further details are given. In this only character, however, the *Ophidium dentatum* differs from my fish, which has four large hooked teeth in the upper and two in the under jaw."

" Although when this fish first came into my possession, I saw that it might be classed under the *Malacopterygii Apodes*, and be placed near *Ophidium*, I considered that in a natural arrangement it would best constitute a new genus of the family *Tænioidea* (Riband-shaped). In being apodal it was not excluded from this family, as two genera belonging to it are destitute of ventral fins. I did not hesitate to place it under the *Acanthopterygii*, as some genera which are included in this order are, like it, strictly Malacopterygian, their natural connexion with genera having fins with spinous rays being considered—and in my opinion most philosophically—to outweigh this character; and further, I felt less reluctance in thus placing it, in consequence of *Cepola rubescens*, which it assimilates in some respects, having but one spinous ray, and that in the ventral fin. At the suggestion of John Edward Gray, Esq. F.R.S. I have, however, reconsidered the subject, and have come to the conclusion here advanced."

As a difference of opinion may still exist with regard to the position of this genus, I think it due to Mr. Thompson to subjoin the observations originally made.

" Like certain other genera which are comprehended under *Acanthopterygii*, the first order of the osseous fishes, its fins are altogether destitute of spinous rays; but, like those alluded to, such as *Zoarcus*, &c. its other characters seem to point out the *Tænioides* as the family to which it belongs. Of the eight genera of *Tænioides* already known, viz. *Lepidopus*, *Trichurus*, *Gymnetrus*, *Stylephorus*, *Cepola*, *Lophotes*, *Trachypterus*, and *Alepisaurus*, the specimen under consideration agrees with *Trichiurus* and *Stylephorus* in being apodal, or wanting ventral fins, but in this character only is there any generic accordance. Though considerably more elongated, from the head posteriorly it approaches most nearly to *Cepola rubescens* in the form of the body, and in the forward commencement of the anal fin, which, with the dorsal, is prolonged until it joins the caudal; but it is only in the continuity of these fins until this junction is effected that the resemblance holds, as in my specimen, the dorsal rays, the five foremost of which are very short, increase in length posteriorly, and near the caudal fin are about three times as long as the depth of the body beneath them; in the anal fin, which is throughout much deeper than the dorsal, the rays likewise increase posteriorly; and near the caudal are in length four times greater than the depth of the body at the same place. The length of the posterior rays of these fins causes the dorsal, anal, and caudal, to appear as one; whilst, though they do join in *Cepola rubescens*, the last ray of the dorsal and anal being much shorter than the outer rays of the caudal, may at the same time be said to mark distinctly the termination of each fin. In my specimen the anal fin originates two lines in advance of the dorsal fin."

In the form of the head, and in dentition, it differs so remarkably from all the other genera as to render a comparison with them unnecessary. Its absolute characters must suffice for distinction.

Description.—"Total length eleven inches; greatest depth at one inch four lines from the snout, six lines, thence posteriorly gradually narrowing; greatest breadth of body anteriorly three lines; at the middle of the entire length one line, and thence to the tail becoming gradually more compressed. Head one inch two lines long, or rather more than one-ninth of the entire length; profile sloping forward equally on both sides to the snout, which is truncated, and projects one line beyond the lower jaw; narrow, increasing in breadth very gradually from the snout, its breadth compared to its length as one to three and a half; height half its length, compressed at the sides, and rather flat above from the eyes backward; from the eyes forward a central bony ridge; snout viewed from above somewhat bifid, in consequence of the forward position of the large teeth on each side. A few large punctures extend from the snout below the eye, and are continued just behind it; a series of small ones closely arranged extend from the upper portion of the eye in a curved form posteriorly to near the edge of the preopercle, and thence a double row extends downwards. Nostrils very large, placed just in advance of, and before the centre of, the eye, and in form a somewhat oval transverse aperture. Eye large, occupying the entire half of the depth of the head; its width greater than its height; in the length of the head occupying the place of one in four and a half; its distance from the snout three lines, or equal to its diameter, consequently two and a half of its diameters are contained between it and the edge of the operculum. Operculum rounded at the base, terminating in a minute point directed backwards, strongly radiated, striæ distant; preoperculum ascending vertically; cheeks smooth and soft. Mouth rather obliquely cleft. Teeth, two large strong ones, placed close together, and curving inwards at each side the extremity of the upper jaw, the two inner one-sixteenth of an inch apart. In the lower jaw one slender rounded tooth, nearly one line long on each side, curving outwards at the base, and inwards at the point. Entire upper and under jaw and vomer densely studded with small bluntish teeth, somewhat uniform in size; vomer extending far forward, and very much developed, forming a cavity in the lower jaw, and in advance of the tongue when the mouth is closed; a series of rows of teeth similar to those last described on the palatal bones: all the teeth of the upper jaw exposed to view when the mouth is closed. Tongue short, not reaching within two lines and a half of the extremity of the lower jaw, and apparently toothless. On the dorsal ridge, one inch from the snout, or two lines and a half behind the cranium, is a short, stout, bony spine, not very conspicuous, and, excepting at its extreme point, covered with skin: it is six lines in advance of the first ray of the dorsal fin. Scales none, but it may have been divested of them during its short exposure on the beach. Lateral line inconspicuous, being a slight depression extending in a straight line along the middle of the sides posteriorly, or throughout the greater portion of its length, but anteriorly nearer to the dorsal than the ventral profile. Vent one inch three lines from the extremity of the lower jaw. Branchiostegous membrane opens forward rather before the extremity of the gape. Dorsal fin commencing one inch six lines from the snout, low at its origin, but gradually increasing in height to near the caudal fin, which it joins, the two or three anterior rays, which are very short, flexible and simple, the remainder articulated. Anal fin originates just behind the vent, or at one inch three lines from the point of the lower jaw, joins the caudal fin, near to which it increases in depth posteriorly from its origin, deeper than the dorsal fin throughout; at about one inch and a half from the caudal fin the rays are in length four times greater than the depth of the body at the same place, the rays of the dorsal fin opposite being three times the depth of the body; the first and second anterior rays flexible and simple, the remainder articulated. Pectoral fins originate one line behind the head, and are equal to half its length, central rays longest, all very flexible, placed below the middle of the sides. Caudal fin, central rays longest. Articulations very long on the rays of all the fins; no branched rays in any one of them.

B. 7 : D. 180 : P. 16 : A. 180 : C. 12.

The number of the fin-rays were reckoned with the greatest care; but without injury to the specimen they could not be ascertained with certainty to a single ray. The vertebræ, which distinctly seen through the skin can be reckoned with accuracy, ninety-eight. Colours, anterior half a dull flesh colour, similar to specimens of *Cepola rubescens* preserved in spirits, hence it is presumed to have been originally red; behind this portion reddish-brown markings appear on the body at the base of the dorsal and anal fins, and suddenly increase in number, until from an inch behind the middle, the whole sides are closely marked and spotted over; the entire top and the sides of the head before the hinder line of the eye are similarly spotted; just behind the cranium a few spots also appear; the posterior rays of the dorsal and anal, and the entire caudal fin, blackish. Irides, operculum, and under surface, a short way beyond the vent, bright silver."

" The two large teeth, resembling serpent's fangs, which terminate the upper jaw on each side, have suggested the generic appellation of *Echiodon ;* and the specific name of Drummondii is proposed in honour of its discoverer."

The figures below represent a side view of the head, the mouth open to show the form and situation of the teeth, enlarged ; and a front view of the anterior terminal teeth, also enlarged. The illustrations here used are derived from Mr. Thompson's paper in the Transactions of the Zoological Society already quoted ; and I with pleasure avail myself of the opportunity in this instance afforded me of recording my obligations to Mr. Thompson for his kind and zealous cooperation in zoology, and particularly for the loan of this rare specimen, and many other Irish fishes, for examination.

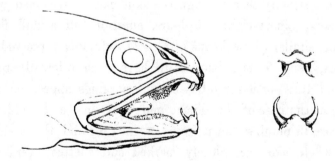

LOPHOBRANCHII. *SYNGNATHIDÆ.*

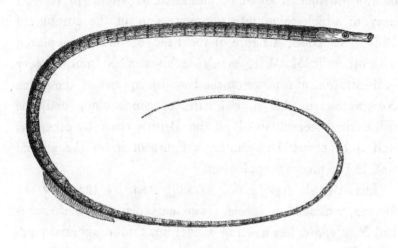

THE STRAIGHT-NOSED PIPE-FISH.

Syngnathus ophidion, Linnæus Syst. Nat. t. i. p. 417, sp. 5.
      „   „   „   Faun. Suec. p. 131, sp. 1.

It is only within a few years, I believe, that writers on the Natural History of European Fishes have become aware that in quoting, as was almost invariably the case, the figure of the *Syngnathus ophidion* of Bloch, tab. 91, fig. 3, as the true *ophidion,* they were not referring to, because that figure does not represent, the true *Syngnathus ophidion* of Artedi and Linnæus. The fish, as represented by Bloch, does not exhibit any appearance of a caudal fin, but if the species there figured from be examined, it will be found to possess a rudimentary caudal fin,* and could not therefore be considered as referred to by Linnæus in the short but expressive description, *S. pinnis caudæ ani pectoralibusque nullis, corpore tereti.*

The first good figure of the true *S. ophidion* of Linnæus

* British Fishes, vol. ii. p. 339, vignette.

that became known to me appeared in an octavo volume by
M. C. U. Ekström, on the Fishes of Morko, in Sudermann-
land, a province in Sweden, published at Berlin in 1835, a
copy of which came into my possession in the autumn of
1836. In 1838, a figure of the head of this fish appeared
with others in M. Wiegmann's Archives of Natural History
in illustration of a paper on the Swedish species of the genus
Syngnathus by M. B. Fr. Fries of Stockholm ; and this
fish having been obtained on the British coast by others as
well as by myself I now insert a figure of it, of the natural
size, in the present supplement.

The British *Syngnathi*, as suggested by the Rev. L.
Jenyns, consist of six species ; two marsupial pipe-fish *S. acus*
and *S. Typhle,* having true caudal fins : four ophidial pipe-
fish, which may be again divided into two sections, the first
of which contains two species, *S. æquoreus* and *S. anguineus,**
having each a rudimentary caudal fin ; † the second section,
also containing two species, *S. ophidion* and *S. lumbrici-
formis,* in which there is no rudimentary caudal fin, the
round tail ending in a fine point.

To this last division belongs the true *S. ophidion* of Artedi
and Linnæus, the males of which in the season of reproduction carry the eggs, after deposition by the female, in three
or four rows of hemispheric depressions on the under surface
of their bodies. This species, which lives among the sea-
weed on our coast, is more rare than some others. It was
found in Cornwall long ago by our countryman and naturalist
John Ray, has been recently described by Mr. Jenyns in his
" Manual of British Vertebrate Animals," from specimens
obtained at Weymouth, and I also possess several specimens
obtained on the Dorsetshire coast.

* A specific name proposed by Mr. Jenyns for that species which we had
previously called, in error, S. *ophidion.*

† See British Fishes, vol. ii. pp. 337 and 339, vignettes.

This little pipe-fish is long, slender, and nearly cylindrical, but slightly compressed from the head to the anal aperture; from thence to the end of the tail round and tapering very gradually to a fine point; the head is short, the length of it only half an inch in a specimen of nine inches; the length of the head therefore, as compared to the whole length of the fish, is as one to eighteen; the nose is straight, rather compressed, a section forming a hexagon slightly elongated, of which the upper and under angles are the most produced; the distance from the point of the nose to the eye, and from thence to the hinder edge of the operculum, equal; no pectoral, anal, or caudal fin; the anal aperture is near the middle of the whole length of the fish, with a delicately-formed dorsal fin in a line over it, nearly one inch in length at its base, with about one-third of the fin, which contains from thirty-five to forty very slender rays, in advance of the vertical line of the anal aperture. Between the head and the anal orifice there are on the body of the fish about thirty sculptured plates or segments, and nearly sixty on the tail, diminishing gradually in size as they approach the tip.

Colour.—Some specimens are uniform olive green, others are tinged with yellowish brown, and both are occasionally varied with darker shades of colour on the body.

The largest specimens seldom exceed nine inches in length. The figure at the head of this subject is the exact size of the specimen from which it was drawn.

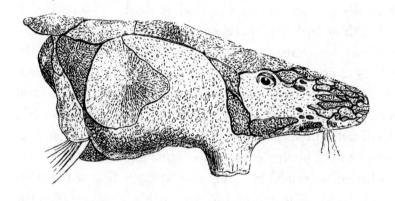

THE BROAD-NOSED STURGEON.

Acipenser latirostris, Broad-nosed Sturgeon, Parnell, Trans. R. S. E. vol. xiv.
pl. 4.
„   „   „   „   „   Fish. of the Forth, Wern.
Mem. vol. vii. p. 405, pl. 39.

In the papers here referred to, Dr. Parnell observes, that but one species of Sturgeon has hitherto been recorded by the different writers on British Ichthyology, but from the observations of practical fishermen, as well as his own, Dr. Parnell adds, I think there is little doubt that two species, at least, will in future be recognised as inhabiting the British coast.

" It has long been noticed by the fishermen of the Solway Frith, that two species of Sturgeon are occasionally entangled in their Salmon-nets, the one with a blunt nose, and the other with a sharp one ; the latter species being the most common of the two.

" A fine specimen of the Blunt-nosed Sturgeon was taken in the Frith of Forth in the month of July 1835, and

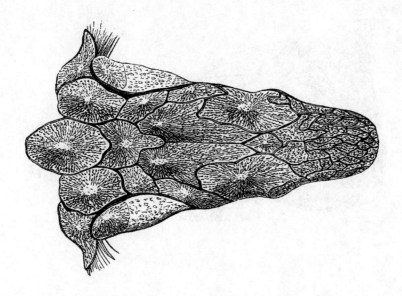

brought to the Edinburgh market for sale, the head of which I preserved. A few weeks after, another was taken in the Tay, which differed in no respect from the former, except in sexual distinction."

" Length seven feet nine inches ; weight eight stone, or one hundred and twelve pounds. The colour of the back and sides is of a light grey, with a shade of olive ; the belly dirty white. The body is armed with five rows of osseous shields, running from the head to the tail. The first row commences behind the head, and runs down the central ridge of the back ; the two next rows arise one on each side of the former. Immediately on the lower margin of the pectorals the other two rows commence. The skin is rough, with a number of small angular osseous plates intermixed with very minute spicula. The first free shield on the dorsal ridge is nearly circular, and very slightly carinated ; all the rest in that row are of an oval form. The snout is wide and depressed, much broader than the diameter of the mouth. On the under surface, placed nearer to the tip of the snout

E 2

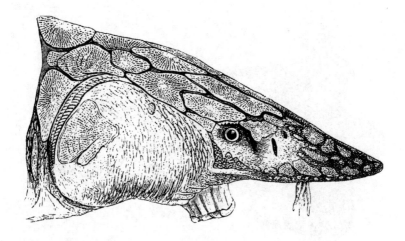

than to the mouth, are four cirri arranged in an irregular line.
The summit of the head is rough, with the central plates
beautifully radiated, and of a fibrous appearance.   The position of the fins is the same as in other Sturgeons."

"This fish differs from the Common Sturgeon, *Acipenser
sturio*, in having the tip of the snout much broader than the
mouth, in the keel of the dorsal plates being but slightly
elevated, and having the cirri placed nearer to the tip of the
snout than to the mouth."

"The Sturgeons are all much allied to each other; and
not being able as yet to find the right synonym for the present one, I have proposed, in the mean time, the name
latirostris, as characteristic of the species."

"In the stomach of the one from the Tay was found an
entire specimen of the Sea-mouse, *Aphrodita aculeata*."

Dr. Parnell has presented the preserved head of this specimen to the Museum of the Zoological Society; but, like
Dr. Parnell, I have been unable to identify it with any described Sturgeon.   It does not agree with either of the nine

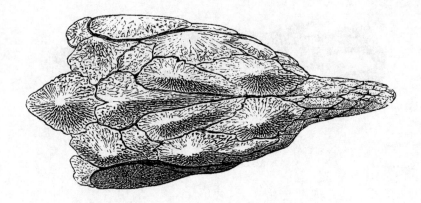

species found in the various waters of the Russian empire, figured and briefly described by M. A. Lovetski, in the third volume of the Transactions of the Imperial Society of Naturalists at Moscow; nor am I able to say that it agrees with either of the eleven species figured and described by Messrs. Brandt and Ratzburg in their Medical Zoology.

Baron Cuvier has observed in his *Regne Animal*, t. ii. p. 379, note, that the species of this genus are not yet well determined by naturalists, nor their comparative characters sufficiently defined. Supposing that the bony plates of the head by their form, size, and relative situation might afford specific characters, I have given two views of these parts in our two British Sturgeons, not without some suspicion, like Dr. Parnell, that we may have even more than two.

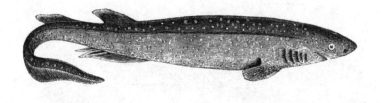

THE SPINOUS SHARK.

Echinorhinus spinosus,	Blainville, Faun. Franç. Poiss. p. 66, sp. 6.
,, ,,	Musignano, Faun. Ital. pt. xiii.
,, *obesus,*	Dr. A. Smith, Zool. South. Afr. No. 1.
Squalis spinosus,	Gmelin, Syst. Nat. I. p. 1500, sp. 27.
,, ,,	Lacepede, Hist. Nat. Poiss. 4to. t. i. p. 30, tab. 3, fig. 2, 8vo. t. 5, p. 354, pl. 22.
,, ,,	Schneider, p. 136, sp. 31.
,, ,,	Risso, Ichth. p. 42, sp. 18.
Scymnus ,,	,, Hist. t. iii. p. 136, sp. 21.
,, ,,	Cuvier, Règne An. t. ii. 1829, p. 393.
Gonoidus ,,	Agassiz, Recherches sur les Poiss. Foss.

Generic Characters. *Echinorhinus,* Blainville. *Gonoidus,* Agassiz.—The first dorsal fin opposite to the abdominal ones. Teeth in both jaws, broad and low, the edge nearly horizontal ; the lateral edges have one or two transverse denticles. (1 species.)*

Soon after the publication of that part of the British Fishes which contained the Sharks, I received a communication from Mr. John Hey, then Honorary Curator to the Leeds Philosophical Society, with a coloured drawing of the well known Spinous Shark of authors, a specimen of which

---

* Müller and Henle. Generic characters of Cartilaginous Fishes. Mag. Nat. Hist. for 1838, p. 89.

had been taken in Filey Bay, on the Yorkshire coast, in the summer of 1830, and therefore entitled to a place among British Fishes; but the whole of the then remaining portion of the work being at that time printed for publication on the 1st of August, 1836, I was unable to avail myself of this interesting information, which came to my hands on the 7th of July.

On the 30th of the same month I was favoured with a letter from Dr. H. S. Boase, of Penzance, containing an account of the capture of a Spinous Shark on the 23rd of that month, near the Land's End; and Dr. Boase also very kindly sent me in his letter pen-and-ink sketches of two views of this Shark, made to a scale of one inch to a foot, with representations and specimens of the teeth and spines.

In November 1837, the Rev. Robert Holdsworth sent me notice by letter of the capture of a Spinous Shark, taken in a trawl-net off Brixham, with pen-and-ink sketches of the form of the body, with a small portion of its spine-studded skin, and some of its teeth.

At the meeting of the British Association at Newcastle-upon-Tyne, in August 1838, Arthur Strickland, Esq. of Bridlington, exhibited in the section devoted to Natural History a drawing, and read a short description, of a Spinous Shark, which had been recently found on the Yorkshire coast, and was evidently of this species, Mr. Gray referring to the figure of it lately published by Dr. Andrew Smith in the first number of his "Illustrations of the Zoology of South Africa," which the drawing exhibited by Mr. Strickland very closely resembled.

Lastly, I may add that on the 9th of November 1838, the Rev. Robert Holdsworth sent me word that another specimen of the Spinous Shark had been caught on a fisherman's line off Berry Head on the previous Tuesday. I soon afterwards received a notice of this last capture from my

friend Mr. Couch, of Polperro, and also from Mr. Heggerty, of Torquay, to which place, as I understood, this last specimen had been brought for preservation.

Four examples of this Shark are therefore known to have been obtained on our coasts within the last three years, and one in the summer of 1830.

This very remarkable Shark was first described by Broussonnet under the name of *Le chien de mer bouclé*, in the "Memoires de l'Académie des Sciences pour 1780," and, as may be seen by the numerous synonymes at the head of this subject, is a species that is exceedingly well known, having a wide geographical range, extending from the North Sea to the Cape of Good Hope in one direction, and from the Shores of Italy into the Atlantic in another.

The specimen described by Broussonnet measured only about four feet in length; but it has been taken upwards of seven feet long on the Cornish coast; and M. Risso mentions that one of four hundred pounds' weight, and therefore probably still longer than the Cornish specimen, was caught by the Mandrague, or Tonnaro fishermen of Nice, in the horizontal nets set up by them to catch Tunnies.

Some differences will be observed in the comparative length and thickness of the figures here given, the first of which is taken from the drawing sent me by Mr. John Hey of the Filey Bay specimen ; the second representing, on the other side, a more bulky fish, is taken from Dr. A. Smith's illustrations. The figures given by Lacépède and the Prince of Musignano are rather long and slender, and were probably taken from specimens of small comparative size ; the figure sent me by Dr. Boase from a fish more than seven feet long, and the drawing exhibited by Mr. Strickland at Newcastle, more resembled the figure by Dr. Smith. Some specimens are described as being intermediate, and all these differences in the same species may be referred to age or sex, or both, a young male and an old female presenting the greatest contrast. The decided similarity in the teeth, which are very peculiar, and which only differ in size, with the particular character of the skin and its spines, with their radiated bases, leave no room to doubt that these various examples belong to one and the same species.

We become a little acquainted with some of the habits of this Shark by noticing the circumstances under which it has been captured. Of the first Cornish specimen, Dr. Boase says, this Shark was caught on the 23rd of July, 1836, west of the Long Slips, Land's End. Just before the moon set the fishermen had been very successful, but all at once lost their sport, or as they expressed it, " the Congers suddenly sheered off to a man." When hooked, it was not more troublesome than a Conger ; but when brought to the water's edge, it gave battle, and was secured with great difficulty. The first specimen noticed by the Rev. Robert Holdsworth as caught in a trawl-net off Brixham, had a portion of a Gurnard in its stomach. Of the third specimen, caught on the southern coast, near Berry Head, Mr. Holdsworth says, this Shark was taken near the bottom on a hook baited with cuttle. The men were fishing for Conger Eel, and other large fish, when this Shark was hooked. They describe his action in the water as most powerful, and were obliged to let him run with the line four times to the bottom before they could hamper him with a sliding noose let down over the line to his tail. These lines and the trawl-net only do their work at the bottom, and we may, therefore, conclude that this species is a Ground Shark. As such Cuvier had arranged it in his genus *Scymnus*, and Dr. Andrew Smith, who from his extensive acquaintance with this division of the cartilaginous fishes is an admitted authority, confirms this opinion. Of this Spinous Shark, Dr. Smith says, " This species is comparatively rare at the Cape of Good Hope. It is described by the fishermen as sluggish and unwieldy in its movements, and but seldom to be observed towards the surface of the water. When they obtain specimens, it is generally at a time when they are fishing in deep water, and when the bait with which the hooks are armed is near to the bottom. In this respect it resembles the *Scyllia*, or Ground Sharks; and, if we were to regard only its internal organisation, we should be disposed to consider it as closely allied to that genus."

Never having seen a specimen of this Shark, the following description of its colour and form is derived from Dr. Smith's work.

Colour :—The head and back, as far as the first dorsal fin, dark leaden grey; the rest of the back, the sides, and the belly, pale coppery yellow, clouded with purple and brownish tints; and the belly besides is marked with blotches of light vermilion red; the fins towards their bases reddish brown, tinged with dull grey, towards their extremities a lighter shade of the same colour; chin, sides of muzzle, and sometimes a spot behind the eye, dull white; eyes coppery green.

Form, &c.—Body very thick in proportion to its length, with only a slight diminution in size towards the tail; the back in front of the first dorsal fin nearly straight; the head flat above, and slightly sloping to the muzzle, which is rounded ; nostrils transverse, and each partially divided by a narrow membranous lobule, which projects backwards from its anterior margin ; their position is nearly over the most projecting, or central portion of the upper jaw, considerably nearer to the eyes than the tip of the snout, and about half way between the latter, and the angle of the mouth. Eyes rather nearer to a line raised from the angle of the mouth than to the nostrils ; pupil circular and small ; postocular spiracle scarcely visible. Gape wide and arched, having at each corner a triangular fold of skin formed by the union of the upper and lower lips. Teeth regularly placed upon each jaw, only one row in use at a time, the rest reclined ; they are large, compressed, and somewhat quadrangular, the cutting edges nearly horizontal, and both of their sides are generally bicuspidate, as will be seen by the figures here inserted, representing from both specimens the teeth of both jaws as opposed to each other.

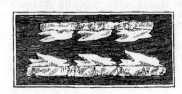 

Branchial openings all in front of pectoral fins ; the first not more than half the length of the fifth. Pectoral fins rather small, the hinder edges nearly square ; the dorsal fins are small, the first narrower at its base than at its extremity, which is slightly rounded ; the second nearly throughout of equal breadth, the hinder edge almost square ; the ventral fins short, broader behind than at their bases, and their posterior edges slightly undulated ; the caudal fin entire, some-

what triangular, and slightly falciform; the upper portion
high above the line of the back, the lower scarcely below the
line of the body immediately in front of it. Lateral line
distinct, commencing above the branchial openings, and extending nearly without curve or undulation to the commencement of the caudal fin, from thence it ascends the latter, and
extends along it, nearer to its anterior than posterior edge,
until it reaches its upper extremity; at its origin this line is
nearer to the middle of the back than the base of the pectoral
fin; to the touch it feels slightly rough, which arises from its
being beset with a number of minute prickles, which are
most distinctly seen in preserved specimens. The surface of
the skin both on the body and fins is more or less sprinkled
with strong bony-looking spines, with large circular and flattened bases, which are striated from the centre towards the
circumference. These spines vary in size as well as form,
some being hooked, others quite straight; in some places
they are disposed in clusters, in others they are solitary, and
on the extremity of the muzzle are nearly wanting. The
appendages to the ventral fins in the male seldom extend
much beyond their posterior margins."

According to M. Risso, the females of this species have a
smaller number of these spines than the males.

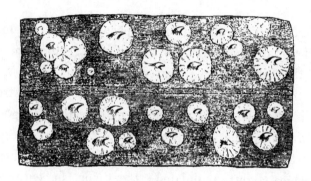

CHONDROPTERYGII. *SQUALIDÆ.*

THE HAMMER-HEADED SHARK.

Zygæna malleus. Val.

Zygæna,	Belon, p. 61.
,,	Rondelet, 1554, p. 389.
Marteau,	,, 1558, p. 304.
Zygæna,	Salvianus, tab. 40.
,, *Salviani,*	Willoughby, p. 55, B. 1.
Squalus zygæna,	Linn. Syst. Nat. t. i. p. 399, sp. 5.
,, ,,	Duhamel, sect. IX. pl. XXI. fig. 3.
Squale marteau,	Lacepede, t. i. p. 257, 4to. edit.
,, ,,	,, t. v. p. 443, 8vo. edit.
,, ,,	Risso, Icht. p. 34.
Zygæna malleus,	,, Hist. p. 125.
,, ,,	Val. Mem. du Mus. t. ix. p. 222.

Generic Characters.—Head depressed, more or less truncated in front, the sides extended horizontally to a considerable length, with the eyes at the external lateral extremity. Teeth of the same shape in the upper and lower jaw, viz. the points directed towards the corner of the mouth, with a smooth edge when young, but distinctly serrated in adult specimens. Branchial openings five. Two dorsal fins, the first in a line close behind the pectorals; the second over the anal fin.

In the sketch of the Natural History of Yarmouth and its Vicinity, by C. J. and James Paget, which I have frequently had the pleasure to refer to in the History of the British Birds, and also in the British Fishes, it is stated at page 17 that a specimen of the *Squalus zygœna*, or Hammer-headed Shark, was taken there in October 1829, and deposited in the Norwich Museum; and by the kindness and influence of J. H. Gurney, Esq. of Norwich, I have had the loan of drawings that were made from this Shark sent to London for my use in this work.

Among the numerous species included in the genus *Squalus* of Linnæus, — and I might say, indeed, in the whole class of Fishes,—there is no form more extraordinary than that of the Hammer-headed Sharks, four species of which are noticed in the memoir by M. Valenciennes here quoted, where they are considered as a sub-genus, under the name of *Zygœna*.

The Hammer-headed Shark taken on the coast of Norfolk, being also a native of the Mediterranean Sea, has been long known, and is figured in the works of Belon, Rondelet, and Salvianus, as already quoted. Its greatest singularity consists in the extraordinary form of the head; but its habits, as far as they are known, afford no physiological illustration of this very remarkable structure. In other respects it is very like the Sharks in general. This species is said to be ferocious, to frequent deep water, and measures from seven to eight feet in length. Baron Cuvier states that it has been known to attain the length of twelve feet. The female produces ten or twelve young ones in spring, which acquire considerable size by the end of autumn. In some countries the flesh of several species of Sharks is eaten, but that of the Hammer-headed Shark is said to be not only hard, but very unpleasant both in smell and flavour.

The head of this Shark, — representations of the upper and under surface of which, on a small scale, are given below, — measured from one eye to the other, is very large and wide; the eyes are furnished with eye-lids, which arise from the internal part of the orbits, the irides are golden yellow, the pupils black; the nostrils are elongated, and open immediately underneath the depression, or notch, in the anterior margin of the laterally expanded portions of the head; the mouth semicircular, and furnished with three, four, or five rows of teeth, depending upon the age of the specimen; these teeth are large, sharp, somewhat triangular and curved, with smooth cutting edges when the Shark is young, but serrated afterwards; the teeth in the upper jaw having their points directed towards the angle of the mouth; those of the lower jaw have the same direction, but they are narrower.

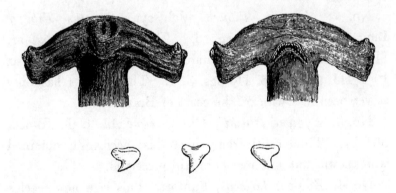

The body is elongated, covered with a skin slightly granulated; the colour greyish brown above, nearly white beneath: branchial openings five, all before the base of the

pectoral fin; the pectoral fins nearly triangular; the first
dorsal fin large; the second small, and placed just in advance
of the commencement of the tail; the inferior lobe of the
tail small, the superior portion as long as the head of the
fish is wide; the anal fin is under the second dorsal.

This species is found in the Mediterranean, on the shores
of the various countries of Europe, in the Ocean, and on the
coast of Brazil.

To make this subject as complete as my means will allow,
and afford an opportunity of identifying any other species
of *Zygæna* that might wander to our shores, I here add,
as a vignette, representations of the heads of the other
known species, of which No. 1 is *Zygæna tudes*, Val. the
synonymes being, according to M. Valenciennes, *Le Squale
pantouflier* of Lacépède, t. i. p. 260, pl. VII. fig. 3. Duhamel, sect. IX. pt. ii. pl. XXI. fig. 4 to 7. Koma
Sora Russel, pl. XII. This species has been found in
the Mediterranean, on the coast of Coromandel, and at
Cayenne, S. America.

No. 2. *Zygæna Tiburo*, Val. syn. *Squalus Tiburo*,
Linn. tom. i. p. 399, sp. 6. *Tiburonis species minor*,
Marcg. 181. Willoughby, tab. B. 9, fig. 3. Klein Misc.
Pisc. III. p. 13, tab. II. figs. 3, 4. This species has only
as yet been met with on the coast of Brazil.

No. 3. *Zygæna Blochii*, Cuv. *Règne An.* t. ii. Bloch,
pl. 117. The locality from which this species was obtained
is unknown, but specimens are still preserved.

No. 4. *Zygæna laticeps*, Cantor. This is a new species
lately described and figured by Dr. Theodore Cantor,
who obtained it in the Bay of Bengal, and in which the
head is still wider than in either of the other known species; a straight line drawn from the one eye to the other
is equal to about one half of the total length of the fish.

In shape the fins are like those of the four species already known ; the only difference I have observed, says Dr. Cantor, is the situation of the anal fin, which in the present species is somewhat anterior to the second dorsal, while these fins in the others are opposite.

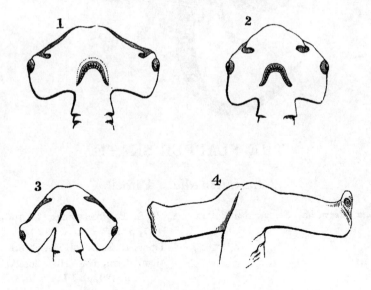

THE FLAPPER SKATE.

Raia intermedia. Parnell.

Raia intermedia, Flapper Skate, Parnell, R. S. E. Proceedings, 17 April,
                     1837, p. 166.
,,   ,,   ,,   ,,   ,,   Trans. R. S. E. vol. xiv. pl. 6.
,,   ,,   ,,   ,,   ,,   Mem. Wern. Nat. Hist. Soc. vol.
                     vii. p. 429, pl. XL.

" This fish," says Dr. Parnell, " which was obtained in
the Frith of Forth in the month of May, seems to be a
new species of Skate, since I am not aware of its having been
previously described. It appears to be the connecting link
between *Raia batis* and *Raia oxyrhynchus,* to both of which
it is closely allied, and it is from this circumstance that I
suggest the specific name of *intermedia.*"

" It is distinguished from *Raia batis*, in the upper surface of the body being perfectly smooth, without granulations, and of a dark olive colour spotted with white ; in the anterior part of each orbit being furnished with a strong spine pointing backwards ; in the dorsal fins being more remote from each other, and in the anterior margins of the pectorals being rather more concave, giving the snout a sharper appearance ; whereas, in *Raia batis*, the upper surface of the body is rough to the touch, of a uniform dusky grey without spots ; the orbits without spines ; the dorsal fins nearly approximate, and the anterior margins of the pectorals nearly straight."

" It is likewise removed from *Raia oxyrhynchus*, in the snout being conic ; the under surface of the body dark grey ; a spine in front of each orbit, and the back of a dark olive-green, spotted with white ; whereas in the *Raia oxyrhynchus*, the snout is sharp and long, with the lateral margins parallel near the tip ; the under surface of the body pure white, and the back of a plain brown without spots."

This species is not uncommon in the Frith of Forth, and I have met, observes Dr. Parnell, " with two examples of a variety of this fish which were taken in the salmon-nets at Queensferry. They were both of small size, about eighteen inches in length. The back was of a uniform dark olive green without spots of any description, covered with a thick mucus ; under surface of a dark grey ; body very thin ; snout sharp, conical ; pectorals at their anterior margin rather sinuous, passing off somewhat suddenly at that part, in a line with the temporal orifices, giving the outline of the anterior part quite a different appearance to that observed in *Raia intermedia ;* the anterior part of each orbit is furnished with a spine ; back perfectly smooth ; tail with one row of spines on the dorsal ridge ; fins, and in all other respects, similar to *Raia intermedia*."

A female specimen of this fish, about two feet in length,

tail included, is thus described by Dr. Parnell :—" Body
rhomboidal, the transverse diameter equalling the distance
between the point of the snout and the last tubercle but
three on the central ridge of the tail; from the point of the
snout to the temporal orifice, rather more than one third the
length as far as the end of the anal fin, and one fourth the
length as far as the termination of the first dorsal. Body
very thin; snout pointed, conical; pectorals large, somewhat
of a triangular form, uniting in front at the snout, and terminating at the base of the ventrals; the anterior margin
rather concave, the posterior margin rounded; ventrals
about three times the length of their breadth; anals
commencing close behind the ventrals, and terminating in
a free point; rounded at the outer margins. Tail short
and firm, being no longer than the distance from the
base of the anal fin to the anterior margin of the orbit;
along the mesial line is a line of tubercles with sharp points
directed downwards, about eighteen in number, commencing
at the base of the anal, and terminating at the commencement of the first dorsal fin; no lateral spines visible. First
dorsal fin small, rounded at the free extremity; situated
about one third of the length of the tail from the tip: the
base of the fin about equalling the length: second dorsal
rather smaller than the first, and about the same form, placed
about half-way between the termination of the first and the
tip of the tail; caudal fin rudimentary. Colour of the upper
surface of the body of a dark olive green, with numerous
white spots; on the under surface dark grey, with minute
specks of a deeper colour. Eyes rather small, flattened
above, placed in front of the temporal orifices; skin both
above and below perfectly smooth; a strong, sharp, bent
spine in front of each orbit; no spine or tubercles of any
description on the back. Mouth large, placed beneath;
teeth small, not so large or so sharp as those in *Raia batis*.

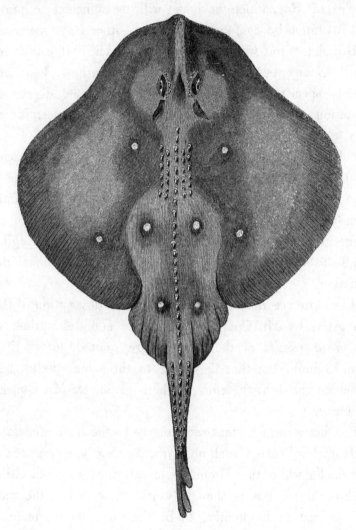

THE SANDY RAY.

Raia radula,	Delar. Mém. Poiss. Ivic. in An. Must. Hist. Nat. t. xiii. p. 321.		
,, ,,	*Raie râpe,*	Risso, Hist. t. iii. p. 151, sp. 38.
,, ,,	,, *ratissoire,*	Blainv. Faun. Franc. p. 25.
,, ,,	*Razza scuffina,*	C. L. Bonap. Faun. Ital. pt. xiii.
	The Sandy Ray,	Couch, Mag. Nat. Hist. vol. xi. p. 71.

In the second volume of the New Series of the Magazine of Natural History, and the eleventh volume of the whole

work, Mr. Couch has given a figure and description of a
species of Ray, which he hopes will be sufficient to prove
that it cannot be confounded with any other Ray recognised
as British; " but whether," says Mr. Couch, " it can be referred to any species described by other authors, I am not
able to specify, except that I have with some degree of
hesitation, supposed it to be possibly the *Raia asterias* of
Ray, Syn. Pisc. p. 27."

" I cannot, however, persuade myself but that this species
has been described by some authors, to whose writings I
have no opportunity of obtaining access; I therefore refrain
from assigning to it a trivial name, that I may be in no
danger of adding to science a useless synonyme. Its English
name of Sandy Ray, will be sufficient as a provisional de-
signation."

The close accordance of the figure and description of this
fish given by Mr. Couch, to the figure and descriptions of
the *Raia radula* of the authors here quoted, leaves little
room to doubt but that they refer to the same species, and
I include the fish, therefore, as here given, on Mr. Couch's
authority.

" It bears but a distant resemblance to the *Raia maculata*,
or Homelyn," Mr. Couch observes, " either in appearance or
value; for while the Homelyn is esteemed as food, either
fresh or salted, this is thought worthy only to bait the crab-
pot, or, just as frequently, to be thrown aside for manure.
It is of frequent occurrence in moderately deep water, from
spring to the end of autumn. In winter, however, it is not
often seen, chiefly, perhaps, because at that season the boats
do not venture quite so far from land; but, perhaps, also,
from the fish having changed its quarters. It seems to be
an indiscriminate feeder, living on small fishes, and different
kinds of crustacea."

" The specimen described, which was of the ordinary size,

measured three feet eight inches in length, of which the tail
was nineteen inches ; the breadth two feet four inches and a
half. The snout projected three-quarters of an inch, prominent and elevated; the mouth three inches and a half
wide, six inches from the snout. Under jaw peaked in the
middle; the teeth slender, sharp, in rows not very closely
placed. The body passes off circularly from the snout, the
greatest breadth opposite the centre of the disk, and of a
rounded form. From the snout the ridge is elevated to the
eyes, a distance of five inches and three-quarters ; eyes two
inches asunder ; temporal orifices large. Body thickest posteriorly; the tail stout at its origin, rounded above, tapering;
a groove along the body and tail ; two fins on the latter
close together. A few spines near the end of the snout ; a
semicircle of them behind each eye ; four short parallel rows
on the centre of the back, and a middle one continued along
the groove to the tail, which is covered with stout hooks,
scarcely in regular order. The remainder of the body
smooth. Colour above a uniform dusky brown, white below.
On the back a variable number of ocellated spots, the size of
the section of a large pea ; the centre pale yellow, the margin
a deeper impression, of the colour of the skin. I have
counted from eight to sixteen of these spots in different
specimens, and believe they have no determinate number ;
but they are always placed, on each side, with corresponding
regularity."

" Besides this description and figure, which I hope will
enable those who visit our fishing vessels to .ascertain this
species, I will further observe, as marks of distinction from
the other British species of this genus, that in addition to
the form of the teeth, which are crooked and slender, resembling a bird's claw in miniature, but which still are less long,
slender, sharp, or crooked, than in young specimens of the
Raia oxyrhynchus, it may be distinguished by a great ten-

dency to circularity in the disk, formed chiefly by a rounding off of the pectoral fins, by a flatness of the anterior portion, by the uniformity of its colour, the regularity of the spots, and the comparatively short and tapering tail."

The vignette below represents the late Hall of the Company of Fishmongers of London. The present new Hall is represented in the British Fishes as the final vignette to Volume II.

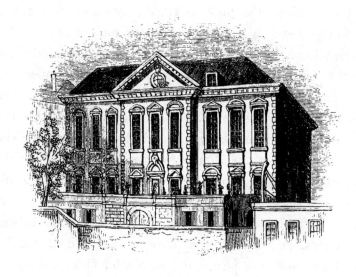

END OF THE SUPPLEMENT TO THE SECOND VOLUME.

London : Printed by Samuel Bentley, Bangor House, Shoe Lane.

By the Author of the " HISTORY OF BRITISH FISHES,"
and to be had of the same Publisher,

A PAPER ON THE

GROWTH OF THE SALMON IN FRESH WATER.

With Six coloured Illustrations of the Fish of the natural size, exhibiting its character and exact appearance at various stages during the first two years.

A GENERAL OUTLINE OF

THE ANIMAL KINGDOM,

AND MANUAL OF COMPARATIVE ANATOMY.

BY THOMAS RYMER JONES, F.Z.S.

PROFESSOR OF COMPARATIVE ANATOMY, IN KING'S COLLEGE, LONDON.

THIS work is intended to comprise a general view of the Animal Creation, exhibiting the structure and internal economy of every class of living beings, and their adaptation to the circumstances in which they are severally destined to exist.

Six parts, at 2s. 6d. each, containing 129 illustrations, are now published—the work will be completed in fifteen; the whole will then form a manual of comparative anatomy and animal physiology, equally adapted to the man of letters, the zoologist, or the anatomical student. In order to render the work as intelligible as possible to unscientific readers, a glossary of technical and scientific terms will be given in the concluding part. About three hundred illustrations will be embodied in illustration of the text.

To accommodate the possessors of the large paper copies of the other works on Natural History issued by the same Publisher, a few copies of this work will be printed to correspond.

A HISTORY OF BRITISH QUADRUPEDS,

INCLUDING THE CETACEA.

BY THOMAS BELL, F.R.S., F.L.S., V.P.Z.S.

PROFESSOR OF ZOOLOGY IN KING'S COLLEGE, LONDON.

The letter-press of this volume contains an account of their habits, utility in food, manufactures, agriculture, or domestic economy, and the noxious qualities of such as are in any way injurious to man; and an attempt has been made to define the characters of many of the species with more accuracy than had been done by previous authors. The illustrations comprise a figure of every species, and of many varieties, with numerous pictorial tail-pieces and anatomical diagrams, illustrative of the text, amounting, in all, to 200.

Price of the work, in demy 8vo. 28s. A few copies are also printed in royal 8vo, price 2l. 16s., and a very limited number in imperial 8vo, price 4l. 4s.

A HISTORY OF BRITISH FISHES;

BY WILLIAM YARRELL, F.L.S., V.P.Z.S.

This work is illustrated with 240 figures of Fishes, mostly taken from the objects themselves, and 145 vignettes, drawn and engraved by the most eminent artists. No pains have been spared to render it worthy of public estimation.

In two vols. demy 8vo, illustrated by nearly 400 beautiful wood-cuts, price 2l. 8s. The royal 8vo, or intermediate size, is out of print, and, of the imperial 8vo, price 7l. 4s. of which fifty only were printed, very few remain.

A HISTORY OF BRITISH REPTILES.

BY THOMAS BELL, F.R.S., F.L.S., V.P.Z.S.

PROFESSOR OF ZOOLOGY IN KING'S COLLEGE, LONDON.

The Reptiles of this country, although few in number, are not devoid of considerable interest; their habits are popularly much misunderstood, and several innocent and useful species are shunned and destroyed, from a mistaken notion that they are directly or indirectly noxious to man. The elucidation of their habits, the distinctive description of the species, their geographical distribution, and the history of the transformation of all the amphibious forms, are amongst the subjects discussed.

In addition to a figure of each species, and of some of the most important varieties, the Illustrations comprise many of structure, developement, and transformation.

In one volume 8vo, of 166 pages, and containing above 40 illustrations, price 8s. 6d. demy 8vo, 17s. royal 8vo, or 1l. 5s. 6d. imperial 8vo.

A HISTORY OF BRITISH BIRDS,

BY WILLIAM YARRELL, F.L.S., V.P.Z.S.

The first volume is now before the public; and, so far as the work has proceeded, the publisher refers to it with pleasure, as a fulfilment of the promises made in the original prospectus. This volume of the History contains descriptions of 105 species, their synonymes, generic and specific characters, geographical range, habits, food, nidification, sometimes with nests, eggs, and other interesting particulars. The illustrations include one representation of each species, and frequently, of male and female: the distinctive difference between the young and adult bird is sometimes given in a third figure, and, occasionally, the variation from summer to winter plumage is shown. Other illustrations, comprising modes of capture, anatomical distinctions, or the most interesting features of internal or external structure, are introduced the more fully to illustrate the descriptions.

Price of the volume 28s. demy 8vo, or in parts, published each alternate month, 2s. 6d. A limited number is also printed on royal 8vo, price 5s. each part, and fifty only on imperial 8vo. The latter will not be delivered until the work is complete.

DR. AIKIN'S CALENDAR OF NATURE;

OR,

NATURAL HISTORY OF EACH MONTH OF THE YEAR;

WITH A FEW ADDITIONS TO THE TEXT, BY A FELLOW OF THE LINNÆAN AND ZOOLOGICAL SOCIETIES.

With eighteen designs. Price 2s. 6d. cloth lettered.

" CATTERMOLE'S ILLUSTRATED EDITION" SHOULD BE PARTICULARLY EXPRESSED IN ORDERING THIS LITTLE VOLUME.

A GEOGRAPHICAL AND COMPARATIVE LIST OF

THE BIRDS OF EUROPE AND NORTH AMERICA.

BY CHARLES LUCIAN BONAPARTE,

PRINCE OF MUSIGNANO.

8vo, 5s. cloth.

THE NATURAL HISTORY OF THE SPERM WHALE,

AND

A SKETCH OF A SOUTH SEA WHALING VOYAGE.

BY THOMAS BEALE.

This is the only work on a subject of much national importance, and the only account of Whaling as practised in the South Seas. Just published, price 12s. post 8vo.

AN ANGLER'S RAMBLES.

BY EDWARD JESSE, F.L.S.

AUTHOR OF "GLEANINGS IN NATURAL HISTORY."

CONTENTS:—Thames Fishing.—Trolling in Staffordshire.—Perch Fishing-club.—Two Days' Fly-fishing on the Test.—Luckford Fishing-club.—Grayling Fishing.—A Visit to Oxford.—The Country Clergyman. Post 8vo, price 10s. 6d. cloth.

THE HONEY BEE,

ITS NATURAL HISTORY, PHYSIOLOGY, AND MANAGEMENT.

BY EDWARD BEVAN, M.D.

A new edition, considerably extended and carefully revised by the Author, one volume, 12mo, with many Illustrations, 10s. 6d. cloth.

A FLORA OF SHROPSHIRE.

BY W. A. LEIGHTON, B.A., F.R.S.E., &c.

This work will comprise the flowering plants indigenous to the county, arranged on the Linnæan system, and will be completed in three Parts.
Parts I. and II. 8vo, sewed, price 4s. each. Part III. is in preparation.

A FLORA OF THE
NEIGHBOURHOOD OF REIGATE, SURREY.

CONTAINING THE FLOWERING PLANTS AND FERNS.

BY GEORGE LUXFORD, A.L.S., F.R.S.E.

12mo, with a map of the district, 5s. cloth.

IN PREPARATION,

and to be published in Parts periodically,

A HISTORY OF BRITISH FOREST TREES, INDIGENOUS AND INTRODUCED. By PRIDEAUX JOHN SELBY, F.R.S.E. F.L.S. &c. With Illustrations.

A HISTORY OF THE FOSSIL FRUITS AND SEEDS OF THE LONDON CLAY. By JAMES SCOTT BOWERBANK, F.G.S. With Illustrations.

A HISTORY OF THE FISHES OF MADEIRA. By RICHARD THOMAS LOWE, M.A. British Chaplain. With Figures by the Hon. C. E. Norton and M. Young.

A HISTORY OF BRITISH CRUSTACEA. By PROFESSOR BELL. The Figures by J. O. Westwood, Sec. E.S.

A HISTORY OF BRITISH FERNS. By EDWARD NEWMAN, F.L.S. Illustrated with a Wood Engraving of every Species and named Variety, drawn on wood by the Author, and showing the figure of the Frond, the Fructification, and Venation of each.